THE EUGENICS MOVEMENT

THE EUGENICS MOVEMENT

An International Perspective

Edited by

Pauline M. H. Mazumdar

Volume I

Routledge
Taylor & Francis Group
LONDON AND NEW YORK

AP
Athena Press

First published 2007
by Routledge
2 Park Square, Milton Park, Abingdon, Oxon, OX14 4RN

Simultaneously published in the USA and Canada
by Routledge
52 Vanderbilt Avenue, New York, NY 10017

Routledge is an imprint of the Taylor & Francis Group, an informa business

Simultaneously published in Japan
by Athena Press
4–33–18 Sengoku, Bunkyo-ku, Tokyo 112–0011
ISBN 978-4-902708-39-4 (Athena Press Set)

Typeset in Times by TechBooks

British Library Cataloguing in Publication Data
A catalogue record for this book is available from the British Library

Library of Congress Cataloging in Publication Data
A catalog record for this book has been requested

ISBN 13: 978-0-415-36871-1 (Set)
ISBN 13: 978-0-415-36872-8 (hbk) (Volume I)

Publisher's note
References within each chapter are as they appear in the original complete work.
The publisher has gone to great lengths to ensure the quality of this reprint, but points out that some imperfections in the original may be apparent.

DISCLAIMER

The publisher has made every effort to contact authors/copyright holders of works reprinted in *The Eugenics Movement*. This has not been possible in every case, however, and we would welcome correspondence from those individuals/companies whom we have been unable to trace.

CONTENTS

VOLUME II

PART 3
The animal model: pedigrees and breeders

Part 3.1 The Geneticists Explain Mendelism to the Breeders

Part 3.2 Animal Breeding and Racism

PART 4
Pauperism, feeblemindedness and differential fertility: the Eugenics Education Society in Britain

Part 4.1 The Founders

VOLUME III

PART 5
Eugenics in the United States: immigration, sterilisation and the law

PART 6
Eugenics and genetics

VOLUME IV

PART 7
Eugenik and *Rassenhygiene*: eugenics in Germany

VOLUME V

PART 9
Eugenics and genetics in the Soviet Union

Part 9.1 The Lysenko Controversy: Contemporary Voices

PART 10
South America: eugenics, *puériculture* and racelessness

Part 10.1 On *Puériculture* and Women

VOLUME VI

ACKNOWLEDGEMENTS

To the students of the eugenics seminar of the spring of 2003: Melissa Nelson, Donald Parker, Boris Petrov, Kiran van Rijn, and Annette Wilkins, who were properly registered, and Erica Michiko Charters, Robert Foo, Akemi Shimomura and Emily Cowall Farrell, who were there because of the compelling nature of the subject. Almost every one of us knew that if we had lived in the heyday of the movement, because of some tare or other in all of our families, eugenics would have profoundly affected our personal lives.

Pauline M. H. Mazumdar
Toronto, May 2006.

GENERAL INTRODUCTION

The problematic that came together as the international eugenics movement had some features that were already in circulation long before the movement proper began. They include the idea that humanity was degenerating, rather than making progress, due to the influence of urban race-poisons, especially alcohol and syphilis, and to the emancipation of women from their duties within the family. In the early twentieth century, these were joined by a focus, both social and scientific, on the mind and body of the urban poor, and a fear that falling birth rates were sapping the military strength of nations. Worse, it was the respectable upper class whose fertility was dropping, while the worthless element, the so-called residuum, was unaffected. The introduction of Mendelian theory, around 1900, gave further scientific support to the fears of the elite. If social worth, and especially intelligence, were inherited, society was doomed.

The movement began in Britain, but quickly spread across the globe, making itself at home with some local modifications everywhere in the Western world. This collection of writing by the eugenists and their opponents comes from Britain, Canada, France, Germany, South America, the USA and the USSR. Recent work has now included China, Sweden and Switzerland. There was also an Indian Eugenics Society, but the movement as a whole was essentially Western.

There is much in the writing of this period that we now find rebarbative: its outspoken elitism, its disgust at the handicapped and its open scorn of the lower classes and of all peoples other than the writer's own. However, it is important that we do not write off such stuff as pseudo-anything. This was a movement that persuaded the best minds of its time. As a modern, science-based ideology, it was attractive to progressives, particularly women, who saw in it scientific validation of women's concerns with reproduction and family. Its opponents were of two kinds, the Catholics and the Left. Strangely, in spite of their profound ideological objections, both were willing to allow it some good points. It was too important to dismiss completely.

PART 1 THE NINETEENTH CENTURY: DEGENERATION

Degeneration had several meanings for nineteenth-century thinkers. In literature, it was taken to mean a decline from the dignified norms of the classical tradition, the very reverse of progress. It meant romanticism, subversion and despair, as well as a concern with the lower depths of society. In medicine, it meant the effects of diseases, stimulants and poisons, passed down to the children of sufferers. The French psychiatrist Bénédict-Augustin Morel, in his book *A Treatise on Physical, Intellectual and Moral Degeneration in the Human Species* (1857) included even the debilitating effects of town life and factory work on both the workers and their children. As the generations succeeded each other, the effects became worse and worse, ending inevitably in insanity and crime. It was an idea that attracted both psychiatrists and that most romantic and pessimistic of novelists, Émile Zola.

The image of the insane and the artist as similarly degenerate was to reappear in Nazi Germany. Adolf Hitler and his Minister of Propaganda, Joseph Goebbels, arranged to clear out from museums and galleries in Germany anything they saw as degenerate and diseased and organised a series of exhibitions of such works. They included works by the German expressionists, such as Emil Nolde, Ernst Ludwig Kirchner and Karl Schmidt-Rotluff (Peter Adam, *Art of the Third Reich*, New York, NY: Abrams, 1992, pp. 120–127). As Hitler wrote in 1939 at the opening of an exhibition of what he regarded as the real stuff, 'As the Reich grows, so grows its art . . . the whole fakery of a fashionable decadent diseased or untruthful art has been brushed aside. A proper standard has been reached. . . ' (ibid., p. 62).

* * *

Max Nordau, 'Fin-de-siècle', Book I, *Entartung* (1892), anonymously translated as *Degeneration*, New York: Fertig, 1968, pp. 1–44.

> Max Nordau was the pseudonym of Max Simon Südfeld, born in Budapest. He studied medicine in Paris under the neurologist J. M. Charcot, an expert on hysteria. As George Bernard Shaw summed up Nordau's views, 'all characteristically modern works of art are symptoms of disease in the artists, and these diseased artists are themselves symptoms of the nervous exhaustion of the race' (Shaw, *The Sanity of Art: An Exposure of the Current Nonsense about Artists Being Degenerate*, London: New Age, 1908, p. 17). Zola was one of the degenerates despised for his 'realism or naturalism', which Nordau found 'obscene and nasty', washed up now as a school, though still popular.

Emile Zola, *Germinal* (1885), translated by Havelock Ellis (1894), London: Dent/Everyman, 1933, Part 7, pp. 349–422.

> Zola's family, the Rougon-Macquarts, is the subject of a series of 20 novels: this is the 13th. Before beginning to write, Zola sketched out his

intentions at length. He writes that their stories demonstrate the action of two basic laws, a more general heredity which becomes darker and darker, and *innéité* or individual nature. Heredity determines moral qualities: *métissage* or mixing, the result of adultery, allows for the introduction of 'bad effects'. Of the two sons of the founding generation, one is legitimate and one illegitimate. The author uses this device to follow the 'natural and social history' of the family in different classes in French society under the Second Empire (see 'Documents et plans préparatoires des Rougon-Macquart', in Henri Mitterand (ed.), Emile Zola, *Les Rougon-Macquart: Histoire naturelle et sociale d'une famille sous le Second Empire*, 5 vols, Paris: Pléiade, 1967, v. 5, pp. 1674–1776; p. 1700).

Cesare Lombroso, *Crime: Its Causes and Remedies* (1906), translated by H. P. Horton, Montclair, NJ: Patterson-Smith, 1968, pp. 365–384.

Lombroso, an Italian psychiatrist, dedicates his book to Nordau, who had dedicated his *Entartung* (1892) to him. According to Lombroso, criminals were born degenerates, recognisable by their depraved physiognomy, and doomed to live out the lives their faces indicated.

Charles Wicksteed Armstrong, 'Is Degeneration a Fact?', *The Survival of the Unfittest*, London: Daniel, 1927, pp. 17–32.

PART 2 THE NINETEENTH CENTURY: SOCIAL ACTIVISM AND SOCIAL CONTROL

In the industrialising and urbanising society of the nineteenth century, the problems of the cause and control of pauperism and crime in the lower classes were a focus of attention for middle-class social activists: in Britain, class, rather than race, was a central part of the eugenics problematic. This was the background against which the eugenics movement got its start.

[Margaret] Fison, *Handbook of the National Association for the Promotion of Social Science*, London: Longman, 1859, pp. 9–34, 67–78, 163–220.

The Social Science Association saw itself as the social equivalent of the British Association for the Advancement of Science. Its *Handbook* lays out the various causes of pauperism that the SSA was interested in. It includes sections on education, especially education for the very poor, on sanitation and on crime and alcoholism. Leaders in all these areas attended the SSA's meetings. As Ms Fison says (p. 20) such societies and their visitors were regarded with hostility and suspicion by the working class who were the purported beneficiaries. However, the members of the SSA were deeply convinced that they were doing good.

Helen Dendy, 'The Industrial Residuum', *Economic Journal* 3 (1863) 600–616, reprinted in Bernard Bosanquet (ed.), *Aspects of the Social Problem by Various Authors*, London: Macmillan, 1895, pp. 82–102.

> Helen Dendy was a Charity Organisation Society activist. In this paper, she distinguishes the 'true industrials' from a 'residuum', whose lack of foresight and animal-like instincts fit them only to occupy the lowest level of society. Interestingly, the group she is describing is what is currently known as the 'informal sector', nowadays regarded as highly important economically, especially in less developed countries.

Great Britain, Inter-Departmental Committee on Physical Deterioration, 'V. Alleged Tendency of Superior Stocks in All Classes towards a Diminished Rate of Reproduction', 'VI. Food', 'VII. Conditions Attending Life of Juvenile Population', 'Appendix 1', 'Appendix 3', *Report of the Inter-Departmental Committee on Physical Deterioration*, Volume I (Report and Appendix), London: HMSO, 1904 ('the FitzRoy Report'), pp. 38–39, 39–44, 44–76, 95–97, 104–111.

> The original mandate of this inquiry was to look into the problem of the increasing numbers of potential recruits who fell below acceptable standards for the British Army (see Appendix 1). Karl Pearson, Galton Professor of Eugenics at University College London claimed (p. 38) that Britain was ceasing to breed intelligence: the mentally better stock is not reproducing itself, while the less able are making up the difference, with obvious consequences for the eugenist. Differential fertility thus stated was to be one of the cornerstones of eugenics. The conclusions of the Committee, however, underline bad nutrition, want of cleanliness and indifference to slight ailments, rather than a degenerate heredity.

PART 3 THE ANIMAL MODEL: PEDIGREES AND BREEDERS

The American Breeders' Association opened its journal in 1910 with a programmatic statement:

> The objects of the American Breeders' Association are the advancement of the discovery of the basic facts concerning heredity, the devising of new plans for creative breeding, and the organization of those projects which led toward improved plants, animals and men . . . [It] promotes and helps organize cooperation among those interested in studying the laws of heredity, those devising improved methods of breeding the new respective species of animals and plants . . . and those interested in eugenics or heredity in man.

However, as Margaret Derry has pointed out in her *Bred for Perfection: Shorthorn Cattle, Collies and Arabian Horses since 1800* (2003), the new genetics did not turn out to be of much help to animal breeders. Textbooks for students of agriculture explained the new Mendelian theory, but then usually admitted it had no clear application on the farm, where practical breeders bred to a type, with no help from the concept of a segregating unit character. The American Breeders' Association gradually became a eugenics society.

On the other hand, as Harriet Ritvo has cleverly observed, the insistence on pedigree in animals had an element of aristocratic anthropomorphism; views on animals tracked views on human aristocracy.

Part 3.1 The Geneticists Explain Mendelism to the Breeders

W. E. Castle, 'Some Biological Principles of Animal Breeding', *American Breeders Magazine* 3 (1912): 270–282.

> Castle explains to the breeders that his rat coat-colours are inherited as 'unit-characters': Irish or black is dominant, hooded is recessive. But he found that by selection he was able to modify the unit-character: it was a quantitative variation, not an unmodifiable entity. It was not until 1919 that he came round to admitting that the changing coat colours could have been due to selection of modifying genes, and that the character might not have been due to a single unit.
>
> Elof Axel Carlson has discussed the history of this concept in *The Gene: A Critical History* (1966). He thought that it did not survive the demonstration that Castle was mistaken. 'World War I had ended,' says Carlson, 'but for Castle it was not armistice, it was unconditional surrender' (Carlson, p. 38). Another historian of classical genetics, Leslie C. Dunn, in his *Short History of Genetics* (1965), felt that the concept had gone down even earlier, with the demonstration in 1909 that the red colour of wheat grains depended on several paired alleles (Dunn, p. 100). However, it was to remain a mainstay of eugenics for many years to come.

Sewall Wright, 'Mendelian Analysis of the Pure Breeds of Livestock: I. The Measurement of Inbreeding and Relationship', *Journal of Heredity* 14 (1923): 339–348.

Sewall Wright, 'Mendelian Analysis of the Pure Breeds of Livestock: II. The Duchess Family of Shorthorns as Bred by Thomas Bates', *Journal of Heredity* 14 (1923): 405–422.

Harry H. Laughlin, 'Racing Capacity in the Thoroughbred Horse: Part I. The Measure of Racing Capacity, Part II. The Inheritance of Racing Capacity', Carnegie Institution of Washington, Supplementary Publications, No. 7, Washington, DC: Carnegie Institution, 1934: 3–14, 15–27.

> Laughlin, one of the most enthusiastic promoters of American eugenics, admits by 1934 (p. 16) that it is not possible to apply genetic analysis (sc., Mendelian prediction) to most of the qualities which are useful in plant and animal breeding . . .

Part 3.2 Animal Breeding and Racism

W. E. Castle, 'Biological and Social Consequences of Race-crossing', *Journal of Heredity* 15 (1924): 363–369.

Castle reviews Norwegian Jon Alfred Mjoen's thesis on 'Harmonic and disharmonic race crossings', from *Eugenics in Race and State* (1923), pp. 41–61. Mjoen also published a version of this paper in the popular Nazi journal *Volk und Rasse* (1929) 4: 72–77.

PART 4 PAUPERISM, FEEBLEMINDEDNESS AND DIFFERENTIAL FERTILITY: THE EUGENICS EDUCATION SOCIETY IN BRITAIN

The Eugenics Education Society was founded in 1907, with Francis Galton, a cousin of Charles Darwin, as its figurehead. Its membership was gathered from a number of different societies of middle-class activists interested in tackling the problems of the urban poor, if possible through legislation. Galton had been interested in the inheritance of talent, however the Society focused on the other end of the scale; feeblemindedness was also hereditary, and was the single cause that underlay most pauperism.

In the early years of the century, when birth rates were falling all over Europe, eugenists fastened on the frightening implications of differential fertility; the pauper class was far more prolific that their betters, the 'prudential classes'. The problematic within which the British group worked was bounded by these linked concepts, all involving class; the biological inheritance of feeblemindedness, and with it pauperism, the differential birth rate, and the possibility of control by legislation.

Part 4.1 The Founders

Francis Galton, 'Hereditary Talent and Character', Part I, *Macmillan's Magazine* 12 (1865): 157–166, 318–327.

Many more of Galton's papers can be found on the web. An interesting approach to Galton's thought is included in Hermann Muckermann's *Eugenics and Catholicism* (Berlin, 1933), Part I; Muckermann's book is in the present collection.

Charles Darwin, 'Natural Selection as Affecting Civilized Nations', *The Descent of Man and Selection in Relation to Sex*, 2nd revised edn, London: Murray, 1909, pp. 944–947.

* * *

All five of the workers chosen, Elderton, Heron, Lidbetter, Pearson and Slater show the kind of class hostility that we met above in Mary Dendy's paper. Elderton and Heron were co-workers of Karl Pearson, Galton Professor of Eugenics and Statistics at University College London. Lidbetter was Relieving Officer of a London Workhouse under the Poor Law. He was trained in the Eugenics Education Society's courses and played an important part in the Society's long-running study of pauper family pedigrees. All five focus on the lower class and its fearsome fertility. The feeling is that classes are hereditary, inbreeding groups, in which high value is related to low fertility. If the less valuable elements can outbreed their betters, the quality of the population is bound to decline. Slater's edited account of the conclusions of the Royal Commission proposes that segregation of the feeble-minded can be used to control their dangerous fertility.

David Heron, 'On the Relation of Fertility in Man to Social Status, and on the Changes in This Relation That Have Taken Place in the Last Fifty Years', Department of Applied Mathematics, University College, University of London, in the series 'Draper's Company Research Memoirs: Studies in National Deterioration', London: Dulau, 1906, pp. 3–22.

A. C. Pigou and Mary Dendy, 'The Economic Aspect of the Problem', 'Can the Feebleminded Be Made Happy in Confinement', in [Mrs Walter Slater] *The Problem of the Feebleminded: An Abstract of the Report of the Royal Commission on the Care and Control of the Feebleminded*, London: King, 1909, 97–101 (A. C. Pigou), pp. 105–113 (Mary Dendy).

> This report was a popular summary of the Royal Commission's official report, put together by a committee of members of the Eugenics Education Society and the National Association for the Care and Control of the Feeble Minded.

E. J. Lidbetter, 'Some Examples of Poor Law Eugenics', *Eugenics Review* 2 (1910–1911): 204–228.

Ethel M. Elderton, Amy Barrington, Gertrude Jones, Edith M. M. de G. Lamotte, H. J. Laski and K. Pearson, *On the Correlation of Fertility with Social Value: A Cooperative Study*, University of London, Francis Galton for National Eugenics, Eugenics Laboratory Memoirs XVIII, London: Dulau, 1913, pp. 1–72.

Karl Pearson, *Social Problems: Their Treatment, Past, Present and Future*, a lecture delivered at the Galton Laboratory for National Eugenics, 19 March 1912, in the series 'Questions of the Day and of the Fray', No. V, London: Dulau, 1912, pp. 3–40.

> Pearson presents himself as a socialist, who has the best interests of society at heart. Any legislation, he says (p. 38), which renders the fertility of the unfit dominant is dangerous: we have suspended Nature's ways of keeping up the quality of our stock, and we must find an alternative. We must check the fertility of the unfit, and encourage that of the fit.

PART 5 EUGENICS IN THE UNITED STATES: IMMIGRATION, STERILISATION AND THE LAW

In the US, the eugenics movement tended to conflate class, race and immigration. It had two practical consequences: the State Sterilization Laws initiated between 1905 and 1922, and the Immigration Acts of 1921 and 1924. A most effective advocate of both sterilisation and immigration restriction was Harry H. Laughlin, a protégé of C. B. Davenport and, from 1910, Superintendent of the Eugenics Record Office at Cold Spring Harbor under Davenport. Laughlin testified before the House Committee on Immigration and Naturalization in 1921, 1922 and 1924, the Committee's only scientific witness.

Jonas Robitscher in his *Eugenic Sterilization* (1973) calculates that about 104, 648 sterilisations were carried out, most of them between 1930 and 1970, when the laws were repealed state by state.

Charles Benedict Davenport, 'Migrations and Their Eugenic Significance', *Heredity in Relation to Eugenics*, New York, NY: Holt, 1911, pp. 204–224.

On Mendelian heredity in man, and on immigration.

Harry H. Laughlin, 'Analysis of America's Modern Melting Pot', *Hearings before the Committee on Immigration and Naturalization*, House of Representatives, 67th Congress, 3rd Session, Washington, DC: GPO, 1923, pp. 724–815, 818–831.

Madison Grant, 'The Nordic Fatherland', *The Passing of the Great Race, or, the Racial Basis of European History*, New York, NY: Scribners, 1922, pp. 213–222.

On the 'Nordic fatherland' of Western culture, and the fall of the Roman Empire. A further argument for the restriction of immigration to those of Northern European stock. Elsewhere, Grant demands the extermination of the insane as the scum of society. According to Paul Weindling, this text was cited by the defendants at the Nuremberg Trial to point to the fact that the US had subscribed to many of the same principles as the Nazis (see his *Nazi Medicine and the Nuremberg Trials: From Medical War Crimes to Informed Consent*, Basingstoke: Palgrave-Macmillan, 2004, pp. 225–249; p. 229).

Paul Popenoe and Roswell Hill Johnson, 'The Need for Negative Eugenics', 'Selective Segregation', 'Selective Sterilization' and 'Selective Contraception', *Applied Eugenics*, New York, NY: Macmillan, 1933, pp. 123–180.

On negative eugenics: controlling the fertility of the undesirables by segregation, sterilisation and contraception.

PART 6 EUGENICS AND GENETICS

Apart from the Catholic Church, which we shall look at here and under Canada and Germany, there were few voices raised against the eugenist's viewpoint until about 1930. This was the time when the British Eugenics Society, founded in 1907 as the Eugenics Education Society, began its ultimately unsuccessful campaign to legalise voluntary sterilisation. From 1931 on, the geneticists of the Left, Lancelot Hogben and J. B. S. Haldane, and the Quaker L. S. Penrose, mounted their attack.

It is notable, however, that though they took every opportunity of badmouthing the eugenists for their snobbery of race and class, they never completely lost hope that sometime in the future, progress in genetics would in fact allow it to be of some use to the human race. Even if Mendelism currently was not very helpful, the progress of science might change that at some future time.

David Heron, *Mendelism and the Problem of Mental Defect: A Criticism of Recent American Work*, No. 7 in the series 'Questions of the Day and Fray', London: Dulau, 1913, pp. 3–62.

> A biometrician from the group around the statistician Karl Pearson at University College London trashes Davenport's vulgar Mendelism.

Sewall Wright, 'Correlation and Causation', *Journal of Agricultural Research* 20 (1921): 557–585.

* * *

Lancelot Hogben, 'The Serological Data for the Study of Gene Localisation', *Genetic Principles in Medicine and Social Science*, London: Williams and Wilkins, 1931; New York: Alfred A. Knopf, 1932, pp. 68–90, 200–220.

> Hogben, a Marxist, hates the Eugenics Society for its position on working-class intelligence and social value. He is a geneticist too, though: Mendelian analysis may fail to predict complex behaviour, but it is not useless. He has discovered in the inheritance of the blood groups the only instance to date of value-free science in the field of human genetics. If something could be shown to be linked to a blood group, then that something would have to be genetically determined. But there was no example so far.

Lionel Penrose, *Mental Defect*, from the series 'Textbooks of Social Biology', Lancelot Hogben (ed.), London: Sidgwick, 1933, pp. 1–13, 85–95.

> Penrose was a Quaker, who at this time was engaged in research on mental defect. He states that:
>
>> Mental defect is fundamentally a social, and not simply a biological or psychological, problem. The social and economic aspect of it is liable to obscure all others, and public opinion varies in its attitude towards mental defect precisely in accordance with its attitude towards other social and economic problems. To investigate the subject from the biological or psychological point of view we must throw aside all social and moral prejudices. (p. 13)

J. B. S. Haldane, 'Evolution and Our Weak Points', *Science Advances*, London: Allen and Unwin, 1947, pp. 118–120.

J. B. S. Haldane, 'The Biology of Inequality', *Heredity and Politics*, New York, NY: Norton, 1938, pp. 13–41.

These three, Haldane, Hogben and Penrose, beginning in 1931 mounted the first explicit attack on the science of the eugenists. Hogben was Professor of Social Biology at the London School of Economics; Haldane was Weldon Professor of Biometry at University College London; both of them were Marxists. Haldane wrote many editorials for the London Marxist newspaper, *The Daily Worker*; 'Evolution and our Weak Points,' is one of them. Their argument is that we cannot say whether success in life is genetic or environmental, until the environment is the same for everyone. As Marxists, they know it is not: the eugenist's attitude to the working class can never have scientific support. On the attacks on eugenics from the Left, see P. M. H. Mazumdar, *Eugenics, Human Genetics and Human Failings – the Eugenics Society: Its Sources and Its Critics in Britain* (London: Routledge, 1992), pp. 146–195.

1930 was also the date of the Pope's Encyclical *Casti connubii* in which he spoke out directly against eugenics. At this point, Marxists sided with Catholics.

Letitia Fairfield, *The Case against Sterilization*, 2nd edn, London: Catholic Truth Society, 1935, pp. 3–32.

Fairfield was the *bête noir* of the British Eugenics Society in its push to legalise voluntary sterilisation. 'Dr Fairfield's book' was frequently quoted by the Society's Catholic opponents. She was cited almost as much as the papal Encyclical itself.

* * *

Fritz Lenz, 'Methods for the Study of Human Heredity', in Erwin Baur, Eugen Fischer and Fritz Lenz, *Human Heredity*, translated by Eden and Cedar Paul, London: Unwin, 1931, pp. 495–561.

Lenz was among the founders of the German eugenics movement. 'Baur-Fischer-Lenz' was the key textbook of the period. The 3rd edition (1931) was translated into English.

R. A. Fisher, 'The Causes of Human Variability', *Eugenics Review* 10 (1918): 213–220; 'The Elimination of Mental Defect', and 'The Biometrical Study of Heredity', *Eugenics Review* 16 (1924): 114–116, 189–210; 'Eugenics: Can It Solve the Problem of Decay of Civilizations?', *Eugenics Review* 18 (1926): 128–136.

The statistician R. A. Fisher was a protégé of the Eugenics Society from his early days as a student at Cambridge, where he started a branch of the Society. In these papers from the *Eugenics Review* between the wars, he

> deploys the heavy artillery of genetics both Mendelian and biometrical. It is veiled in a chatty style, but nonetheless, it is a very serious defence of the eugenic cause.

J. B. S. Haldane, 'A Provisional Map of a Human Chromosome', *Nature* 137 (1936): 398–400.

PART 7 *EUGENIK* AND *RASSENHYGIENE*: EUGENICS IN GERMANY

The science of eugenics often seems to be somewhat naïve, as in the case of C. B. Davenport's 'vulgar Mendelism'. In Germany, however, it was anything but. At a time when British and American eugenists were content to point to a shocking Lidbetter-style pedigree as evidence of degeneration, the German geneticists had already mathematised their Mendelism. The German eugenics movement was one of the earliest to organise; its leaders were Alfred Ploetz, Fritz Lenz and Ernst Rüdin. The geneticists began by trying to fit the genetics of psychoses, such as schizophrenia, into a Mendelian pattern with the help of the mathematical Mendelism of Wilhelm Weinberg. However, the calculations were difficult and tended to minimise the numbers of affected persons in the family of a given propositus.

Empirical prognosis, however, introduced by Rüdin and his group, was looser, easier to explain politically, and maximised the numbers of potentially affected family members, making the argument for sterilisation far more convincing. The empirical school developed during the twenties, and reached its apotheosis in 1933 with the Nazi accession, when it became a part of official policy. It worked through the National Centre for Psychiatric Research in Munich, under Ernst Rüdin. On the transition from Mendelism to empirical prognosis, see Mazumdar, 'Two Models for Human Genetics: Blood Grouping and Psychiatry in Germany between the World Wars', *Bulletin of the History of Medicine* 70 (1996): 609–657. Rüdin's draft of a sterilisation bill (based, it is said, on the Indiana Sterilization Act) was enacted into law as the Gesetz zur Verhütung erbkranken Nachwuchses of 1933. The numbers sterilised under this law vastly exceeded those in America. From 1940, sterilisation was replaced by so-called euthanasia.

Ernst Rüdin, 'Praktische Ergebnisse der psychiatrischen Erblichkeitsforschung', *Archiv für Rassen- und Gesellschaftsbiologie* 24 (1930): 228–237.

> Translated by P. M. H. Mazumdar as 'Practical Results of Research in Psychiatric Genetics'. Rüdin was Director of the Genealogical-Demographical Section of the Centre for Psychiatric Research in Munich. The Centre introduced a new method, *empirische Erbprognose* or empirical genetic prognosis, for research on the inheritance of mental problems. They abandoned a complex mathematical Mendelism for a simplified probability estimate of the likelihood of disease in the relatives of patients, based on the collection of data on thousands of cases. The

practical upshot of this work was to work out general indications for sterilisation.

Fritz Lenz, 'Die Stellung des Nationalsozialismus zur Rassenhygiene', *Archiv für Rassen- und Gesellschaftsbiologie* 25 (1931): 300–308.

Translated by P. M. H. Mazumdar as 'The Position of National Socialism on Race Hygiene'. Lenz was one of the founders of the German movement. He was the 'Lenz' in 'Baur-Fischer-Lenz', the massive and authoritative textbook of human genetics first published in 1921, *Menschliche Erblichkeitslehre und Rassenhygiene*. He wrote Volume II, on *Rassenhygiene.* In this paper, he says with pride that Adolf Hitler read his book while he was in gaol following the Munich *Putsch*, and that much of the racist phraseology of Hitler's *Mein Kampf* comes from it.

Hermann Muckermann, *Eugenik und Katholizismus*, Berlin: Metzner Verlag, 1933.

Translated and annotated by P. M. H. Mazumdar as 'Eugenics and Catholicism'. This text is interesting enough to include the whole of it. Part I discusses the sources and more recent scientific findings of eugenics; Part II discusses the Encyclical *Casti connubii* of 1930 and its sources. Muckermann, a geneticist and Catholic intellectual, tries to reconcile eugenics and Catholicism, or at least to justify his failure to come out publicly against eugenics.

Pope Pius XI, *Casti connubii*, Encyclical of 31 December 1930 on Christian marriage.

The Encyclical is important not only for the stand it took against eugenics, but also for the dialogue it created among Catholics both for and against eugenics. It is combed through at length by Hermann Muckermann and by the Canadian Hervé Blais, both of whom are included in the collection. The Encyclical is clear in its rejection of eugenics, birth control and sterilisation, but it allows a priest to counsel against marriage if he feels it to be inappropriate.

A further Encyclical of the same Pope, Pius XI, *Mit brennender Sorge* of 1937, was smuggled into Nazi Germany, and read aloud from the pulpits on Palm Sunday. It denounces Nazi racial theory as 'an idolatrous cult of race and blood', and underlines the Jewish roots of Christianity. These and other Encyclicals can be found on the Vatican website, http://www.vatican.va/holy_father/pius_xi/encyclicals/index.

PART 8 FRANCE: EUGENICS AND LAMARCKISM

The eugenics of France was a little different from that in other countries. Since it was led by a well-known paediatrician, Adolphe Pinard, it emphasised *puériculture* or infant management, including care of the pregnant mother and infant nutrition.

In addition, there was a constant fear in France that Germany was becoming more and more powerful and populous, while France's birth rate fell lower and lower. French eugenics therefore concentrated less on sterilisation and birth control than on pro-natalism – quality and quantity, as William Schneider puts it in his *Quality and Quantity: The Quest for Biological Regeneration in Twentieth-Century France*, Cambridge: Cambridge University Press, 1990. In France, he says, the call was more for having healthier babies, than for sterilising the unfit (p. 82).

Part 8.1 First International Eugenics Congress, London, 24–30 July 1912

The International Congresses met first in London in 1912, as a respectful nod to Galton's homeland. They broke off during the 1914–1918 War, and continued in New York in 1921, 1932 and 1940. Another series of international meetings, the International Eugenics Conferences, met in 1913, then roughly every two years from 1919 to 1936 in a different European city each time. Like the Socialist International, they served to make eugenics an international movement – linking, for example, American and German eugenics. Stefan Kühl calls them the 'racist international' (see his *Die Internationale der Rassisten: Aufstieg und Niedergang der internationalen Bewegung für Eugenik und Rassenhygiene im 20. Jahrhundert*, Frankfurt: Campus Verlag, 1997).

H. Hallopeau, 'Sur la prophylaxie de la syphilis héréditaire et son action eugénique', *Problems in Eugenics*, London: Eugenics Education Society, 1912, pp. 343–347.

> Translated by the Eugenics Education Society as 'On the Prophylaxis of Hereditary Syphilis and Its Effect on Eugenics', *Problems in Engenics*, pp. 347–351. There was a large French delegation to the London Congress,which served to introduce the French style to London. Hallopeau was a professor at the École de Médecine, Paris. His contribution underlines the very clinical nature of French eugenics.

Adolphe Pinard, 'Considérations générales sur "La puériculture avant la procréation"', *Problems in Eugenics*, London: Eugenics Education Society, 1912, pp. 457–458.

> Translated by the Eugenics Education Society as 'General Considerations upon "Education before Procreation"', *Problems in Eugenics*, pp. 458–459. Dr Pinard of the Baudeloque Clinic was probably the most famous obstetrician in France. It was his claim that both spouses, but especially the mother, should retreat and rest before procreation. Pre-natal care, post-natal care and feeding of the newborn under supervision were also in his territory. It was his contacts with March and Richet that introduced him to hereditarianism and eugenics, but the origin of French thinking in clinical paediatrics left a permanent Lamarckian stamp on it.

Part 8.2 Congress of the Royal Institute of Public Health, Paris, 15–19 May 1913

[Anon.], 'Eugenics and Public Health', *Eugenics Review* 5 (1913): 157–165.

Part 8.3 The First World War

Lucien March, 'Some Attempts towards Race Hygiene in France during the War', *Eugenics Review* 10 (1918): 195–212.

> March was Director of the Statistique générale de France. Along with Pinard, he was one of eight Vice-Presidents of the Consultative Committee for the First Eugenics Congress of 1912. As a pro-natalist, he argued against the concern with differential fertility that so frightened British and other eugenists. Here he joins a widespread debate about the dysgenic effects of war.

Charles Robert Richet, 'Le problème ou le préjugé des races', *Revue générale des sciences* 16 (1906): 883–891.

> Translated by P. M. H. Mazumdar as 'Race Problem or Race Prejudice'. Richet was a well-known physiologist, and the discoverer of anaphylaxis.

Charles Robert Richet, 'Conclusions générales', *La sélection humaine*, Paris: Félix Alcan, 1912–1919, pp. 224–235.

> Translated by P. M. H. Mazumdar as 'Human Selection'. Richet's book sketches a full eugenics programme along Darwinian lines. As William Schneider points out, Richet's *Sélection humaine* added the element of social biology to Pinard's Lamarckian *puériculture.* Pinard, it seems, found its argument new and compelling (see Schneider, loc. cit., above, p. 81).

Part 8.4 The Population Problem

Hon. H. Onslow, 'The French Commission on Depopulation', *Eugenics Review* 5 (1913): 130–152.

D. V. Glass, 'Population and Policy in Germany', 'Family Allowances and the Birth-rate in France and Belgium' and 'Conclusion', *The Struggle for Population*, Oxford: Oxford University Press, 1936, pp. 16–32, 72–86, 87–91.

* * *

Alexis Carrel, 'The Remaking of Man', *Man, the Unknown* (1935), London: Hamish Hamilton, 1938, pp. 257–298.

Carrel was a French immunologist, who worked for most of his life at the Rockefeller Institute in New York. After the German takeover, he returned to France to work for the Vichy Government. His book was a bestseller: it appeared in French and English simultaneously. He concludes with a paean to eugenics, tossing off rather lightly the suggestion that petty criminals should be beaten with whips and hospitalised, while the more serious might be executed in 'small euthanasic institutions supplied with proper gases' (p. 296). Like Madison Grant (above) he was cited by the defence in the Nuremberg Trials (Paul Weindling, *Nazi Medicine and the Nuremberg Trials*, 2004, p. 254).

PART 9 EUGENICS AND GENETICS IN THE SOVIET UNION

Eugenics in the Soviet Union of 1924 seemed perfectly compatible with Marxism. Joravsky cites a source: 'Only at the communist stage will it become possible to bring health to society as a whole. Hygiene, the study of the health of man and society will be converted into eugenics... Eugenics will place the interests of the whole society, of the collective, first, above the interests of individual persons' (David Joravsky, *The Lysenko Affair*, Cambridge, MA: Harvard University Press, 1970, p. 259). Nikolai Konstantinov Kol'tsov of the Institute of Experimental Biology founded the Russian Eugenics Society in 1920, with official approval. However, it was not long before eugenics soon showed itself to be linked to an intellectual caste system, incompatible with an ideology that cherished the lower classes, rather than damning them as dangerously fertile. Arguments for enforced sterilisation on the German or American model were unacceptable. By 1930, the movement was over. The eugenists drew in their horns and turned to the genetics of wheat or farm stock. By 1938, medical genetics too, and Kol'tsov with it, had been branded as racist and fascist; his earlier eugenic articles were compared with those of Fritz Lenz, 'a fascist thug' (Zhores A. Medvedev, *The Rise and Fall of T. D. Lysenko*, New York, NY: Columbia University Press, 1969, pp. 78–85). The domination of genetics, breeding and agronomy by Trofim Denisovich Lysenko dated from the late thirties.

G. K. Batkis, 'Evgenika', *Bol'shaia sovetskaia entsiklopediia* 23 (1931), pp. 812–819.

Translated by G. I. Kichigina as 'Eugenics'. This article defines eugenics as the 'bourgeois doctrine of the biological improvement of the race'. Kol'tsov's ideas were fascistic and eugenics was 'Menshevising idealism'. As Loren Graham has said in a famous article, the end of eugenics meant the end of studies of human heredity as a whole (Graham, 'Science and Values: The Eugenics Movement in Germany and Russia in the 1920s', *American Historical Review* (1978) 82: 1135–1164).

J. B. S. Haldane, 'Vavilov' and 'Genetics in the Soviet Union', *Science Advances*, London: Allen, 1947, pp. 210–212, 220–226.

> Haldane was Weldon Professor of Biometry, one of the successors to Karl Pearson, at University College London. He was a Marxist, who wrote frequently for the *Daily Worker*, where these short pieces were first published. He was one of the left-wing geneticists who attacked the Eugenics Society in the thirties, without, however, entirely giving up hope that, some day, genetics might be of help to humanity.

Raissa L. Berg, *Sukhovei Vospominiania Genetika* (1983), translated by David Lowe as *Acquired Traits: Memoirs of a Geneticist from the Soviet Union*, New York, NY: Viking-Penguin 1988, pp. 149–157, 158–163, 270–292, 293–318.

> Berg was a geneticist who suffered under the Stalinist regime, when Mendelian genetics or Mendelism-Morganism was proscribed as bourgeois science and its practitioners either hounded out or imprisoned.

* * *

Part 9.1 The Lysenko Controversy: Contemporary Voices

Winter wheats need a spell of cold weather to make them sprout and ripen; that could be replaced by cooling the seed artificially. Lysenko's Lamarckian claim was that a winter wheat could be changed genetically into spring wheat by growing it in a warm greenhouse; it would not then need the usual cold spell to 'vernalise' it. The environment could be assimilated into the plant and change its heredity. Plants could be conditioned to grow in any environment – even Siberia. It was a hope that was not to be realised, in spite of the political support it received under Stalin.

Lysenkoism has usually been presented as a political aberration, and obviously unscientific. However, it is striking that wheat-growing countries worldwide – Canada, for example – shared Soviet hopes that Lysenko's techniques would allow wheat to be grown in areas where it had never grown before.

Trofim Denisovich Lysenko, 'The Situation in the Science of Biology', *Soviet Biology: Report to the Lenin Academy of Agricultural Sciences*, London: Birch, 1948, pp. 5–51.

P. S. Hudson and R. H. Richens, *The New Genetics in the Soviet Union*, Cambridge: Imperial Bureau of Plant Breeding and Genetics, 1946, 'The Evidence', pp. 32–51, 'Interpretation', pp. 52–70, 'Anti-Mendelism', pp. 70–76.

> This small book gives a very good contemporary critique of Lysenkoism from the geneticists' point of view. They are clearly not rejecting it out of hand. Lysenko's hopeful claim that he could change the genetics of wheat by environmental means was taken very seriously. Environmentalism, of course, fitted neatly into an ideology that rejected class and was to reject classical genetics in its entirety.

PART 10 SOUTH AMERICA: EUGENICS, *PUÉRICULTURE* AND RACELESSNESS

In our example, Brazilians are shown to pride themselves on their lack of race-consciousness and on the mixing of peoples of European and African origin in their population. South American eugenics had a scientific linkage to France, rather than to its nearest neighbour, the US. It played down the separation of nature from nurture, and tended to be Lamarckian, rather than Mendelian. There was no suggestion that the differential fertility could pose any problem to the population, although its focus on alcoholism and the diseases of poor black Brazilians introduced an element of race. Race mixture, however, was part of the national self-image.

Nancy Stepan's study, *The Hour of Eugenics: Race, Gender and Nation in Latin America* (Ithaca, NY: Cornell University Press, 1991) is presently the best secondary source for Latin America.

Part 10.1 On *Puériculture* and Women

D. F. Ramos, 'Homiculture in Its Relations to Eugenics in Cuba', *Eugenics in Race and State: Scientific Papers of the Second International Congress of Eugenics, New York, 1921*, Baltimore, MD: Williams, 1923, pp. 432–434.

Renato Kehl, 'Eugenics Abroad: III. In Brazil', *Eugenics Review* 23 (1931–1932): 234–237.

> Kehl's life was devoted to the eugenics movement. He started the Eugenics Society of São Paulo in 1917, after studying Galton's work, and acted as its Secretary; the members were mainly doctors, as in France, with strong sanitary connections. The Society only lasted two years, but Renato Kehl went on to publish large numbers of articles and pamphlets arguing for a specifically Brazilian eugenics. By 1947, he had published 26 books on the subject, as well as many theses from the Medical School of Rio de Janeiro. He was in contact with the British society, as these publications show, but British eugenists found the Brazilian link between sanitation and eugenics an example of muddled thinking (K. E. Trounson, 'The Literature Reviewed', *Eugenics Review* (1931) 13: 236).

Part 10.2 On Race

Gilberto Freyre, 'Social Life in Brazil in the Middle of the Nineteenth Century' (1922), *The Hispano-American Review: Readings in Latin-American History*, Vol. 2, New York, NY: Crowell, 1966, pp. 215–236.

> Freyre was a student of the American anthropologist Franz Boas, and shared Boas's anti-racism. His work offered Brazilians a self-concept that included black, whites and mulattoes in a mixed-race 'luso-tropical' civilisation. He defended neo-Lamarckianism.

PART 11 EUGENICS IN CANADA: 'OUR OWN MASTER RACE'

Eugenics in Canada started later than elsewhere, but persisted well into the 1970s, as the legal documents cited here show. The Alberta Sexual Sterilization Act was passed in 1928, and repealed in 1971; the British Colombia and Manitoba Acts were passed in 1933 and repealed in 1973. Eugenic sterilisation was essentially confined to 'progressive' Anglophone Canada, since the Catholic Church in Quebec, as elsewhere, was clear that the purpose of man's life on earth was to prepare his soul for the next life; the physical body was irrelevant to this goal. The current *locus classicus* for the history of Canadian eugenics is Angus MacLaren's *Our Own Master Race: Eugenics in Canada 1885–1945*, Toronto: McLellan and Stewart, 1990.

Part 11.1 The Advocates

It is no accident that both our Canadian examples are Anglophone women. Progressive women saw in eugenics a means of promoting their concerns of motherhood and family into scientific prominence. MacMurchy was a school doctor and hygienist, concerned with the genetics of feeble-mindedness and degeneration among the school-children. Macklin was a geneticist, who, although well-known in her field, and the holder of a university appointment in anatomy at Western Ontario Medical School, was never permitted as a woman to teach genetics as long as she stayed in Canada.

Helen MacMurchy, 'The History of the Movement', 'Sterilization?' and 'Results of Sterilization', *Sterilization? Birth Control? A Book for Family Welfare and Safety*, Toronto: Macmillan, 1934, pp. 9–34, 35–46, 68–84.

Madge Thurlow Macklin, 'The Value of Medical Genetics to the Clinician', in Charles B. Davenport, Clyde E. Keeler, Maud Slye and Madge Thurlow Macklin, *Medical Genetics and Eugenics*, Philadelphia, PA: Women's Medical College, 1940, pp. 128–141.

Part 11.2 The Church

Hostility to eugenics was concentrated in Catholic, francophone Quebec. Hervé Blais was a member of the Order of Friars Minor, from the Seminaire des Saints Apôtres, Sherbrooke, Quebec. On Blais as a representative of Catholic thinking, see Sebastien Normandin, 'Eugenics, McGill and the Catholic Church in Montreal', *Canadian Bulletin of the History of Medicine/Bulletin canadien de l'histoire de la médecine* 15 (1998): 59–86. LaPierre was Professor of Paediatrics at the Université de Montréal. He was a Catholic, conservative and traditionalist who rejected progressivism and the emancipation of women, which was what he saw in eugenics.

These two French-Canadian excerpts argue a very general Catholic position. However, not all Catholics were against eugenics. Muckermann (see above) managed to reconcile the two.

Hervé Blais, 'L'eugénique devant la pensée catholique', *Les Tendances eugénistes au Canada*, Montréal, Québec: L'Institute familiale, 1942, pp. v–vii, 133–172.

Translated by P. M. H. Mazumdar as, 'The Eugenic Tendency in Canada'. See also the Encyclical, *Casti connubii* (1930) and the commentary on it by the German Catholic eugenist, Abbé Hermann Muckermann (1933), also in this collection. Blais is writing in part to answer Muckermann: his arguments have the *Imprimatur* of authority behind them.

Gaston LaPierre, 'Les campagnes internationales actuelles d'eugénisme', *Revue trimestrielle canadienne* 21 (1935): 356–372.

Translated by P. M. H. Mazumdar as 'International Eugenics Today'. This paper is interesting because it touches on a broad range of anti-progressivist Catholic writings concerning women and the family. I have tried to identify his references: they are often cited from secondary sources that have themselves cited them without identification.

Part 11.3 Law Reform

These final legal documents show the dismantling of legal support for forced sterilisation in Canada, and some of its consequences.

Law Reform Commission of Canada, 'Approaches to Sterilization', *Sterilization: Implications for Mentally Retarded and Mentally Ill Persons*, Working Paper 24 (1979): 23–73.

Institute of Law Research and Reform, Edmonton, Alberta, *Sterilization Decisions: Minors and Mentally Incompetent Adults*, Report for Discussion 6 (March 1988): 31–83.

Institute of Law Research and Reform, Edmonton, Alberta, *Competence and Human Reproduction*, Report 52 (February 1989): 31–45.

Anne Marie Owen, 'Alberta Woman Fights to Control Money, *National Post*, 7 April 1999, A9.

CONCLUSION

The repeal of the Canadian and US laws in the seventies would seem to be the end of the eugenics movement. The problematic has disintegrated: no-one now fears differential fertility on class lines, though its natural successor, population control in the third world, is still discussed. The British Eugenics Society's *Eugenics Review* was succeeded in 1968 by its *Journal of Biosocial Science*, where such papers still appear. There is still debate about race and intelligence, especially in the US. Richard Herrnstein and Charles Murray's *The Bell Curve: Intelligence and Class Structure in American Life*, New York, NY: Free Press, 1994, has reignited a critical discourse that in many ways is looking to rebut the racist ideology of American eugenics all over again.

Yet the expectations of the geneticists of the thirties may have been realised; genetics is indeed being used by the human race. Assisted reproduction has opened a door to a new, and different, means of expunging abnormalities, on an individual rather than a communitarian basis. Genetic counselling may be discreetly doing one-on-one what segregation and sterilisation could not. Counselling needs no legislation.

PART 1

The nineteenth century: degeneration

DEGENERATION

BOOK I.

FIN-DE-SIÈCLE.

CHAPTER I.

THE DUSK OF THE NATIONS.

FIN-DE-SIÈCLE is a name covering both what is characteristic of many modern phenomena, and also the underlying mood which in them finds expression. Experience has long shown that an idea usually derives its designation from the language of the nation which first formed it. This, indeed, is a law of constant application when historians of manners and customs inquire into language, for the purpose of obtaining some notion, through the origins of some verbal root, respecting the home of the earliest inventions and the line of evolution in different human races. *Fin-de-siècle* is French, for it was in France that the mental state so entitled was first consciously realized. The word has flown from one hemisphere to the other, and found its way into all civilized languages. A proof this that the need of it existed. The *fin-de-siècle* state of mind is to-day everywhere to be met with; nevertheless, it is in many cases a mere imitation of a foreign fashion gaining vogue, and not an organic evolution. It is in the land of its birth that it appears in its most genuine form, and Paris is the right place in which to observe its manifold expressions.

No proof is needed of the extreme silliness of the term. Only the brain of a child or of a savage could form the clumsy idea that the century is a kind of living being, born like a beast or a man, passing through all the stages of existence, gradually ageing and declining after blooming childhood, joyous youth, and vigorous maturity, to die with the expiration of the hundredth year, after being afflicted in its last decade with all the infirmities of mournful senility. Such a childish anthropomorphism or zoomorphism never stops to consider that the

I

arbitrary division of time, rolling ever continuously along, is not identical amongst all civilized beings, and that while this nineteenth century of Christendom is held to be a creature reeling to its death presumptively in dire exhaustion, the fourteenth century of the Mahommedan world is tripping along in the baby-shoes of its first decade, and the fiftieth century of the Jews strides gallantly by in the full maturity of its fifty-second year. Every day on our globe 130,000 human beings are born, for whom the world begins with this same day, and the young citizen of the world is neither feebler nor fresher for leaping into life in the midst of the death-throes of 1900, nor on the birthday of the twentieth century. But it is a habit of the human mind to project externally its own subjective states. And it is in accordance with this naïvely egoistic tendency that the French ascribe their own senility to the century, and speak of *fin-de-siècle* when they ought correctly to say *fin-de-race*.*

But however silly a term *fin-de-siècle* may be, the mental constitution which it indicates is actually present in influential circles. The disposition of the times is curiously confused, a compound of feverish restlessness and blunted discouragement, of fearful presage and hang-dog renunciation. The prevalent feeling is that of imminent perdition and extinction. *Fin-de-siècle* is at once a confession and a complaint. The old Northern faith contained the fearsome doctrine of the Dusk of the Gods. In our days there have arisen in more highly-developed minds vague qualms of a Dusk of the Nations, in which all suns and all stars are gradually waning, and mankind with all its institutions and creations is perishing in the midst of a dying world.

It is not for the first time in the course of history that the horror of world-annihilation has laid hold of men's minds. A similar sentiment took possession of the Christian peoples at the approach of the year 1000. But there is an essential difference between chiliastic panic and *fin-de-siècle* excitement. The despair at the turn of the first millennium of Christian chronology proceeded from a feeling of fulness of life and joy of life. Men were aware of throbbing pulses, they were conscious of unweakened capacity for enjoyment, and found it unmitigatedly appalling to perish together with the world, when

* This passage has been misunderstood. It has been taken to mean that all the French nation had degenerated, and their race was approaching its end. However, from the concluding paragraph of this chapter, it may be clearly seen that I had in my eye only the upper ten thousand. The peasant population, and a part of the working classes and the *bourgeoisie*, are sound. I assert only the decay of the rich inhabitants of great cities and the leading classes. It is they who have discovered *fin-de-siècle*, and it is to them also that *fin-de-race* applies.

there were yet so many flagons to drain and so many lips to kiss, and when they could yet rejoice so vigorously in both love and wine. Of all this in the *fin-de-siècle* feeling there is nothing. Neither has it anything in common with the impressive twilight-melancholy of an aged Faust, surveying the work of a lifetime, and who, proud of what has been achieved, and contemplating what is begun but not completed, is seized with vehement desire to finish his work, and, awakened from sleep by haunting unrest, leaps up with the cry: 'Was ich gedacht, ich eil' es zu vollbringen.'*

Quite otherwise is the *fin-de-siècle* mood. It is the impotent despair of a sick man, who feels himself dying by inches in the midst of an eternally living nature blooming insolently for ever. It is the envy of a rich, hoary voluptuary, who sees a pair of young lovers making for a sequestered forest nook; it is the mortification of the exhausted and impotent refugee from a Florentine plague, seeking in an enchanted garden the experiences of a Decamerone, but striving in vain to snatch one more pleasure of sense from the uncertain hour. The reader of Turgenieff's *A Nest of Nobles* will remember the end of that beautiful work. The hero, Lavretzky, comes as a man advanced in years to visit at the house where, in his young days, he had lived his romance of love. All is unchanged. The garden is fragrant with flowers. In the great trees the happy birds are chirping; on the fresh turf the children romp and shout. Lavretzky alone has grown old, and contemplates, in mournful exclusion, a scene where nature holds on its joyous way, caring nought that Lisa the beloved is vanished, and Lavretzky, a broken-down man, weary of life. Lavretzky's admission that, amidst all this ever-young, ever-blooming nature, for him alone there comes no morrow; Alving's dying cry for 'The sun—the sun!' in Ibsen's *Ghosts*—these express rightly the *fin-de-siècle* attitude of to-day.

This fashionable term has the necessary vagueness which fits it to convey all the half-conscious and indistinct drift of current ideas. Just as the words 'freedom,' 'ideal,' 'progress' seem to express notions, but actually are only sounds, so in itself *fin-de-siècle* means nothing, and receives a varying signification according to the diverse mental horizons of those who use it.

The surest way of knowing what *fin-de-siècle* implies, is to consider a series of particular instances where the word has been applied. Those which I shall adduce are drawn from French books and periodicals of the last two years.†

* 'My thought I hasten to fulfil.'

† A four-act comedy, by H. Micard and F. de Jouvenot, named *Fin-de-Siècle*, which was played in Paris in 1890, hardly avails to determine the

A king abdicates, leaves his country, and takes up his residence in Paris, having reserved certain political rights. One day he loses much money at play, and is in a dilemma. He therefore makes an agreement with the Government of his country, by which, on receipt of a million francs, he renounces for ever every title, official position and privilege remaining to him. *Fin-de-siècle* king.

A bishop is prosecuted for insulting the minister of public worship. The proceedings terminated, his attendant canons distribute amongst the reporters in court a defence, copies of which he has prepared beforehand. When condemned to pay a fine, he gets up a public collection, which brings in tenfold the amount of the penalty. He publishes a justificatory volume containing all the expressions of support which have reached him. He makes a tour through the country, exhibits himself in every cathedral to the mob curious to see the celebrity of the hour, and takes the opportunity of sending round the plate. *Fin-de-siècle* bishop.

The corpse of the murderer Pranzini after execution underwent autopsy. The head of the secret police cuts off a large piece of skin, has it tanned, and the leather made into cigar-cases and card-cases for himself and some of his friends. *Fin-de-siècle* official.

An American weds his bride in a gas-factory, then gets with her into a balloon held in readiness, and enters on a honeymoon in the clouds. *Fin-de-siècle* wedding.

An *attaché* of the Chinese Embassy publishes high-class works in French under his own name. He negotiates with banks respecting a large loan for his Government, and draws large advances for himself on the unfinished contract. Later it comes out that the books were composed by his French secretary, and that he has swindled the banks. *Fin-de-siècle* diplomatist.

A public schoolboy walking with a chum passes the gaol where his father, a rich banker, has repeatedly been imprisoned for fraudulent bankruptcy, embezzlement and similar lucrative misdemeanours. Pointing to the building, he tells his friend with a smile: 'Look, that's the governor's school.' *Fin-de-siècle* son.

Two young ladies of good family, and school friends, are chatting together. One heaves a sigh. 'What's the matter?' asks the other. 'I'm in love with Raoul, and he with me.' 'Oh, that's lovely! He's handsome, young, elegant: and yet you're sad?' 'Yes, but he has nothing, and is nothing, and

sense of the word as the French use it. The authors were concerned, not to depict a phase of the age or a psychological state, but only to give an attractive title to their piece.

my parents want me to marry the baron, who is fat, bald, and ugly, but has a huge lot of money.' 'Well, marry the baron without any fuss, and make Raoul acquainted with him, you goose.' *Fin-de-siècle* girls.

Such test-cases show how the word is understood in the land of its birth. Germans who ape Paris fashions, and apply *fin-de-siècle* almost exclusively to mean what is indecent and improper, misuse the word in their coarse ignorance as much as, in a previous generation, they vulgarized the expression *demi-monde*, misunderstanding its proper meaning, and giving it the sense of *fille de joie*, whereas its creator Dumas intended it to denote persons whose lives contained some dark period, for which they were excluded from the circle to which they belong by birth, education, or profession, but who do not by their manner betray, at least to the inexperienced, that they are no longer acknowledged as members of their own caste.

Prima facie, a king who sells his sovereign rights for a big cheque seems to have little in common with a newly-wedded pair who make their wedding-trip in a balloon, nor is the connection at once obvious between an episcopal Barnum and a well-brought-up young lady who advises her friend to a wealthy marriage mitigated by a *cicisbeo*. All these *fin-de-siècle* cases have, nevertheless, a common feature, to wit, a contempt for traditional views of custom and morality.

Such is the notion underlying the word *fin-de-siècle*. It means a practical emancipation from traditional discipline, which theoretically is still in force. To the voluptuary this means unbridled lewdness, the unchaining of the beast in man; to the withered heart of the egoist, disdain of all consideration for his fellow-men, the trampling under foot of all barriers which enclose brutal greed of lucre and lust of pleasure; to the contemner of the world it means the shameless ascendency of base impulses and motives, which were, if not virtuously suppressed, at least hypocritically hidden; to the believer it means the repudiation of dogma, the negation of a super-sensuous world, the descent into flat phenomenalism; to the sensitive nature yearning for æsthetic thrills, it means the vanishing of ideals in art, and no more power in its accepted forms to arouse emotion. And to all, it means the end of an established order, which for thousands of years has satisfied logic, fettered depravity, and in every art matured something of beauty.

One epoch of history is unmistakably in its decline, and another is announcing its approach. There is a sound of rending in every tradition, and it is as though the morrow would not link itself with to-day. Things as they are totter and plunge, and they are suffered to reel and fall, because man is weary, and there is no faith that it is worth an effort to uphold them. Views that have hitherto governed minds are dead or driven hence like disenthroned kings, and for their inheritance they that hold the titles and they that would usurp are locked in struggle. Meanwhile interregnum in all its terrors prevails; there is confusion among the powers that be; the million, robbed of its leaders, knows not where to turn; the strong work their will; false prophets arise, and dominion is divided amongst those whose rod is the heavier because their time is short. Men look with longing for whatever new things are at hand, without presage whence they will come or what they will be. They have hope that in the chaos of thought, art may yield revelations of the order that is to follow on this tangled web. The poet, the musician, is to announce, or divine, or at least suggest in what forms civilization will further be evolved. What shall be considered good to-morrow—what shall be beautiful? What shall we know to-morrow—what believe in? What shall inspire us? How shall we enjoy? So rings the question from the thousand voices of the people, and where a market-vendor sets up his booth and claims to give an answer, where a fool or a knave suddenly begins to prophesy in verse or prose, in sound or colour, or professes to practise his art otherwise than his predecessors and competitors, there gathers a great concourse, crowding around him to seek in what he has wrought, as in oracles of the Pythia, some meaning to be divined and interpreted. And the more vague and insignificant they are, the more they seem to convey of the future to the poor gaping souls gasping for revelations, and the more greedily and passionately are they expounded.

Such is the spectacle presented by the doings of men in the reddened light of the Dusk of the Nations. Massed in the sky the clouds are aflame in the weirdly beautiful glow which was observed for the space of years after the eruption of Krakatoa. Over the earth the shadows creep with deepening gloom, wrapping all objects in a mysterious dimness, in which all certainty is destroyed and any guess seems plausible. Forms lose their outlines, and are dissolved in floating mist. The day is over, the night draws on. The old anxiously watch its approach, fearing they will not live to see the end. A few amongst the young and strong are conscious of the vigour of life in all their veins and nerves, and rejoice in the coming sunrise. Dreams, which fill up the hours of darkness till the breaking of the new day, bring to the former comfortless memories, to the latter high-souled hopes. And in the artistic products of the age we see the form in which these dreams become sensible.

Here is the place to forestall a possible misunderstanding.

The great majority of the middle and lower classes is naturally not *fin-de-siècle*. It is true that the spirit of the times is stirring the nations down to their lowest depths, and awaking even in the most inchoate and rudimentary human being a wondrous feeling of stir and upheaval. But this more or less slight touch of moral sea-sickness does not excite in him the cravings of travailing women, nor express itself in new æsthetic needs. The Philistine or the Proletarian still finds undiluted satisfaction in the old and oldest forms of art and poetry, if he knows himself unwatched by the scornful eye of the votary of fashion, and is free to yield to his own inclinations. He prefers Ohnet's novels to all the symbolists, and Mascagni's *Cavalleria Rusticana* to all Wagnerians and to Wagner himself; he enjoys himself royally over slap-dash farces and music-hall melodies, and yawns or is angered at Ibsen; he contemplates gladly chromos of paintings depicting Munich beer-houses and rustic taverns, and passes the open-air painters without a glance. It is only a very small minority who honestly find pleasure in the new tendencies, and announce them with genuine conviction as that which alone is sound, a sure guide for the future, a pledge of pleasure and of moral benefit. But this minority has the gift of covering the whole visible surface of society, as a little oil extends over a large area of the surface of the sea. It consists chiefly of rich educated people, or of fanatics. The former give the *ton* to all the snobs, the fools, and the blockheads; the latter make an impression upon the weak and dependent, and intimidate the nervous. All snobs affect to have the same taste as the select and exclusive minority, who pass by everything that once was considered beautiful with an air of the greatest contempt. And thus it appears as if the whole of civilized humanity were converted to the æsthetics of the Dusk of the Nations.

CHAPTER II.

THE SYMPTOMS.

LET us follow in the train frequenting the palaces of European capitals, the highways of fashionable watering-places, the receptions of the rich, and observe the figures of which it is composed.

Amongst the women, one wears her hair combed smoothly back and down like Rafael's Maddalena Doni in the Uffizi at Florence; another wears it drawn up high over the temples like Julia, daughter of Titus, or Plotina, wife of Trajan, in the busts in the Louvre; a third has hers cut short in front on the brow and long in the nape, waved and lightly puffed, after the fashion of the fifteenth century, as may be seen in the pages and young knights of Gentile Bellini, Botticelli and Mantegna. Many have their hair dyed, and in such a fashion as to be startling in its revolt against the law of organic harmony, and the effect of a studied discord, only to be resolved into the higher polyphony of the toilet taken as a whole. This swarthy, dark-eyed woman snaps her fingers at nature by framing the brown tones of her face in copper-red or golden-yellow; yonder blue-eyed fair, with a complexion of milk and roses, intensifies the brightness of her cheeks by a setting of artificially blue-black tresses. Here is one who covers her head with a huge heavy felt hat, an obvious imitation, in its brim turned up at the back, and its trimming of large plush balls, of the sombrero of the Spanish bull-fighters, who were displaying their skill in Paris at the exhibition of 1889, and giving all kinds of *motifs* to modistes. There is another who has stuck on her hair the emerald-green or ruby-red biretta of the mediæval travelling student. The costume is in keeping with the bizarre coiffure. Here is a mantle reaching to the waist, slit up on one side, draping the breast like a *portière,* and trimmed round the hem with little silken bells, by the incessant clicking of which a sensitive spectator would in a very short time either be hypnotized or driven to take frantic fright. There is a Greek peplos, of which the tailors speak as glibly as any venerable philologist. Next to the stiff monumental trim of Catharine de Medicis, and the high ruff of Mary, Queen of Scots, goes the flowing white raiment of the angel of the Annunciation in Memling's pictures, and, by way of antithesis, that caricature of masculine array, the fitting cloth coat, with widely opened lapels, waistcoat, stiffened shirt-front, small stand-up collar, and necktie. The majority, anxious to be inconspicuous in unimaginative mediocrity, seems to have for its leading style a laboured rococo, with bewildering oblique lines, incomprehensible swellings, puffings, expansions and contractions, folds with irrational beginning and aimless ending, in which all the outlines of the human figure are lost, and which cause women's bodies to resemble now a beast of the Apocalypse, now an armchair, now a triptych, or some other ornament.

The children, strolling beside their mothers thus bedecked, are embodiments of one of the most afflicting aberrations into which the imagination of a spinster ever lapsed. They are living copies of the pictures of Kate Greenaway, whose love of children, diverted from its natural outlet, has sought gratification in the most affected style of drawing, wherein the sacredness of childhood is profaned under absurd disguises. Here is an imp dressed from head to foot in the blood-red costume of a mediæval executioner; there a four-year-old girl wears a cabriolet bonnet of her great-grandmother's days and sweeps after her a court mantle of loud-hued velvet. Another wee dot, just able to keep on her tottering legs, has been arrayed in the long dress of a lady of the First Empire, with puffed sleeves and short waist.

The men complete the picture. They are preserved from excessive oddity through fear of the Philistine's laugh, or through some remains of sanity in taste, and, with the exception of the red dress-coat with metal buttons, and knee-breeches with silk stockings, with which some idiots in eye-glass and gardenia try to rival burlesque actors, present little deviation from the ruling canon of the masculine attire of the day. But fancy plays the more freely among their hair. One displays the short curls and the wavy double-pointed beard of Lucius Verus, another looks like the whiskered cat in a Japanese kakemono. His neighbour has the *barbiche* of Henri IV., another the fierce moustache of a lansquenet by F. Brun, or the chin-tuft of the city-watch in Rembrandt's 'Ronde de Nuit.'

The common feature in all these male specimens is that they do not express their real idiosyncrasies, but try to present something that they are not. They are not content to show their natural figure, nor even to supplement it by legitimate accessories, in harmony with the type to which they approximate, but they seek to model themselves after some artistic pattern which has no affinity with their own nature, or is even antithetical to it. Nor do they for the most part limit themselves to one pattern, but copy several at once, which jar one with another. Thus we get heads set on shoulders not belonging to them, costumes the elements of which are as disconnected as though they belonged to a dream, colours that seem to have been matched in the dark. The impression is that of a masked festival, where all are in disguises, and with heads too in character. There are several occasions, such as the varnishing day at the Paris Champs de Mars salon, or the opening of the Exhibition of the Royal Academy in London, where this impression is so weirdly intensified, that one seems to be moving amongst dummies patched together at haphazard, in a mythical mortuary, from fragments of bodies, heads, trunks, limbs, just as they came to hand, and which the designer, in heedless pell-mell, clothed at random in the garments of all epochs and countries. Every single figure strives visibly by some singularity in outline, set, cut, or colour, to startle attention violently, and imperiously to detain it. Each one wishes to create a strong nervous excitement, no matter whether agreeably or disagreeably. The fixed idea is to produce an effect at any price.

Let us follow these folk in masquerade and with heads in character to their dwellings. Here are at once stage properties and lumber-rooms, rag-shops and museums. The study of the master of the house is a Gothic hall of chivalry, with cuirasses, shields and crusading banners on the walls; or the shop of an Oriental bazaar with Kurd carpets, Bedouin chests, Circassian narghilehs and Indian lacquered caskets. By the mirror on the mantelpiece are fierce or funny Japanese masks. Between the windows are staring trophies of swords, daggers, clubs and old wheel-trigger pistols. Daylight filters in through painted glass, where lean saints kneel in rapture. In the drawing-room the walls are either hung with worm-eaten Gobelin tapestry, discoloured by the sun of two centuries (or it may be by a deftly mixed chemical bath), or covered with Morris draperies, on which strange birds flit amongst crazily ramping branches, and blowzy flowers coquet with vain butterflies. Amongst armchairs and padded seats, such as the cockered bodies of our contemporaries know and expect, there are Renaissance stools, the heart or shell-shaped bottoms of which would attract none but the toughened hide of a rough hero of the jousting lists. Startling is the effect of a gilt-painted couch between buhl-work cabinets and a puckered Chinese table, next an inlaid writing-table of graceful rococo. On all the tables and in all the cabinets is a display of antiquities or articles of vertù, big or small, and for the most part warranted not genuine; a figure of Tanagra near a broken jade snuff-box, a Limoges plate beside a long-necked Persian waterpot of brass, a *bonbonnière* between a breviary bound in carved ivory, and snuffers of chiselled copper. Pictures stand on easels draped with velvet, the frames made conspicuous by some oddity, such as a spider in her web, a metal bunch of thistle-heads, and the like. In a corner a sort of temple is erected to a squatting or a standing Buddha. The boudoir of the mistress of the house partakes of the nature of a chapel and of a harem. The toilet-table is designed and decorated like an altar, a *prie-Dieu* is a pledge for the piety of the inmate, and a broad divan, with an orgiastic *abandon* about the cushions, gives reassurance that things are not so bad. In the dining-room the walls are hung with the whole stock-in-trade of a porcelain shop, costly silver is displayed in an old farmhouse dresser, and on the table bloom aristocratic orchids, and proud silver vessels shine between rustic stone-ware plates and ewers. In the evening, lamps of the stature of a man illumine these rooms with light both subdued and tinted by sprawling shades, red, yellow or green of hue, and even covered by black lace. Hence the inmates appear, now bathed in variegated diaphanous mist, now suffused with coloured radiance, while the corners and backgrounds are shrouded in depths of artfully-effected *clair-obscur*, and the furniture and bric-à-brac are dyed in unreal chords of colour. Unreal, too, are the studied postures, by assuming which the inmates are enabled to reproduce on their faces the light effects of Rembrandt or Schalcken. Everything in these houses aims at exciting the nerves and dazzling the senses. The disconnected and antithetical effects in all arrangements, the constant contradiction between form and purpose, the outlandishness of most objects, is intended to be bewildering. There must be no sentiment of repose, such as is felt at any composition, the plan of which is easily taken in, nor of the comfort attending a prompt comprehension of all the details of one's environment. He who enters here must not doze, but be thrilled. If the master of the house roams about these rooms clothed after the example of Balzac in a white monk's cowl, or after the model of Richepin in the red cloak of the robber-chieftain of an operetta, he only gives expression to the admission that in such a comedy theatre a clown is in place. All is discrepant, indiscriminate jumble. The unity of abiding by one definite historic style counts as old-fashioned, provincial, Philistine, and the time has not yet produced a style of its own. An approach is, perhaps, made to one in the furniture of Carabin, exhibited in the salon of the Champs de Mars. But these balusters, down which naked furies and possessed creatures are rolling in mad riot, these bookcases, where base and pilaster consist of a pile of guillotined heads, and even this table, representing a gigantic open book borne by gnomes, make up a style that is feverish and infernal. If the director-general of Dante's 'Inferno' had an audience-chamber, it might well be furnished with such as these. Carabin's creations may be intended to equip a house, but they are a nightmare.

We have seen how society dresses and where it dwells. We shall now observe how it enjoys itself, and where it seeks stimulation and distraction. In the art exhibition it crowds, with proper little cries of admiration, round Besnard's women, with their grass-green hair, faces of sulphur-yellow or fiery-red, and arms spotted in violet and pink, dressed in a shining blue cloud resembling faintly a sort of nightdress; that is to say, it has a fondness for bold, revolutionary debauch of colour. But not exclusively so. Next to Besnard it worships with equal or greater rapture the works of Puvis de Chavannes, wan, and as though blotted out with a half-transparent wash of lime; or those of Carriere, suffused in a problematical vapour, reeking as if with a cloud of incense; or those of Roll, shimmering in a soft and silvery sheen. The purple of the Manet school, steeping the whole visible creation in bluish glamour, the half-tones, or, rather, phantom-colours of the 'Archaists,' that seem to have risen, faded and nebulous, out of some primeval tomb, and all these palettes of 'dead leaves,' 'old ivory,' evaporating yellows, smothered purple, attract on the whole more rapturous glances than the voluptuous 'orchestration' of the Besnard section. The subject of the picture leaves these select gazers apparently indifferent; it is only seamstresses and country-folk, the grateful *clientèle* of the chromo, who linger over the 'story.' And yet these as they pass stop by preference before Henry Martin's 'Every Man and his Chimæra,' in which bloated figures, in an atmosphere of yellow broth, are doing incomprehensible things that need profound explanation; or before Jean Béraud's 'Christ and the Adulteress,' where, in a Parisian dining-room, in the midst of a company in dress-coats, and before a woman in ball-dress, a Christ robed in correct Oriental gear, and with an orthodox halo, acts a scene out of the Gospel; or before Raffaelli's topers and cut-throats of the purlieus of Paris, drawn in high relief, but painted with ditch-water and dissolved clay. Steering in the wake of 'society' through a picture-gallery, one will be unalterably convinced that they turn up their eyes and fold their hands before pictures at which the commoner sort burst out laughing or pull the grimace of a man who believes he is made a fool of; and that they shrug their shoulders and hasten with scornful exchange of looks past such as the latter pause at in grateful enjoyment.

At opera and concert the rounded forms of ancient melody are coldly listened to. The translucent thematic treatment of classic masters, their conscientious observance of the laws of counterpoint, are reckoned flat and tedious. A coda graceful in cadence, serene in its 'dying fall,' a pedal-base with correct harmonization, provoke yawns. Applause and wreaths are reserved for Wagner's *Tristan and Isolde*, and especially the mystic *Parsifal*, for the religious music in Bruneau's *Dream*, or the symphonies of César Franck. Music in order to please must either counterfeit religious devotion, or agitate the mind by its form. The musical listener is accustomed involuntarily to develop a little in his mind every motive occurring in a piece. The mode in which the composer carries out his *motif* is bound, accordingly, to differ entirely from this anticipated development. It must not admit of being guessed. A dissonant interval must appear where a consonant interval was expected; if the hearer is hoping that a phrase in what is an obvious final cadence will be spun out to its natural end, it must be sharply interrupted in the middle of a bar. Keys and pitch must change suddenly. In the orchestra a vigorous polyphony must summon the attention in several directions at once; particular instruments, or groups of instruments, must address the listener simultaneously without heeding each other, till he gets as nervously excited as the man who vainly endeavours to understand what is being said in the jangle of a dozen voices. The theme, even if in the first instance it has a distinct outline, must become ever more indefinite, ever more dissolving into a mist, in which the imagination can see any forms it likes, as in driving clouds of night. The tide of sound must flow on without any perceptible limit or goal, surging up and down in endless chromatic passages of triplets. If now and then it delude the listener, borne along by it, and straining his eyes to see land with glimpses of a distant shore, this is soon discovered to be a fleeting mirage. The music must continually promise, but never perform; must seem about to tell some great secret, and grow dumb or break away ere to throbbing hearts it tells the word they wait for. The audience go to their concert-room in quest of Tantalus moods, and leave it with all the nervous exhaustion of a young pair of lovers, who for hours at the nightly tryst have sought to exchange caresses through a closely-barred window.

The books in which the public here depicted finds its delight or edification diffuse a curious perfume yielding distinguishable odours of incense, eau de Lubin and refuse, one or the other preponderating alternately. Mere sewage exhalations are played out. The filth of Zola's art and of his disciples in literary canal-dredging has been got over, and nothing remains for it but to turn to submerged peoples and social strata. The vanguard of civilization holds its nose at the pit of undiluted naturalism, and can only be brought to bend over it with sympathy and curiosity when, by cunning engineering, a drain from the boudoir and the sacristy has been turned into it. Mere sensuality passes as commonplace, and only finds admission when disguised as something unnatural and degenerate. Books treating of the relations between the sexes, with no matter how little reserve, seem too dully moral. Elegant titillation only begins where normal sexual relations leave off. Priapus has become a symbol of virtue. Vice looks to Sodom and Lesbos, to Bluebeard's castle and the servants' hall of the 'divine' Marquis de Sade's *Justine*, for its embodiments.

The book that would be fashionable must, above all, be obscure. The intelligible is cheap goods for the million only. It must further discourse in a certain pulpit tone—mildly unctuous, not too insistent; and it must follow up risky scenes by tearful outpourings of love for the lowly and the suffering, or glowing transports of piety. Ghost-stories are very popular, but they must come on in scientific disguise, as hypnotism, telepathy, somnambulism. So are marionette-plays, in which seemingly naïve but knowing rogues make used-up old ballad dummies babble like babies or idiots. So are esoteric novels, in which the author hints that he could say a deal about magic, kabbala, fakirism, astrology and other white and black arts if he only chose. Readers intoxicate themselves in the hazy word-sequences of symbolic poetry. Ibsen dethrones Goethe; Maeterlinck ranks with Shakespeare; Nietzsche is pronounced by German and even by French critics to be the leading German writer of the day; the *Kreutzer Sonata* is the Bible of ladies, who are amateurs in love, but bereft of lovers; dainty gentlemen find the street ballads and gaol-bird songs of Jules Jouy, Bruant, MacNab and Xanroff very *distingué* on account of 'the warm sympathy pulsing in them,' as the stock phrase runs; and society persons, whose creed is limited to baccarat and the money market, make pilgrimages to the Oberammergau Passion-play, and wipe away a tear over Paul Verlaine's invocations to the Virgin.

But art exhibitions, concerts, plays and books, however extraordinary, do not suffice for the æsthetic needs of elegant society. Novel sensations alone can satisfy it. It demands more intense stimulus, and hopes for it in spectacles, where different arts strive in new combinations to affect all the senses at once. Poets and artists strain every nerve incessantly to satisfy this craving. A painter, who for that matter is less occupied with new impressions than with old puffs, paints a picture indifferently well of the dying Mozart working at his *Requiem*, and exhibits it of an evening in a darkened room, while a dazzling ray of skilfully directed electric light falls on the painting, and an invisible orchestra softly plays the *Requiem*. A musician goes one step further. Developing to the utmost a Bayreuth usage, he arranges a concert in a totally darkened hall, and thus delights those of the audience who find opportunity, by happily chosen juxtapositions, to augment their musical sensations by hidden enjoyment of another sort. Haraucourt, the poet, has his paraphrase of the Gospel, written in spirited verse, recited on the stage by Sarah Bernhardt, while, as in the old-fashioned melodrama, soft music in unending melody accompanies the actress. Even the nose, hitherto basely ignored by the fine arts, attracts the pioneers, and is by them invited to take part in æsthetic delights. A hose is set up in the theatre, by which the spectators are sprayed with perfumes. On the stage a poem in approximately dramatic form is recited. In every division, act, scene, or however the thing is called, a different vowel-sound is made to preponderate; during each the theatre is illuminated with a differently tinted light, the orchestra discourses music in a different key, and the jet gives out a different perfume. This idea of accompanying verses with odours was thrown out years ago, half in jest, by Ernest Eckstein. Paris has carried it out in sacred earnest. The new school fetch the puppet theatre out of the nursery, and enact pieces for adults which, with artificial simplicity, pretend to hide or reveal a profound meaning, and with great talent and ingenuity execute a magic-lantern of prettily drawn and painted figures moving across surprisingly luminous backgrounds; and these living pictures make visible the process of thought in the mind of the author who recites his accompanying poem, while a piano endeavours to illustrate the leading emotion. And to enjoy such exhibitions as these society crowds into a suburban circus, the loft of a back tenement, a second-hand costumier's shop, or a fantastic artist's restaurant, where the performances, in some room consecrated to beery potations, bring together the greasy *habitué* and the dainty aristocratic fledgling.

CHAPTER III.

DIAGNOSIS.

THE manifestations described in the preceding chapter must be patent enough to everyone, be he never so narrow a Philistine. The Philistine, however, regards them as a passing fashion and nothing more; for him the current terms, caprice, eccentricity, affectation of novelty, imitation, instinct, afford a sufficient explanation. The purely literary mind, whose merely æsthetic culture does not enable him to understand the connections of things, and to seize their real meaning, deceives himself and others as to his ignorance by means of sounding phrases, and loftily talks of a 'restless quest of a new ideal by the modern spirit,' 'the richer vibrations of the refined nervous system of the present day,' 'the unknown sensations of an elect mind.' But the physician, especially if he have devoted himself to the special study of nervous and mental maladies, recognises at a glance, in the *fin-de-siècle* disposition, in the tendencies of contemporary art and poetry, in the life and conduct of the men who write mystic, symbolic and 'decadent' works, and the attitude taken by their admirers in the tastes and æsthetic instincts of fashionable society, the confluence of two well-defined conditions of disease, with which he is quite familiar, viz. degeneration (degeneracy) and hysteria, of which the minor stages are designated as neurasthenia. These two conditions of the organism differ from each other, yet have many features in common, and frequently occur together; so that it is easier to observe them in their composite forms, than each in isolation.

The conception of degeneracy, which, at this time, obtains throughout the science of mental disease, was first clearly grasped and formulated by Morel. In his principal work—often quoted, but, unfortunately, not sufficiently read*—the following definition of what he wishes to be understood by 'degeneracy' is given by this distinguished expert in mental pathology, who was, for a short time, famous in Germany, even outside professional circles.†

'The clearest notion we can form of degeneracy is to regard it as *a morbid deviation from an original type.* This deviation, even if, at the outset, it was ever so slight, contained transmissible elements of such a nature that anyone bearing in him the germs becomes more and more incapable of fulfilling his functions in the world; and mental progress, already checked in his own person, finds itself menaced also in his descendants.'

When under any kind of noxious influences an organism becomes debilitated, its successors will not resemble the healthy, normal type of the species, with capacities for development, but will form a new sub-species, which, like all others, possesses the capacity of transmitting to its offspring, in a continuously increasing degree, its peculiarities, these being morbid deviations from the normal form—gaps in development, malformations and infirmities. That which distinguishes degeneracy from the formation of new species (phylogeny) is, that the morbid variation does not continuously subsist and propagate itself, like one that is healthy, but, fortunately, is soon rendered sterile, and after a few generations often dies out before it reaches the lowest grade of organic degradation.‡

Degeneracy betrays itself among men in certain physical

* *Traité des Dégénérescences physiques, intellectuelles et morales de l'Espèce humaine et des Causes qui produisent ces Variétés maladives.* Par le Dr. B. A. Morel. Paris, 1857, p. 5.

† At the instigation of his mistress Ebergenyi, Count Chorinsky had poisoned his wife, previously an actress. The murderer was an epileptic, and a 'degenerate,' in the Morelian sense. His family summoned Morel from Normandy to Munich, for the purpose of proving to the jury, before whom the case (1868) was tried, that the accused was irresponsible. The latter was singularly indignant at this; and the Attorney-General also contradicted, in the most emphatic manner, the evidence of the French alienist, and supported himself by the approbation of the most prominent alienists in Munich. Chorinsky was pronounced guilty. Nevertheless, only a short time after his conviction, insanity developed itself in him, and a few months later he died, in the deepest mental darkness, thus justifying all the previous assertions of the French physician, who had, in the German tongue, demonstrated to a German jury the incompetence of his professional confrères in Munich.

‡ Morel, *op. cit.*, p. 683.

characteristics, which are denominated 'stigmata,' or brand-marks—an unfortunate term derived from a false idea, as if degeneracy were necessarily the consequence of a fault, and the indication of it a punishment. Such stigmata consist of deformities, multiple and stunted growths in the first line of asymmetry, the unequal development of the two halves of the face and cranium; then imperfection in the development of the external ear, which is conspicuous for its enormous size, or protrudes from the head, like a handle, and the lobe of which is either lacking or adhering to the head, and the helix of which is not involuted; further, squint-eyes, hare-lips, irregularities in the form and position of the teeth; pointed or flat palates, webbed or supernumerary fingers (syn- and poly-dactylia), etc. In the book from which I have quoted, Morel gives a list of the anatomical phenomena of degeneracy, which later observers have largely extended. In particular, Lombroso* has conspicuously broadened our knowledge of stigmata, but he apportions them merely to his 'born criminals'—a limitation which from the very scientific standpoint of Lombroso himself cannot be justified, his 'born criminals' being nothing but a subdivision of degenerates. Féré† expresses this very emphatically when he says, 'Vice, crime and madness are only distinguished from each other by social prejudices.'

There might be a sure means of proving that the application of the term 'degenerates' to the originators of all the *fin-de-siècle* movements in art and literature is not arbitrary, that it is no baseless conceit, but a fact; and that would be a careful physical examination of the persons concerned, and an inquiry into their pedigree. In almost all cases, relatives would be met with who were undoubtedly degenerate, and one or more stigmata discovered which would indisputably establish the diagnosis of 'Degeneration.' Of course, from human consideration, the result of such an inquiry could often not be made public; and he alone would be convinced who should be able to undertake it himself.

Science, however, has found, together with these physical stigmata, others of a mental order, which betoken degeneracy quite as clearly as the former; and they allow of an easy demonstration from all the vital manifestations, and, in particular, from all the works of degenerates, so that it is not necessary to measure the cranium of an author, or to see the lobe of a painter's ear, in order to recognise the fact that he belongs to the class of degenerates.

* *L'Uomo delinquente in rapporto all' Antropologia, Giurisprudenza e alle Discipline carcerarie.* 3a edizione. Torino, 1884, p. 147 *et seq.* See also Dr. Ch. Fére, 'La Famille nevropathique.' Paris, 1894, pp. 176-212.

† 'La Famille nevropathique,' *Archives de Nevrologie*, 1884, *Nos.* 19 *et* 20.

Quite a number of different designations have been found for these persons. Maudsley and Ball call them 'Borderland dwellers'—that is to say, dwellers on the borderland between reason and pronounced madness. Magnan gives to them the name of 'higher degenerates' (*dégénérés supérieurs*), and Lombroso speaks of 'mattoids' (from *matto*, the Italian for insane), and 'graphomaniacs,' under which he classifies those semi-insane persons who feel a strong impulse to write. In spite, however, of this variety of nomenclature, it is a question simply of one single species of individuals, who betray their fellowship by the similarity of their mental physiognomy.

In the mental development of degenerates, we meet with the same irregularity that we have observed in their physical growth. The asymmetry of face and cranium finds, as it were, its counterpart in their mental faculties. Some of the latter are completely stunted, others morbidly exaggerated. That which nearly all degenerates lack is the sense of morality and of right and wrong. For them there exists no law, no decency, no modesty. In order to satisfy any momentary impulse, or inclination, or caprice, they commit crimes and trespasses with the greatest calmness and self-complacency, and do not comprehend that other persons take offence thereat. When this phenomenon is present in a high degree, we speak of 'moral insanity' with Maudsley;* there are, nevertheless, lower stages in which the degenerate does not, perhaps, himself commit any act which will bring him into conflict with the criminal code, but at least asserts the theoretical legitimacy of crime; seeks, with philosophically sounding fustian, to prove that 'good' and 'evil,' virtue and vice, are arbitrary distinctions; goes into raptures over evildoers and their deeds; professes to discover beauties in the lowest and most repulsive things; and tries to awaken interest in, and so-called 'comprehension' of, every bestiality. The two psychological roots of moral insanity, in all its degrees of development, are, firstly, unbounded egoism,† and, secondly, impulsiveness‡—*i.e.*, in-

* See, on this subject, in particular, Krafft Ebing, *Die Lehre vom moralischen Wahnsinn*, 1871; H. Maudsley, *Responsibility in Mental Disease*, International Scientific Series; and Ch. Féré, *Dégénérescence et Criminalité*, Paris, 1888.

† J. Roubinovitch, *Hystérie mâle et Dégénérescence*; Paris, 1890, p. 62: 'The society which surrounds him (the degenerate) always remains strange to him. He knows nothing, and takes interest in nothing but himself.'

Legrain, *Du Délire chez les Dégénérés*; Paris, 1886, p. 10: 'The patient is . . . the plaything of his passions; he is carried away by his impulses, and has only one care—to satisfy his appetites.' P. 27: 'They are egoistical, arrogant, conceited, self-infatuated,' etc.

‡ Henry Colin, *Essai sur l'État mental des Hystériques*; Paris, 1890, p. 59: 'Two great facts control the being of the hereditary degenerate: obsession [the tyrannical domination of one thought from which a man cannot free

ability to resist a sudden impulse to any deed; and these characteristics also constitute the chief intellectual stigmata of degenerates. In the following sections of this work, I shall find occasion to show on what organic grounds, and in consequence of what peculiarities of their brain and nervous system, degenerates are necessarily egoistical and impulsive. In these introductory remarks I would wish only to point out the stigma itself.

Another mental stigma of degenerates is their emotionalism. Morel* has even wished to make this peculiarity their chief characteristic—erroneously, it seems to me, for it is present in the same degree among hysterics, and, indeed, is to be found in perfectly healthy persons, who, from any transient cause, such as illness, exhaustion, or any mental shock, have been temporarily weakened. Nevertheless it is a phenomenon rarely absent in a degenerate. He laughs until he sheds tears, or weeps copiously without adequate occasion; a commonplace line of poetry or of prose sends a shudder down his back; he falls into raptures before indifferent pictures or statues; and music especially,† even the most insipid and least commendable, arouses in him the most vehement emotions. He is quite proud of being so vibrant a musical instrument, and boasts that where the Philistine remains completely cold, he feels his inner self confounded, the depths of his being broken up, and the bliss of the Beautiful possessing him to the tips of his fingers. His excitability appears to him a mark of superiority; he believes himself to be possessed by a peculiar insight lacking in other mortals, and he is fain to despise the vulgar herd for the dulness and narrowness of their minds. The unhappy creature does not suspect that he is conceited about a disease and boasting of a derangement of the mind; and certain silly critics, when, through fear of being pronounced deficient in comprehension, they make desperate efforts to share the emotions of a degenerate in regard to some insipid or ridiculous production, or when they praise in exaggerated expressions the beauties which the degenerate asserts he finds therein, are unconsciously simulating one of the stigmata of semi-insanity.

Besides moral insanity and emotionalism, there is to be observed in the degenerate a condition of mental weakness and despondency, which, according to the circumstances of his life, assumes the form of pessimism, a vague fear of all men, and of

himself; Westphal has created for this the good term 'Zwangs-Vorstellung,' *i.e.*, coercive idea] and impulsion—both irresistible.'

* Morel, 'Du Délire émotif,' *Archives générales*, 6 série, vol. vii., pp. 385 and 530. See also Roubinovitch, *op. cit.*, p. 53.

† J. Roubinovitch, *op. cit.*, p. 68: 'Music excites him keenly

the entire phenomenon of the universe, or self-abhorrence. 'These patients,' says Morel,* 'feel perpetually compelled . . . to commiserate themselves, to sob, to repeat with the most desperate monotony the same questions and words. They have delirious presentations of ruin and damnation, and all sorts of imaginary fears.' 'Ennui never quits me,' said a patient of this kind, whose case Roubinovitch† describes, 'ennui of myself.' 'Among moral stigmata,' says the same author,‡ 'there are also to be specified those undefinable apprehensions manifested by degenerates when they see, smell, or touch any object.' And he further§ calls to notice 'their unconscious fear of everything and everyone.' In this picture of the sufferer from melancholia; downcast, sombre, despairing of himself and the world, tortured by fear of the Unknown, menaced by undefined but dreadful dangers, we recognise in every detail the man of the Dusk of the Nations and the *fin-de-siècle* frame of mind, described in the first chapter.

With this characteristic dejectedness of the degenerate, there is combined, as a rule, a disinclination to action of any kind, attaining possibly to abhorrence of activity and powerlessness to will (*aboulia*). Now, it is a peculiarity of the human mind, known to every psychologist, that, inasmuch as the law of causality governs a man's whole thought, he imputes a rational basis to all his own decisions. This was prettily expressed by Spinoza when he said: 'If a stone flung by a human hand could think, it would certainly imagine that it flew because it wished to fly.' Many mental conditions and operations of which we become conscious are the result of causes which do not reach our consciousness. In this case we fabricate causes *a posteriori* for them, satisfying our mental need of distinct causality, and we have no trouble in persuading ourselves that we have now truly explained them. The degenerate who shuns action, and is without will-power, has no suspicion that his incapacity for action is a consequence of his inherited deficiency of brain. He deceives himself into believing that he despises action from free determination, and takes pleasure in inactivity; and, in order to justify himself in his own eyes, he constructs a philosophy of renunciation and of contempt for the world and men, asserts that he has convinced himself of the excellence of Quietism, calls himself with consummate self-consciousness a Buddhist, and praises Nirvana in poetically eloquent phrases as the highest and worthiest ideal of the human mind. The degenerate and insane are the predestined disciples of Schopenhauer and Hartmann,

* Morel, 'Du Délire panophobique des Aliénés gémisseurs,' *Annales médico-psychologiques*, 1871.

† Roubinovitch, *op. cit.*, p. 28. ‡ *Ibid.*, p. 37. § *Ibid.*, p. 66.

and need only to acquire a knowledge of Buddhism to become converts to it.

With the incapacity for action there is connected the predilection for inane reverie. The degenerate is not in a condition to fix his attention long, or indeed at all, on any subject, and is equally incapable of correctly grasping, ordering, or elaborating into ideas and judgments the impressions of the external world conveyed to his distracted consciousness by his defectively operating senses. It is easier and more convenient for him to allow his brain-centres to produce semi-lucid, nebulously blurred ideas and inchoate embryonic thoughts, and to surrender himself to the perpetual obfuscation of a boundless, aimless, and shoreless stream of fugitive ideas; and he rarely rouses himself to the painful attempt to check or counteract the capricious, and, as a rule, purely mechanical associations of ideas and succession of images, and bring under discipline the disorderly tumult of his fluid presentations. On the contrary, he rejoices in his faculty of imagination, which he contrasts with the insipidity of the Philistine, and devotes himself with predilection to all sorts of unlicensed pursuits permitted by the unshackled vagabondage of his mind; while he cannot endure well-ordered civil occupations, requiring attention and constant heed to reality. He calls this 'having an idealist temperament,' ascribes to himself irresistible æsthetic propinquities, and proudly styles himself an artist.*

We will briefly mention some peculiarities frequently manifested by a degenerate. He is tormented by doubts, seeks for the basis of all phenomena, especially those whose first causes are completely inaccessible to us, and is unhappy when his inquiries and ruminations lead, as is natural, to no result.† He is ever supplying new recruits to the army of system-inventing metaphysicians, profound expositors of the riddle of the universe, seekers for the philosopher's stone, the squaring of the circle and perpetual motion.‡ These last three subjects have such a special attraction for him, that the Patent Office at Washington is forced to keep on hand printed replies to the numberless memorials in which patents are constantly

* Charcot, 'Leçons du Mardi à la Salpêtrière,' *Policlinique*, Paris, 1890, 2e partie, p. 392: 'This person [the invalid mentioned] is a performer at fairs; he calls himself "artist." The truth is that his art consists in personating a "wild man" in fair-booths.'

† Legrain, *op. cit.*, p. 73: 'The patients are perpetually tormented by a multitude of questions which invade their minds, and to which they can give no answer; inexpressible moral sufferings result from this incapacity. Doubt envelops every possible subject:—metaphysics, theology, etc.'

‡ Magnan, 'Considérations sur la Folie des Héréditaires ou Dégénerés,' *Progrès médical*, 1886, p. 1110 (in the report of a medical case): 'He also thought of seeking for the philosopher's stone, and of making gold.'

demanded for the solution of these chimerical problems. In view of Lombroso's researches,* it can scarcely be doubted that the writings and acts of revolutionists and anarchists are also attributable to degeneracy. The degenerate is incapable of adapting himself to existing circumstances. This incapacity, indeed, is an indication of morbid variation in every species, and probably a primary cause of their sudden extinction. He therefore rebels against conditions and views of things which he necessarily feels to be painful, chiefly because they impose upon him the duty of self-control, of which he is incapable on account of his organic weakness of will. Thus he becomes an improver of the world, and devises plans for making mankind happy, which, without exception, are conspicuous quite as much by their fervent philanthropy, and often pathetic sincerity, as by their absurdity and monstrous ignorance of all real relations.

Finally, a cardinal mark of degeneration which I have reserved to the last, is mysticism. Colin says:† 'Of all the delirious manifestations peculiar to the hereditarily-afflicted, none indicates the condition more clearly, we think, than mystical delirium, or, when the malady has not reached this point, the being constantly occupied with mystical and religious questions, an exaggerated piety, etc.' I will not here multiply evidence and quotations. In the following books, where the art and poetry of the times are treated of, I shall find occasion to show the reader that no difference exists between these tendencies and the religious manias observed in nearly all degenerates and sufferers from hereditary mental taint.

I have enumerated the most important features characterizing the mental condition of the degenerate. The reader can now judge for himself whether or not the diagnosis 'degeneration' is applicable to the originators of the new æsthetic tendencies. It must not for that matter be supposed that degeneration is synonymous with absence of talent. Nearly all the inquirers who have had degenerates under their observation expressly establish the contrary. 'The degenerate,' says Legrain,‡ 'may be a genius. A badly balanced mind is susceptible of the highest conceptions, while, on the other hand, one meets in the same mind with traits of meanness and pettiness all the more striking from the fact that they co-exist with the most brilliant qualities.' We shall find this reservation in all authors

* Lombroso, 'La Physionomie des Anarchistes,' *Nouvelle Revue*, May 15, 1891, p. 227: 'They [the anarchists] frequently have those characteristics of degeneracy which are common to criminals and lunatics, for they are anomalies, and bear hereditary taints.' See also the same author's *Pazzi ed Anomali*. Turin, 1884.

† Colin, *op. cit.*, p. 154.

‡ Legrain, *op. cit.*, p. 11.

who have contributed to the natural history of the degenerate. 'As regards their intellect, they can,' says Roubinovitch,* 'attain to a high degree of development, but from a moral point of view their existence is completely deranged. . . . A degenerate will employ his brilliant faculties quite as well in the service of some grand object as in the satisfaction of the basest propensities.' Lombroso† has cited a large number of undoubted geniuses who were equally undoubted mattoids, graphomaniacs, or pronounced lunatics; and the utterance of a French savant, Guérinsen, 'Genius is a disease of the nerves,' has become a 'winged word.' This expression was imprudent, for it gave ignorant babblers a pretext, and apparently a right, to talk of exaggeration, and to contemn experts in nervous and mental diseases, because they professedly saw a lunatic in everyone who ventured to be something more than the most ordinary, characterless, average being. Science does not assert that every genius is a lunatic; there are some geniuses of superabundant power whose high privilege consists in the possession of one or other extraordinarily developed faculty, without the rest of their faculties falling short of the average standard. Just as little, naturally, is every lunatic a genius; most of them, even if we disregard idiots of different degrees, are much rather pitiably stupid and incapable; but in many, nay, in abundant cases, the 'higher degenerate' of Magnan, just as he occasionally exhibits gigantic bodily stature or the disproportionate growth of particular parts, has some mental gift exceptionally developed at the cost, it is true, of the remaining faculties, which are wholly or partially atrophied.‡ It is this which enables the well-informed to distinguish at the first glance between the sane genius, and the highly, or even the most highly, gifted degenerate. Take from the former the special capacity through which he becomes a genius, and there still remains a capable, often conspicuously intelligent, clever, moral, and judicious man, who will hold his ground with propriety in our social mechanism. Let the same be tried in the case of a degenerate, and there remains only a criminal or madman, for whom healthy humanity can find no use. If

* Roubinovitch, *op. cit.*, p. 33.

† Lombroso, *Genie und Irrsinn;* German translation by A. Courth. Reclam's *Universal Bibliothek*, Bde. 2313-16. See also in particular, J. F. Nisbet, *The Insanity of Genius.* London, 1891.

‡ Falret, *Annales médico-psychologiques*, 1867, p. 76: 'From their childhood they usually display a very unequal development of their mental faculties, which, weak in their entirety, are remarkable for certain special aptitudes; they have shown an extraordinary gift for drawing, arithmetic, music, sculpture, or mechanics . . . and, together with those specially developed aptitudes, obtaining for them the fame of "infant phenomena," they for the most part give evidence of very great deficiencies in their intelligence, and of a radical debility in the remaining faculties.'

Goethe had never written a line of verse, he would, all the same, have still remained a man of the world, of good principles, a fine art connoisseur, a judicious collector, a keen observer of nature. Let us, on the contrary, imagine a Schopenhauer who had written no astounding books, and we should have before us only a repulsive *lusus naturæ*, whose morals would necessarily exclude him from all respectable society, and whose fixed idea that he was a victim of persecution would point him out as a subject for a madhouse. The lack of harmony, the absence of balance, the singular incapacity of usefully applying, or deriving satisfaction from, their own special faculty among highly-gifted degenerates, strikes every healthy censor who does not allow himself to be prejudiced by the noisy admiration of critics, themselves degenerates: and will always prevent his mistaking the mattoid for the same exceptional man who opens out new paths for humanity and leads it to higher developments. I do not share Lombroso's opinion* that highly-gifted degenerates are an active force in the progress of mankind. They corrupt and delude; they do, alas! frequently exercise a deep influence, but this is always a baneful one. It may not be at once remarked, but it will reveal itself subsequently. If cotemporaries do not recognise it, the historian of morals will point it out *a posteriori*. They, likewise, are leading men along the paths they themselves have found to new goals; but these goals are abysses or waste places. They are guides to swamps like will-o'-the-wisps, or to ruin like the ratcatcher of Hammelin. Observers lay stress on their unnatural sterility. 'They are,' says Tarabaud,† 'cranks; wrongheaded, unbalanced, incapable creatures; they belong to the class of whom it may not be said that they have no mind, but whose mind produces nothing.' 'A common type,' writes Legrain,‡ 'unites them:—weakness of judgment and unequal development of mental powers. . . . Their conceptions are never of a high order. They are incapable of great thoughts and prolific ideas. This fact forms a peculiar contrast to the frequently excessive development of their powers of imagination.' 'If they are painters,' we read in Lombroso,§ 'then their predominant attribute will be the colour-sense; they will be decorative. If they are poets, they will be rich in rhyme, brilliant in style, but barren of thought; sometimes they will be "decadents."'

Such are the qualities of the most gifted of those who are

* *Nouvelle Revue*, July 15, 1891.

† Tarabaud, *Des Rapports de la Dégénérescence mentale et de l'Hystérie.* Paris, 1888, p. 12.

‡ Legrain, *op. cit.*, pp. 24 and 26.

§ Lombroso, *Nouvelles recherches de Psychiatrie et d'Anthropologie criminelle.* Paris, 1892, p. 74.

discovering new paths, and are proclaimed by enthusiastic followers as the guides to the promised land of the future. Among them degenerates and mattoids predominate. The second of the above-mentioned diagnoses, on the contrary, applies for the most part to the multitude who admire these individuals and swear by them, who imitate the fashions they design, and take delight in the extravagances described in the previous chapter. In their case we have to deal chiefly with hysteria, or neurasthenia.

For reasons which will be elucidated in the next chapter, hysteria has hitherto been less studied in Germany than in France, where, more than elsewhere, it has formed a subject of earnest inquiry. We owe what we know of it almost exclusively to French investigators. The copious treatises of Axenfeld,* Richer,† and in particular Gilles de la Tourette,‡ adequately comprise our present knowledge of this malady; and I shall refer to these works when I enumerate the symptoms chiefly indicative of hysteria.

Among the hysterical—and it must not be thought that these are met with exclusively, or even preponderantly, among females. for they are quite as often, perhaps oftener, found among males§ —among the hysterical, as among the degenerate, the first thing which strikes us is an extraordinary emotionalism. 'The leading characteristic of the hysterical,' says Colin,|| 'is the disproportionate impressionability of their psychic centres. . . . They are, above all things, impressionable.' From this primary peculiarity proceeds a second quite as remarkable and important—the exceeding ease with which they can be made to yield to suggestion.¶ The earlier observers always mentioned the boundless mendacity of the hysterical; growing, indeed, quite indignant at it, and making it the most prominent mark of the mental condition of such patients. They were mistaken. The hysterical subject does not consciously lie. He believes in the truth of his craziest inventions. The morbid mobility of his mind, the excessive excitability of his imagination, conveys to his consciousness all sorts of queer and senseless ideas. He suggests to himself that these ideas are founded on true perceptions, and believes in the truth of his foolish inventions until

* Axenfeld, *Des Névroses.* 2 vols., 2e édition, revue et complétée par le Dr. Huchard. Paris, 1879.

† Paul Richer, *Études cliniques sur l'Hystéro-épilepsie ou Grande Hystérie.* Paris, 1891.

‡ Gilles de la Tourette, *Traité clinique et thérapeutique de l'Hystérie.* Paris, 1891.

§ Paul Michaut, *Contribution à l'Étude des Manifestations de l'Hystérie chez l'Homme.* Paris, 1890.

|| Colin, *op. cit.*, p. 14.

¶ Gilles de la Tourette, *op. cit.*, p. 548 *et passim.*

a new suggestion—perhaps his own, perhaps that of another person—has ejected the earlier one. A result of the susceptibility of the hysterical subject to suggestion is his irresistible passion for imitation,* and the eagerness with which he yields to all the suggestions of writers and artists.† When he sees a picture, he wants to become like it in attitude and dress; when he reads a book, he adopts its views blindly. He takes as a pattern the heroes of the novels which he has in his hand at the moment, and infuses himself into the characters moving before him on the stage.

Added to this emotionalism and susceptibility to suggestion is a love of self never met with in a sane person in anything like the same degree. The hysterical person's own 'I' towers up before his inner vision, and so completely fills his mental horizon that it conceals the whole of the remaining universe. He cannot endure that others should ignore him. He desires to be as important to his fellow-men as he is to himself. 'An incessant need pursues and governs the hysterical—to busy those about them with themselves.'‡ A means of satisfying this need is the fabrication of stories by which they become interesting. Hence come the adventurous occurrences which often enough occupy the police and the reports of the daily press. In the busiest thoroughfare the hysterical person is set upon, robbed, maltreated and wounded, dragged to a distant place, and left to die. He picks himself up painfully, and informs the police. He can show the wounds on his body. He gives all the details. And there is not a single word of truth in the whole story; it is all dreamt and imagined. He has himself inflicted his wounds in order for a short time to become the centre of public attention. In the lower stages of hysteria this need of making a sensation assumes more harmless forms. It displays itself in eccentricities of dress and behaviour. 'Other hysterical subjects are passionately fond of glaring colours and extravagant forms; they wish to attract attention and make themselves talked about.'§

It is certainly unnecessary to draw the reader's attention in a special manner to the complete coincidence of this clinical picture of hysteria with the description of the peculiarities of the *fin-de-siècle* public, and to the fact that in the former we meet with all the features made familiar to us by the consideration of contemporary phenomena; in particular with the passion for imitating in externals—in dress, attitude, fashion of the hair and beard—the figures in old and modern pictures, and the feverish effort, through any sort of singularity, to make

* Colin, *op. cit.*, pp. 15 and 16.

† Gilles de la Tourette, *op. cit.*, p. 493.

‡ *Ibid.*, p. 303

§ Legrain, *op. cit.*, p. 39.

themselves talked about. The observation of pronounced cases of degeneration and hysteria, whose condition makes them necessary subjects for medical treatment, gives us also the key to the comprehension of subordinate details in the fashions of the day. The present rage for collecting, the piling up, in dwellings, of aimless bric-à-brac, which does not become any more useful or beautiful by being fondly called *bibelots*, appear to us in a completely new light when we know that Magnan has established the existence of an irresistible desire among the degenerate to accumulate useless trifles. It is so firmly imprinted and so peculiar that Magnan declares it to be a stigma of degeneration, and has invented for it the name 'oniomania,' or 'buying craze.' This is not to be confounded with the desire for buying, which possesses those who are in the first stage of general paralysis. The purchases of these persons are due to their delusion as to their own greatness. They lay in great supplies because they fancy themselves millionaires. The oniomaniac, on the contrary, neither buys enormous quantities of one and the same thing, nor is the price a matter of indifference to him as with the paralytic. He is simply unable to pass by any lumber without feeling an impulse to acquire it.

The curious style of certain recent painters—'impressionists,' 'stipplers,' or 'mosaists,' 'papilloteurs' or 'quiverers,' 'roaring' colourists, dyers in gray and faded tints—becomes at once intelligible to us if we keep in view the researches of the Charcot school into the visual derangements in degeneration and hysteria. The painters who assure us that they are sincere, and reproduce nature as they see it, speak the truth. The degenerate artist who suffers from *nystagmus*, or trembling of the eyeball, will, in fact, perceive the phenomena of nature trembling, restless, devoid of firm outline, and, if he is a conscientious painter, will give us pictures reminding us of the mode practised by the draughtsmen of the *Fliegende Blätter* when they represent a wet dog shaking himself vigorously. If his pictures fail to produce a comic effect, it is only because the attentive beholder reads in them the desperate effort to reproduce fully an impression incapable of reproduction by the expedients of the painter's art as devised by men of normal vision.

There is hardly a hysterical subject whose retina is not partly insensitive.* As a rule the insensitive parts are connected, and include the outer half of the retina. In these cases the field of vision is more or less contracted, and appears to him not as it does to the normal man—as a circle—but as a picture bordered by whimsically zigzag lines. Often, however, the

* Dr. Emile Berger, *Les Maladies des Yeux dans leurs rapports avec la Pathologie général.* Paris, 1892, p. 129 *et seq.*

insensitive parts are not connected, but are scattered in isolated spots over the entire retina. Then the sufferer will have all sorts of gaps in his field of vision, producing strange effects, and if he paints what he sees, he will be inclined to place in juxtaposition larger or smaller points or spots which are completely or partially dissociated. The insensitiveness need not be complete, and may exist only in the case of single colours, or of all. If the sensitiveness is completely lost ('achromatopsy') he then sees everything in a uniform gray, but perceives differences in the degree of lustre. Hence the picture of nature presents itself to him as a copper-plate or a pencil drawing, where the effect of the absent colours is replaced by differences in the intensity of light, by greater or less depth and power of the white and black portions. Painters who are insensitive to colour will naturally have a predilection for neutral-toned painting; and a public suffering from the same malady will find nothing objectionable in falsely-coloured pictures. But if, besides the whitewash of a Puvis de Chavannes, obliterating all colours equally, fanatics are found for the screaming yellow, blue, and red of a Besnard, this also has a cause, revealed to us by clinical science. 'Yellow and blue,' Gilles de la Tourette* teaches us, 'are peripheral colours' (*i.e.*, they are seen with the outermost parts of the retina); 'they are, therefore, the last to be perceived' (if the sensitiveness for the remaining colours is destroyed). 'These are . . . the very two colours the sensations of which in hysterical amblyopia [dulness of vision] endure the longest. In many cases, however, it is the red, and not the blue, which vanishes last.'

Red has also another peculiarity explanatory of the predilection shown for it by the hysterical. The experiments of Binet† have established that the impressions conveyed to the brain by the sensory nerves exercise an important influence on the species and strength of the excitation distributed by the brain to the motor nerves. Many sense-impressions operate enervatingly and inhibitively on the movements; others, on the contrary, make these more powerful, rapid and active; they are 'dynamogenous,' or 'force-producing.' As a feeling of pleasure is always connected with dynamogeny, or the production of force, every living thing, therefore, instinctively seeks for dynamogenous sense-impressions, and avoids enervating and inhibitive ones. Now, red is especially dynamogenous. 'When,' says Binet,‡ in a report of an experiment on a female hysterical subject who was paralyzed in one half of her body, 'we place a dynamo-

* *Traité clinique et thérapeutique de l'Hystérie*, p. 339. See also Drs. A. Marie et J. Bonnet, *La Vision chez les Idiots et les Imbéciles*. Paris, 1892.

† Alfred Binet, 'Recherches sur les Altérations de la Conscience chez les Hystériques,' *Revue philosophique*, 1889, vol. xxvii.

‡ *Op. cit.*, p. 150.

meter in the anæsthetically insensible right hand of Amélie Cle...... the pressure of the hand amounts to 12 kilogrammes. If at the same time she is made to look at a red disc, the number indicating the pressure in kilogrammes is at once doubled.' Hence it is intelligible that hysterical painters revel in red, and that hysterical beholders take special pleasure in pictures operating dynamogenously, and producing feelings of pleasure.

If red is dynamogenous, violet is conversely enervating and inhibitive.* It was not by accident that violet was chosen by many nations as the exclusive colour for mourning, and by us also for half-mourning. The sight of this colour has a depressing effect, and the unpleasant feeling awakened by it induces dejection in a sorrowfully-disposed mind. This suggests that painters suffering from hysteria and neurasthenia will be inclined to cover their pictures uniformly with the colour most in accordance with their condition of lassitude and exhaustion. Thus originate the violet pictures of Manet and his school, which spring from no actually observable aspect of nature, but from a subjective view due to the condition of the nerves. When the entire surface of walls in salons and art exhibitions of the day appears veiled in uniform half-mourning, this predilection for violet is simply an expression of the nervous debility of the painter.

There is yet another phenomenon highly characteristic in some cases of degeneracy, in others of hysteria. This is the formation of close groups or schools uncompromisingly exclusive to outsiders, observable to-day in literature and art. Healthy artists or authors, in possession of minds in a condition of well-regulated equilibrium, will never think of grouping themselves into an association, which may at pleasure be termed a sect or band; of devising a catechism, of binding themselves to definite æsthetic dogmas, and of entering the lists for these with the fanatical intolerance of Spanish inquisitors. If any human activity is individualistic, it is that of the artist. True talent is always personal. In its creations it reproduces itself, its own views and feelings, and not the articles of faith learnt from any æsthetic apostle; it follows its creative impulses, not a theoretical formula preached by the founder of a new artistic or literary church; it constructs its work in the form organically necessary to it, not in that proclaimed by a leader as demanded by the fashion of the day. The mere fact that an artist or author allows himself to be

* Ch. Féré, 'Sensation et Mouvement,' *Revue philosophique*, 1886. See also the same author's *Sensation et Mouvement*, Paris, 1887; *Dégénérescence et criminalité*, Paris, 1888; and 'L'Énergie et la Vitesse des Mouvements volontaires,' *Revue philosophique*, 1889.

sworn in to the party cry of any 'ism,' that he perambulates with jubilations behind a banner and Turkish music, is complete evidence of his lack of individuality—that is, of talent. If the mental movements of a period—even those which are healthy and prolific—range themselves, as a rule, under certain main tendencies, which receive each its distinguishing name, this is the work of historians of civilization or literature, who subsequently survey the combined picture of an epoch, and for their own convenience undertake divisions and classifications, in order that they may more correctly find their way among the multifariousness of the phenomena. These are, however, almost always arbitrary and artificial. Independent minds (we are not here speaking of mere imitators), united by a good critic into a group, may, it is true, have a certain resemblance to each other, but, as a rule, this resemblance will be the consequence, not of actual internal affinity, but of external influences. No one is able completely to withdraw himself from the influences of his time, and under the impression of events which affect all contemporaries alike, as well as of the scientific views prevailing at a given time, certain features develop themselves in all the works of an epoch, which stamp them as of the same date. But the same men who subsequently appear so naturally in each other's company, in historical works, that they seem to form a family, went when they lived their separate ways far asunder, little suspecting that at one time they would be united under one common designation. Quite otherwise it is when authors or artists consciously and intentionally meet together and found an æsthetic school, as a joint-stock bank is founded, with a title for which, if possible, the protection of the law is claimed, with by-laws, joint capital, etc. This may be ordinary speculation, but as a rule it is disease. The predilection for forming societies met with among all the degenerate and hysterical may assume different forms. Criminals unite in bands, as Lombroso expressly establishes.* Among pronounced lunatics it is the *folie à deux*, in which a deranged person completely forces his insane ideas on a companion; among the hysterical it assumes the form of close friendships, causing Charcot to repeat at every opportunity: 'Persons of highly-strung nerves attract each other;'† and finally authors found schools.

The common organic basis of these different forms of one and the same phenomenon—of the *folie à deux*, the association of neuropaths, the founding of æsthetic schools, the banding of criminals—is, with the active part, viz., those who lead and inspire, the predominance of obsessions: with the

* Lombroso, *L'Uomo délinquente*, p. 524.

† 'Les Nerveux se recherchent,' Charcot, *Leçons du Mardi*, *passim*.

associates, the disciples, the submissive part, weakness of will and morbid susceptibility to suggestion.* The possessor of an obsession is an incomparable apostle. There is no rational conviction arrived at by sound labour of intellect, which so completely takes possession of the mind, subjugates so tyrannically its entire activity, and so irresistibly impels it to words and deeds, as delirium. Every proof of the senselesness of his ideas rebounds from the deliriously insane or half-crazy person. No contradiction, no ridicule, no contempt, affects him; the opinion of the majority is to him a matter of indifference; facts which do not please him he does not notice, or so interprets that they seem to support his delirium; obstacles do not discourage him, because even his instinct of self-preservation is unable to cope with the power of his delirium, and for the same reason he is often enough ready, without further ado, to suffer martyrdom. Weak-minded or mentally-unbalanced persons, coming into contact with a man possessed by delirium, are at once conquered by the strength of his diseased ideas, and are converted to them. By separating them from the source of inspiration, it is often possible to cure them of their transmitted delirium, but frequently their acquired derangement outlasts this separation.

This is the natural history of the æsthetic schools. Under the influence of an obsession, a degenerate mind promulgates some doctrine or other—realism, pornography, mysticism, symbolism, diabolism. He does this with vehement penetrating eloquence, with eagerness and fiery heedlessness. Other degenerate, hysterical, neurasthenical minds flock around him, receive from his lips the new doctrine, and live thenceforth only to propagate it.

In this case all the participants are sincere—the founder as well as the disciples. They act as, in consequence of the diseased constitution of their brain and nervous system, they are compelled to act. The picture, however, which from a clinical standpoint is perfectly clear, gets dimmed if the apostle of a craze and his followers succeed in attracting to themselves the attention of wider circles. He then receives a concourse of unbelievers, who are very well able to recognise the insanity of the new doctrine, but who nevertheless accept it, because they hope, as associates of the new sect, to acquire fame and money. In every civilized nation which has a developed art and literature there are numerous intellectual eunuchs, incapable of producing with their own powers a living mental work, but quite able to imitate the process of production. These cripples

* Legrain, *op. cit.*, p. 173: 'The true explanation of the occurrence of *folie à deux* must be sought for, on the one hand, in the predisposition to insanity, and, on the other hand, in the accompanying weakness of mind.' See also Régis, *La Folie à Deux.* Paris, 1880.

form, unfortunately, the majority of professional authors and artists, and their many noxious followers often enough stifle true and original talent. Now it is these who hasten to act as camp-followers for every new tendency which seems to come into fashion. They are naturally the most modern of moderns, for no precept of individuality, no artistic knowledge, hinders them from bunglingly imitating the newest model with all the assiduity of an artisan. Clever in discerning externals, unscrupulous copyists and plagiarists, they crowd round every original phenomenon, be it healthy or unhealthy, and without loss of time set about disseminating counterfeit copies of it. To-day they are symbolists, as yesterday they were realists or pornographists. If they can promise themselves fame and a good sale, they write of mysteries with the same fluency as if they were spinning romances of knights and robbers, tales of adventure, Roman tragedies, and village stories at a time when newspaper critics and the public seemed to demand these things in preference to others. Now these practitioners, who, let it be again asserted, constitute the great majority of the mental workers of the fashionable sects in art and literature, and therefore of the associates of these sects also, are intellectually quite sane, even if they stand at a very low level of development, and were anyone to examine them, he might easily doubt the accuracy of the diagnosis, 'Degeneration' as regards the confessors of the new doctrines. Hence some caution must be exercised in the inquiry, and the sincere originators be always distinguished from the aping intriguers,—the founder of the religion and his apostles from the rabble to whom the Sermon on the Mount is of less concern than the miraculous draught of fishes and the multiplication of loaves.

It has now been shown how schools originate. They arise from the degeneration of their founders and of the imitators they have convinced. That they come into fashion, and for a short time attain a noisy success, is due to the peculiarities of the recipient public, namely, to hysteria. We have seen that hypersusceptibility to suggestion is the distinguishing characteristic of hysteria. The same power of obsession with which the degenerate in mind wins imitators, gathers round him adherents. When a hysterical person is loudly and unceasingly assured that a work is beautiful, deep, pregnant with the future, he believes in it. He believes in everything suggested to him with sufficient impressiveness. When the little cow-girl, Bernadette, saw the vision of the Holy Virgin in the grotto of Lourdes, the women devotees and hysterical males of the surrounding country who flocked thither did not merely believe that the hallucinant maiden had herself seen the vision, but all of them saw the Holy Virgin with their own eyes. M. E. de

Goncourt* relates that in 1870, during the Franco-Prussian War, a multitude of men, numbering tens of thousands, in and before the Bourse in Paris, were convinced that they had themselves seen—indeed, a part of them had read—a telegram announcing French victories fastened to a pillar inside the Exchange, and at which people were pointing with their finger; but as a matter of fact it never existed. It would be possible to cite examples by the dozen, of illusions of the senses suggested to excited crowds. Thus the hysterical allow themselves without more ado to be convinced of the magnificence of a work, and even find in it beauties of the highest kind, unthought of by the authors themselves and the appointed trumpeters of their fame. If the sect is so completely established that, in addition to the founders, the priests of the temple, the paid sacristans and choir-boys, it has a congregation, processions, and far-sounding bells, it then attaches to itself other converts besides the hysterical who have accepted the new belief by way of suggestion. Young persons without judgment, still seeking their way, go whither they see the multitude streaming, and unhesitatingly follow the procession, because they believe it to be marching on the right road. Superficial persons, fearing nothing so much as to be thought behind the times, attach themselves to the procession, shouting 'Hurrah!' and 'All hail!' so as to convince themselves that they also are really dancing along before the latest conqueror and newest celebrity. Decrepit graybeards, filled with a ridiculous dread of betraying their real age, eagerly visit the new temple and mingle their quavering voices in the song of the devout, because they hope to be thought young when seen in an assembly in which young persons predominate.

Thus a regular concourse is established about a victim of degeneration. The fashionable coxcomb, the æsthetic 'gigerl,'† peeps over the shoulder of the hysterical whose admiration has been suggested to him; the intriguer marches at the heel of the dotard, simulating youth; and between all these comes pushing the inquisitive young street-loafer, who must always be in every place where 'something is going on.' And this crowd, because it is driven by disease, self-interest and vanity, makes very much more noise and bustle than a far larger number of sane men, who, without self-seeking after-thought, take quiet enjoyment in works of sane talent, and do not feel obliged to shout out their appreciation in the streets, and to threaten with death harmless passers-by who do not join in their jubilations.

* *Journal des Goncourt.* Dernière série, premier volume, 1870-71. Paris, 1890, p. 17.

† Viennese for 'fop.'—TRANSLATOR.

CHAPTER IV.

ETIOLOGY.

We have recognised the effect of diseases in these *fin-de-siècle* literary and artistic tendencies and fashions, as well as in the susceptibility of the public with regard to them, and we have succeeded in maintaining that these diseases are degeneracy and hysteria. We have now to inquire how these maladies of the day have originated, and why they appear with such extraordinary frequency at the present time.

Morel,* the great investigator of degeneracy, traces this chiefly to poisoning. A race which is regularly addicted, even without excess, to narcotics and stimulants in any form (such as fermented alcoholic drinks, tobacco, opium, hashish, arsenic), which partakes of tainted foods (bread made with bad corn), which absorbs organic poisons (marsh fever, syphilis, tuberculosis, goitre), begets degenerate descendants who, if they remain exposed to the same influences, rapidly descend to the lowest degrees of degeneracy, to idiocy, to dwarfishness, etc. That the poisoning of civilized peoples continues and increases at a very rapid rate is widely attested by statistics.† The consumption of tobacco has risen in France from 0·8 kilogramme per head in 1841 to 1·9 kilogrammes in 1890. The corresponding figures for England are 13 and 26 ounces;‡ for Germany, 0·8 and 1·5 kilogrammes. The consumption of alcohol§ during

* *Traité des Dégénérescences, passim.*

† Personally communicated by the distinguished statistician, Herr Josef Körösi, Head of the Bureau of Statistics at Budapest.

‡ Speech of the Chancellor of the Exchequer, Mr. Goschen, in the House of Commons, April 11, 1892.

§ J. Vavasseur in the *Economiste français* of 1890. See also *Bulletin de Statistique* for 1891. The figures are uncertain, for they have been given differently by every statistician whom I have consulted. The fact of the increase in the consumption of alcohol alone stands out with certainty in all the publications consulted. Besides spirits, fermented drinks are consumed per head of the population, according to J. Körösi :

GREAT BRITAIN.

	Wine. Gall.	Beer and Cider. Gall.
1830—1850	0·2	26
1880—1888	0·4	27

FRANCE.

1840—1842	23	3
1870—1872	25	6

the same period has risen in Germany (1844) from 5·45 quarts to (1867) 6·86 quarts; in England from 2·01 litres to 2·64 litres; in France from 1·33 to 4 litres. The increase in the consumption of opium and hashish is still greater, but we need not concern ourselves about that, since the chief sufferers from them are Eastern peoples, who play no part in the intellectual development of the white races. To these noxious influences, however, one more may be added, which Morel has not known, or has not taken into consideration—residence in large towns. The inhabitant of a large town, even the richest, who is surrounded by the greatest luxury, is continually exposed to unfavourable influences which diminish his vital powers far more than what is inevitable. He breathes an atmosphere charged with organic detritus; he eats stale, contaminated, adulterated food; he feels himself in a state of constant nervous excitement, and one can compare him without exaggeration to the inhabitant of a marshy district. The effect of a large town on the human organism offers the closest analogy to that of the Maremma, and its population falls victim to the same fatality of degeneracy and destruction as the victims of malaria. The death-rate in a large town is more than a quarter greater than the average for the entire population; it is double that of the open country, though in reality it ought to be less, since in a large town the most vigorous ages predominate, during which the mortality is lower than in infancy and old age.* And the children of large towns who are not carried off at an early age suffer from the peculiar arrested development which Morel† has ascertained in the population of fever districts. They develop more or less normally until fourteen or fifteen years of age, are up to that time alert, sometimes brilliantly endowed, and give the highest promise; then suddenly there is a standstill, the mind loses its facility of comprehension, and the boy who, only yesterday, was a model scholar, becomes an obtuse, clumsy dunce, who can only be steered with the greatest

PRUSSIA.

	Quarts.
1839	13·48
1871	17 92

GERMAN EMPIRE.

	Litres.
1872	81·7
1889—1890	90·3

* In France the general mortality was, from 1886 to 1890, 22·21 per 1,000. But in Paris it rose to 23·4; in Marseilles to 34·8; in all towns with more than 100,000 inhabitants to a mean of 28·31; in all places with less than 5,000 inhabitants to 21·74. (*La Médecine moderne*, year 1891.)

† *Traité des Dégénérescences*, pp. 614, 615.

difficulty through his examinations. With these mental changes bodily modifications go hand in hand. The growth of the long bones is extremely slow, or ceases entirely, the legs remain short, the pelvis retains a feminine form, certain other organs cease to develop, and the entire being presents a strange and repulsive mixture of incompleteness and decay.*

Now we know how, in the last generation, the number of the inhabitants of great towns increased† to an extraordinary degree. At the present time an incomparably larger portion of the whole population is subjected to the destructive influences of large towns than was the case fifty years ago; hence the number of victims is proportionately more striking, and continually becomes more remarkable. Parallel with the growth of large towns is the increase in the number of the degenerate of all kinds—criminals, lunatics, and the 'higher degenerates' of Magnan; and it is natural that these last should play an ever more prominent part in endeavouring to introduce an ever greater element of insanity into art and literature.

The enormous increase of hysteria in our days is partly due to the same causes as degeneracy, besides which there is one cause much more general still than the growth of large towns —a cause which perhaps of itself would not be sufficient to bring about degeneracy, but which is unquestionably quite enough to produce hysteria and neurasthenia. This cause is the fatigue of the present generation. That hysteria is in reality a consequence of fatigue Féré has conclusively demonstrated by convincing experiments. In a communication to the Biological Society of Paris, this distinguished investigator says:‡ 'I have recently observed a certain number of facts which have made apparent the analogy existing between fatigue and the chronic condition of the hysterical. One knows that among the hysterical [involuntary!] symmetry of move-

* Brouardel, *La Semaine médicale.* Paris, 1887, p. 254. In this very remarkable study by the Parisian Professor, the following passage appears: 'What will these [those remaining stationary in their development] young Parisians become by-and-by? Incapable of accomplishing a long and conscientious work, they excel, as a rule, in artistic activities. If they are painters they are stronger in colour than in drawing. If they are poets, the flow of their verses assures their success rather than the vigour of the thought.'

† The 26 German towns which to-day have more than 100,000 inhabitants, numbered altogether, in 1891, 6,000,000, and in 1835 1,400,000. The 31 English towns of this category, in 1891, 10,870 000; in 1841, 4,590,000; the 11 French towns, in 1891, 4,180,000; in 1836, 1,710,000. It should be remarked that about a third of these 68 towns had not in 1840 as many generally as 100,000 inhabitants. To-day, in the large towns in Germany, France, and England, there reside 21,050,000 individuals, while in 1840 only 4,800,000 were living under these conditions. (Communicated by Herr Josef Körösi.)

‡ Féré, *La Semaine médicale.* Paris, 1890, p. 192.

ments frequently shows itself in a very characteristic manner. I have proved that in normal subjects this same symmetry of movements is met with under the influence of fatigue. A phenomenon which shows itself in a very marked way in serious hysteria is that peculiar excitability which demonstrates that the energy of the voluntary movements, through peripheral stimulations or mental presentations, suffers rapid and transitory modifications co-existing with parallel modifications of sensibility, and of the functions of nutrition. This excitability can be equally manifested during fatigue. . . . Fatigue constitutes a true temporary experimental hysteria. It establishes a transition between the states which we call normal and the various states which we designate hysteria. One can change a normal into a hysterical individual by tiring him. . . . All these causes (which produce hysteria) can, as far as the pathogenic part they play is concerned, be traced to one simple physiological process—to fatigue, to depression of vitality.'

Now, to this cause—fatigue—which, according to Féré, changes healthy men into hysterical, the whole of civilized humanity has been exposed for half a century. All its conditions of life have, in this period of time, experienced a revolution unexampled in the history of the world. Humanity can point to no century in which the inventions which penetrate so deeply, so tyrannically, into the life of every individual are crowded so thick as in ours. The discovery of America, the Reformation, stirred men's minds powerfully, no doubt, and certainly also destroyed the equilibrium of thousands of brains which lacked staying power. But they did not change the material life of man. He got up and laid down, ate and drank, dressed, amused himself, passed his days and years as he had been always wont to do. In our times, on the contrary, steam and electricity have turned the customs of life of every member of the civilized nations upside down, even of the most obtuse and narrow-minded citizen, who is completely inaccessible to the impelling thoughts of the times.

In an exceptionally remarkable lecture by Professor A. W. von Hofmann, in 1890, before the Congress of German Natural Science held in Bremen, he gave, in concluding, a short description of the life of an inhabitant of a town in the year 1822. He shows us a student of science who at that date is arriving with the coach from Bremen to Leipzig. The journey has lasted four days and four nights, and the traveller is naturally stiff and bruised. His friends receive him, and he wishes to refresh himself a little. But there is yet no Munich beer in Leipzig. After a short interview with his comrades, he goes in search of his inn. This is no easy task, for in the streets an Egyptian darkness reigns, broken only at long distances by the smoky flame of an oil-lamp. He at last finds his quarters, and wishes for a light. As matches do not yet exist, he is reduced to bruising the tips of his fingers with flint and steel, till he succeeds at last in lighting a tallow candle. He expects a letter, but it has not come, and he cannot now receive it till after some days, for the post only runs twice a week between Frankfort and Leipzig.*

But it is unnecessary to go back to the year 1822, chosen by Professor Hofmann. Let us stop, for purposes of comparison, at the year 1840. This year has not been arbitrarily selected. It is about the date when that generation was born which has witnessed the irruption of new discoveries in every relation of life, and thus personally experienced those transformations which are the consequences. This generation reigns and governs to-day; it sets the tone everywhere, and its sons and daughters are the youth of Europe and America, in whom the new æsthetic tendencies gain their fanatical partisans. Let us now compare how things went on in the civilized world in 1840 and a half-century later.†

In 1840 there were in Europe 3,000 kilometres of railway; in 1891 there were 218,000 kilometres. The number of travellers in 1840, in Germany, France and England, amounted to 2½ millions; in 1891 it was 614 millions. In Germany every inhabitant received, in 1840, 85 letters; in 1888, 200 letters. In 1840 the post distributed in France 94 millions of letters; in England, 277 millions; in 1881, 595 and 1,299 millions respectively. The collective postal intercourse between all countries, without including the internal postage of each separate country, amounted, in 1840, to 92 millions; in 1889, to 2,759 millions. In Germany, in 1840, 305 newspapers were published; in 1891, 6,800; in France, 776 and 5,182; in England (1846), 551 and 2,255. The German book trade produced, in 1840, 1,100 new works; in 1891, 18,700. The exports and imports of the world had, in 1840, a value of 28, in 1889 of 74, milliards of marks. The ships which, in 1840, entered all the ports of Great Britain contained 9½, in 1890 74½, millions of tons. The whole British merchant navy measured, in 1840, 3,200,000; in 1890, 9,688,000 tons.

* See, besides the lecture by Hofmann, the excellent book: *Eine deutsche Stadt vor 60 Jahren, Kulturgeschichtliche Skizze*, von Dr. Otto Bähr, 2 Auflage. Leipzig, 1891.

† In order not to make the footnotes too unwieldy, I state here that the following figures are borrowed in part from communications made by Herr Josef Körösi, in part from a remarkable study by M. Charles Richet: 'Dans Cent Ans,' *Revue scientifique*, 1891-92; and in a small degree from private publications (such as *Annuaire de la Presse*, *Press Directory*, etc.) For some of the figures I have also used, with profit, Mulhall, and the speech of Herr von Stephan to the Reichstag, February 4, 1892.

Let us now consider how these formidable figures arise. The 18,000 new publications, the 6,800 newspapers in Germany, desire to be read, although many of them desire in vain; the 2,759 millions of letters must be written; the larger commercial transactions, the numerous journeys, the increased marine intercourse, imply a correspondingly greater activity in individuals. The humblest village inhabitant has to-day a wider geographical horizon, more numerous and complex intellectual interests, than the prime minister of a petty, or even a second-rate state a century ago. If he do but read his paper, let it be the most innocent provincial rag, he takes part, certainly not by active interference and influence, but by a continuous and receptive curiosity, in the thousand events which take place in all parts of the globe, and he interests himself simultaneously in the issue of a revolution in Chili, in a bush-war in East Africa, a massacre in North China, a famine in Russia, a street-row in Spain, and an international exhibition in North America. A cook receives and sends more letters than a university professor did formerly, and a petty tradesman travels more and sees more countries and people than did the reigning prince of other times.

All these activities, however, even the simplest, involve an effort of the nervous system and a wearing of tissue. Every line we read or write, every human face we see, every conversation we carry on, every scene we perceive through the window of the flying express, sets in activity our sensory nerves and our brain centres. Even the little shocks of railway travelling, not perceived by consciousness, the perpetual noises, and the various sights in the streets of a large town, our suspense pending the sequel of progressing events, the constant expectation of the newspaper, of the postman, of visitors, cost our brains wear and tear. In the last fifty years the population of Europe has not doubled, whereas the sum of its labours has increased tenfold, in part even fifty-fold. Every civilized man furnishes, at the present time, from five to twenty-five times as much work as was demanded of him half a century ago.

This enormous increase in organic expenditure has not, and cannot have, a corresponding increase of supply. Europeans now eat a little more and a little better than they did fifty years ago, but by no means in proportion to the increase of effort which to-day is required of them. And even if they had the choicest food in the greatest abundance, it would do nothing towards helping them, for they would be incapable of digesting it. Our stomachs cannot keep pace with the brain and nervous system. The latter demand very much more than the former are able to perform. And so there follows what always happens if great expenses are met by small incomes; first the savings are consumed, then comes bankruptcy.

Its own new discoveries and progress have taken civilized humanity by surprise. It has had no time to adapt itself to its changed conditions of life. We know that our organs acquire by exercise an ever greater functional capacity, that they develop by their own activity, and can respond to nearly every demand made upon them; but only under one condition—that this occurs gradually, that time be allowed them. If they are obliged to fulfil, without transition, a multiple of their usual task, they soon give out entirely. No time was left to our fathers. Between one day and the next, as it were, without preparation, with murderous suddenness, they were obliged to change the comfortable creeping gait of their former existence for the stormy stride of modern life, and their heart and lungs could not bear it. The strongest could keep up, no doubt, and even now, at the most rapid pace, no longer lose their breath, but the less vigorous soon fell out right and left, and fill to-day the ditches on the road of progress.

To speak without metaphor, statistics indicate in what measure the sum of work of civilized humanity has increased during the half-century. It had not quite grown to this increased effort. It grew fatigued and exhausted, and this fatigue and exhaustion showed themselves in the first generation, under the form of acquired hysteria; in the second, as hereditary hysteria.

The new æsthetic schools and their success are a form of this general hysteria; but they are far from being the only one. The malady of the period shows itself in yet many other phenomena which can be measured and counted, and thus are susceptible of being scientifically established. And these positive and unambiguous symptoms of exhaustion are well adapted to enlighten the ignorant, who might believe at first sight that the specialist acts arbitrarily in tracing back fashionable tendencies in art and literature to states of fatigue in civilized humanity.

It has become a commonplace to speak of the constant increase of crime, madness and suicide. In 1840, in Prussia, out of 100,000 persons of criminally responsible age, there were 714 convictions; in 1888, 1,102 (from a letter communicated by the Prussian bureau of statistics). In 1865, in every 10,000 Europeans there were 63 suicides; in 1883, 109; and since that time the number has increased considerably. In the last twenty years a number of new nervous diseases have been discovered and named.* Let it not be believed that they always existed, and were merely overlooked. If they had been

* See G. André, *Les nouvelles maladies nerveuses.* Paris, 1892.

met with anywhere they would have been detected, for even if the theories which prevailed in medicine at various periods were erroneous, there have always been perspicacious and attentive physicians who knew how to observe. If, then, the new nervous diseases were not noticed, it is because they did not formerly appear. And they are exclusively a consequence of the present conditions of civilized life. Many affections of the nervous system already bear a name which implies that they are a direct consequence of certain influences of modern civilization. The terms 'railway-spine' and 'railway-brain,' which the English and American pathologists have given to certain states of these organs, show that they recognise them as due partly to the effects of railway accidents, partly to the constant vibrations undergone in railway travelling. Again, the great increase in the consumption of narcotics and stimulants, which has been shown in the figures above, has its origin unquestionably in the exhausted systems with which the age abounds. There is here a disastrous, vicious circle of reciprocal effects. The drinker (and apparently the smoker also) begets enfeebled children, hereditarily fatigued or degenerated, and these drink and smoke in their turn, because they are fatigued. These crave for a stimulus, for a momentary, artificial invigoration, or an alleviation of their painful excitability, and then, when they recognise that this increases, in the long-run, their exhaustion as well as their excitability, they cannot, through weakness of will, resist those habits.*

Many observers assert that the present generation ages much more rapidly than the preceding one. Sir James Crichton-Browne points out this effect of modern circumstances on contemporaries in his speech at the opening of the winter term, 1891, before the medical faculty of the Victoria University.† From 1859 to 1863 there died in England, of heart-disease, 92,181 persons; from 1884 to 1888, 224,102. Nervous complaints carried off from 1864 to 1868, 196,000 persons; from 1884 to 1888, 260,558. The difference of figures would have been still more striking if Sir James had chosen a more remote period for comparison with the present, for in 1865 the high pressure under which the English worked was already nearly as great as in 1885. The dead carried off by heart and nerve diseases are the victims of civilization. The heart and nervous system first break down under the overstrain. Sir James in his speech says further on: 'Men and women grow

* Legrain, *op. cit.*, p. 251: 'Drinkers are "degenerates";' and p. 258 (after four reports of invalids which serve as a basis to the following summary): 'Hence, at the base of all forms of alcoholism we find mental degeneracy.'

† *Revue scientifique*, year 1892; vol. xlix., p. 168 *et seq.*

old before their time. Old age encroaches upon the period of vigorous manhood. . . . Deaths due exclusively to old age are found reported now between the ages of forty-five and fifty-five. . . .' Mr. Critchett (an eminent oculist) says: 'My own experience, which extends now over a quarter of a century, leads me to believe that men and women, in the present day, seek the aid of spectacles at a less advanced period of life than their ancestors. . . . Previously men had recourse to spectacles at the age of fifty. The average age is now forty-five years.' Dentists assert that teeth decay and fall out at an earlier age than formerly. Dr. Lieving attests the same respecting the hair, and assures us that precocious baldness is to be specially observed 'among persons of nervous temperaments and active mind, but of weak general health.' Everyone who looks round the circle of his friends and acquaintances will remark that the hair begins to turn gray much sooner than in former days. Most men and women show their first white hairs at the beginning of the thirties, many of them at a very much younger age. Formerly white hair was the accompaniment of the fiftieth year.

All the symptoms enumerated are the consequences of states of fatigue and exhaustion, and these, again, are the effect of contemporary civilization, of the vertigo and whirl of our frenzied life, the vastly increased number of sense impressions and organic reactions, and therefore of perceptions, judgments, and motor impulses, which at present are forced into a given unity of time. To this general cause of contemporary pathological phenomena, one may be added special to France. By the frightful loss of blood which the body of the French people suffered during the twenty years of the Napoleonic wars, by the violent moral upheavals to which they were subjected in the great Revolution and during the imperial epic, they found themselves exceedingly ill-prepared for the impact of the great discoveries of the century, and sustained by these a more violent shock than other nations more robust and more capable of resistance. Upon this nation, nervously strained and predestined to morbid derangement, there broke the awful catastrophe of 1870. It had, with a self-satisfaction which almost attained to megalomania, believed itself the first nation in the world; it now saw itself suddenly humiliated and crushed. All its convictions abruptly crumbled to pieces. Every single Frenchman suffered reverses of fortune, lost some members of his family, and felt himself personally robbed of his dearest conceptions, nay, even of his honour. The whole people fell into the condition of a man suddenly visited by a crushing blow of destiny, in his fortune, his position, his family, his reputation, even in his self-respect. Thousands lost their reason. In

Paris a veritable epidemic of mental diseases was observed, for which a special name was found—*la folie obsidionale*, 'siege-madness.' And even those who did not at once succumb to mental derangement, suffered lasting injury to their nervous system. This explains why hysteria and neurasthenia are much more frequent in France, and appear under such a greater variety of forms, and why they can be studied far more closely in this country than anywhere else. But it explains, too, that it is precisely in France that the craziest fashions in art and literature would necessarily arise, and that it is precisely there that the morbid exhaustion of which we have spoken became for the first time sufficiently distinct to consciousness to allow a special name to be coined for it, namely, the designation of *fin-de-siècle*.

The proposition which I set myself to prove may now be taken as demonstrated. In the civilized world there obviously prevails a twilight mood which finds expression, amongst other ways, in all sorts of odd æsthetic fashions. All these new tendencies, realism or naturalism, 'decadentism,' neo-mysticism, and their sub-varieties, are manifestations of degeneration and hysteria, and identical with the mental stigmata which the observations of clinicists have unquestionably established as belonging to these. But both degeneration and hysteria are the consequences of the excessive organic wear and tear suffered by the nations through the immense demands on their activity, and through the rank growth of large towns.

Led by this firmly linked chain of causes and effects, everyone capable of logical thought will recognise that he commits a serious error if, in the æsthetic schools which have sprung up in the last few years, he sees the heralds of a new era. They do not direct us to the future, but point backwards to times past. Their word is no ecstatic prophecy, but the senseless stammering and babbling of deranged minds, and what the ignorant hold to be the outbursts of gushing, youthful vigour and turbulent constructive impulses are really nothing but the convulsions and spasms of exhaustion.

We should not allow ourselves to be deceived by certain catch-words, frequently uttered in the works of these professed innovators. They talk of socialism, of emancipation of the mind, etc., and thereby create the outward show of being deeply imbued with the thoughts and struggles of the times. But this is empty sham. The catch-words in vogue are scattered through the works without internal sequence, and the struggles of the times are merely painted on the outside. It is a phenomenon observed in every kind of mania, that it receives its special colouring from the degree of culture of the invalid, and from the views prevailing at the times in which he lived.

The Catholic who is a prey to megalomania fancies he is the Pope; the Jew, that he is the Messiah; the German, that he is the Emperor or a field-marshal; the Frenchman, that he is the President of the Republic. In the persecution-mania, the invalid of former days complained of the wickedness and knavery of magicians and witches; to-day he grumbles because his imaginary enemies send electric streams through his nerves, and torment him with magnetism. The degenerates of to-day chatter of Socialism and Darwinism, because these words, and, in the best case, the ideas connected with these, are in current use. These so-called socialist and freethinking works of the degenerate as little advance the development of society towards more equitable economic forms, and more rational views of the relations among phenomena, as the complaints and descriptions of an individual suffering from persecution-mania, and who holds electricity responsible for his disagreeable sensations, advance the knowledge of this force of nature. Those obscure or superficially verbose works which pretend to offer solutions for the serious questions of our times, or, at least, to prepare the way thereto, are even impediments and causes of delay, because they bewilder weak or unschooled brains, suggest to them erroneous views, and make them either more inaccessible to rational information or altogether closed to it.

The reader is now placed at those points of view whence he can see the new æsthetic tendencies in their true light and their real shape. It will be the task of the following books to demonstrate the pathological character of each one of these tendencies, and to inquire what particular species of degenerate delirium or hysterical psychological process they are related to or identical with.

PART SEVEN

CHAPTER I

THE shots fired at Montsou had reached as far as Paris with a formidable echo. For four days all the opposition journals had been indignant, displaying atrocious narratives on their front pages: twenty-five wounded, fourteen dead, including three women and two children. And there were prisoners taken as well; Levaque had become a sort of hero, and was credited with a reply of antique sublimity to the examining magistrate. The Empire, hit in mid-career by these few balls, affected the calm of omnipotence, without itself realizing the gravity of its wound. It was simply an unfortunate collision, something lost over there in the black country, very far from the Parisian boulevards which formed public opinion; it would soon be forgotten. The Company had received an official intimation to hush up the affair, and to put an end to a strike which from its irritating duration was becoming a social danger.

So on Wednesday morning three of the directors appeared at Montsou. The little town, sick at heart, which had not dared hitherto to rejoice over the massacre, now breathed again, and tasted the joy of being saved. The weather, too, had become fine; there was a bright sun—one of those first February days which, with their moist warmth, tip the lilac shoots with green. All the shutters had been flung back at the administration building, the vast structure seemed alive again. And cheering rumours were circulating; it was said that the directors, deeply affected by the catastrophe, had rushed down to open their paternal arms to the wanderers from the settlements. Now that the blow had fallen—a more vigorous one doubtless than they had wished for—they were prodigal in their task of relief, and decreed measures that were excellent though tardy. First of all they sent away the Borains, and made much of this extreme concession to their workmen. Then they put an end to the military occupation of the pits, which were no longer threatened by the crushed strikers. They also obtained silence regarding the sentinel who had disappeared from the

Voreux; the district had been searched without finding either the gun or the corpse, and although there was a suspicion of crime, it was decided to consider the soldier a deserter. In every way they thus tried to attenuate matters, judging it dangerous to acknowledge the irresistible savagery of a crowd set free amid the falling structure of the old world. And besides, this work of conciliation did not prevent them from bringing purely administrative affairs to a satisfactory conclusion; for Deneulin had been seen to return to the administration buildings, where he met M. Hennebeau. The negotiations for the purchase of Vandame continued, and it was considered certain that Deneulin would accept the Company's offers.

But what particularly stirred the country were the great yellow posters which the directors had stuck up in profusion on the walls. On them were to be read these few lines, in very large letters: 'Workers of Montsou! We do not wish that the errors of which you have lately seen the sad effects should deprive sensible and willing workmen of their livelihood. We shall therefore reopen all the pits on Monday morning, and when work is resumed we shall examine with care and consideration those cases in which there may be room for improvement. We shall, in fact, do all that is just or possible to do.' In one morning the ten thousand colliers passed before these placards. Not one of them spoke, many shook their heads, others went away with trailing steps, without changing one line in their motionless faces.

Up till now the settlement of the Deux-Cent-Quarante had persisted in its fierce resistance. It seemed that the blood of their mates, which had reddened the mud of the pit, was barricading the road against the others. Scarcely a dozen had gone down, merely Pierron and some sneaks of his sort, whose departure and arrival were gloomily watched without a gesture or a threat. Therefore a deep suspicion greeted the placard stuck on to the church. Nothing was said about the returned certificates in that. Would the Company refuse to take them on again? and the fear of retaliations, the fraternal idea of protesting against the dismissal of the more compromised men, made them all obstinate still. It was dubious; they would see. They would return to the pit when these gentlemen were good enough to put things plainly. Silence crushed the low houses. Hunger itself seemed nothing; all might die now that violent death had passed over their roofs.

But one house, that of the Maheus, remained especially black and mute in its overwhelming grief. Since she had followed her man to the cemetery, Maheude kept her teeth clenched. After the battle, she had allowed Étienne to bring back Catherine, muddy and half dead; and as she was undressing her, before the young man, in order to put her to bed, she thought for a moment that her daughter also had received a ball in the belly. for the chemise was marked with large patches of blood. But she soon understood that it was the flood of puberty, which was at last breaking out in the shock of this abominable day. Ah! another piece of luck, that wound! A fine present, to be able to make children for the gendarmes to kill; and she never spoke to Catherine, nor did she, indeed, talk to Étienne. The latter slept with Jeanlin, at the risk of being arrested, seized by such horror at the idea of going back to the darkness of Réquillart that he would have preferred a prison. A shudder shook him, the horror of the night after all those deaths, an unacknowledged fear of the little soldier who slept down there underneath the rocks. Besides, he dreamed of a prison as of a refuge in the midst of the torment of his defeat; but they did not trouble him, and he dragged on his wretched hours, not knowing how to weary out his body. Only at times Maheude looked at both of them, at him and her daughter, with a spiteful air, as though she were asking them what they were doing in her house.

Once more they were all snoring in a heap. Father Bonnemort occupied the former bed of the two youngsters, who slept with Catherine now that poor Alzire no longer dug her hump into her big sister's ribs. It was when going to bed that the mother felt the emptiness of the house by the coldness of her bed, which was now too large. In vain she took Estelle to fill the vacancy; that did not replace her man, and she wept quietly for hours. Then the days began to pass by as before, always without bread, but yet without the luck to die outright; things picked up here and there rendered to the wretches the poor service of keeping them alive. Nothing had changed in their existence, only her man was gone.

On the afternoon of the fifth day, Étienne, made miserable by the sight of this silent woman, left the room, and walked slowly along the paved street of the settlement. The inaction which weighed on him impelled him to take constant walks, with arms swinging idly and lowered head, always tortured by the same thought. He tramped thus for half an hour, when he felt, by an increase in his discomfort, that his mates were

*M 897

coming to their doors to look at him. His little remaining popularity had been driven to the winds by that fusillade, and he never passed now without meeting fiery looks which pursued him. When he raised his head there were threatening men there, women drawing aside the curtains from their windows; and beneath this still silent accusation and the restrained anger of these eyes, enlarged by hunger and tears, he became awkward and could scarcely walk straight. These dumb reproaches seemed to be always increasing behind him. He became so terrified, lest he should hear the entire settlement come out to shout its wretchedness at him, that he returned shuddering. But at the Maheus' the scene which met him still further agitated him. Old Bonnemort was near the cold fireplace, nailed to his chair ever since two neighbours, on the day of the slaughter, had found him on the ground, with his stick broken, struck down like an old thunder-stricken tree. And while Lénore and Henri, to cajole their hunger, were scraping, with deafening noise, an old saucepan in which cabbages had been boiled the day before, Maheude, after having placed Estelle on the table, was standing up, threatening Catherine with her fist.

'Say that again, by God! Just dare to say that again!'

Catherine had declared her intention to go back to the Voreux. The idea of not gaining her bread, of being thus tolerated in her mother's house, like a useless animal that is in the way, was becoming every day more unbearable; and if it had not been from the fear of Chaval she would have gone down on Tuesday.

She said again, stammering:

'What would you have? We can't go on doing nothing. We should get bread, anyhow.'

Maheude interrupted her.

'Listen to me: the first one of you who goes to work, I 'll do for you. No, that would be too much, to kill the father and go on taking it out of the children! I 've had enough of it; I 'd rather see you all put in your coffins, like him that 's gone already.'

And her long silence broke out into a furious flood of words. A fine sum that Catherine would bring her! hardly thirty sous, to which they might add twenty sous if the bosses were good enough to find work for that brigand Jeanlin. Fifty sous, and seven mouths to feed! The brats were only good to swallow soup. As to the grandfather, he must have broken something in his brain when he fell, for he seemed imbecile; unless it had turned his blood to see the soldiers firing at his mates.

'That 's it, old man, isn't it? They 've quite done for you. It 's no good having your hands still strong; you 're done for.'

Bonnemort looked at her with his dim eyes without understanding. He remained for hours with fixed gaze, having no intelligence now except to spit into a plate filled with ashes, which was put beside him for cleanliness.

'And they 've not settled his pension, either,' she went on. 'And I 'm sure they won't give it, because of our ideas. No! I tell you that we 've had too much to do with those people who bring ill-luck.'

'But,' Catherine ventured to say, 'they promise on the placard——'

'Just let me alone with your damned placard! More birdlime for catching us and eating us. They can be mighty kind now that they have ripped us open.'

'But where shall we go, mother? They won't keep us at the settlement, sure enough.'

Maheude made a vague, terrified gesture. Where should they go to? She did not know at all; she avoided thinking, it made her mad. They would go elsewhere—somewhere. And as the noise of the saucepan was becoming unbearable, she turned round on Lénore and Henri and boxed their ears. The fall of Estelle, who had been crawling on all fours, increased the disturbance. The mother quieted her with a push—a good thing if it had killed her! She spoke of Alzire; she wished the others might have that child's luck. Then suddenly she burst out into loud sobs, with her head against the wall.

Étienne, who was standing by, did not dare to interfere. He no longer counted for anything in the house, and even the children drew back from him suspiciously. But the unfortunate woman's tears went to his heart, and he murmured:

'Come, come! courage! we must try to get out of it.'

She did not seem to hear him, and was bemoaning herself now in a low continuous complaint.

'Ah! the wretchedness! is it possible? Things did go on before these horrors. We ate our bread dry, but we were all together; and what has happened, good God! What have we done, then, that we should have such troubles—some under the earth, and the others with nothing left but to long to get there too? It 's true enough that they harnessed us like horses to work, and it 's not at all a just sharing of things to be always getting the stick and making rich people's fortunes bigger without hope of ever tasting the good things. There 's no pleasure in life when hope goes. Yes, that might have gone on longer; we might have breathed a bit. If we had only known! Is it possible to make oneself so wretched through wanting justice?'

Sighs swelled her breast, and her voice choked with immense sadness.

'Then there are always some clever people there who promise you that everything can be arranged by just taking a little trouble. Then one loses one's head, and one suffers so much from things as they are that one asks for things that can't be. Now, I was dreaming like a fool; I seemed to see a life of good friendship with everybody; I got off into the air, my faith! into the clouds. And then one breaks one's back when one tumbles down into the mud again. It 's not true; there 's nothing over there of the things that people tell of. What there is, is only wretchedness; ah! wretchedness, as much as you like of it, and bullets into the bargain.'

Étienne listened to this lamentation, and every tear struck him with remorse. He knew not what to say to calm Maheude, broken by her terrible fall from the heights of the ideal. She had come back to the middle of the room, and was now looking at him; she addressed him with contemptuous familiarity in a last cry of rage:

'And you, do you talk of going back to the pit, too, after driving us out of the bloody place! I 've nothing to reproach you with; but if I were in your shoes I should be dead of grief by now after causing such harm to the mates.'

He was about to reply, but then shrugged his shoulders in despair. What was the good of explaining, for she would not understand in her grief? And he went away, for he was suffering too much, and resumed his wild walk outside.

There again he found the settlement apparently waiting for him, the men at the doors, the women at the windows. As soon as he appeared growls were heard, and the crowd increased. The breath of gossip, which had been swelling for four days, was breaking out in a universal malediction. Fists were stretched towards him, mothers spitefully pointed him out to their boys, old men spat as they looked at him. It was the change which follows on the morrow of defeat, the fatal reverse of popularity, an execration exasperated by all the suffering endured without result. He had to pay for famine and death.

Zacharie, who came up with Philomène, hustled Étienne as he went out, grinning maliciously.

'Well, he gets fat. It's filling, then, to live on other people's deaths?'

The Levaque woman had already come to her door with Bouteloup. She spoke of Bébert, her youngster, killed by a bullet, and cried:

'Yes, there are cowards who get children murdered! Let him go and look for mine in the earth if he wants to give it me back!'

She was forgetting her man in prison, for the household was going on since Bouteloup remained; but she thought of him, however, and went on in a shrill voice:

'Get along! rascals may walk about while good people are put away!'

In avoiding her, Étienne tumbled on to Pierronne, who was running up across the gardens. She had regarded her mother's death as a deliverance, for the old woman's violence threatened to get them hanged; nor did she weep over Pierron's little girl, that street-walker Lydie—a good riddance. But she joined in with her neighbours with the idea of gaining reconciliation.

'And my mother, eh, and the little girl? You 've been seen; you were hiding yourself behind them when they caught the lead instead of you!'

What was to be done? Strangle Pierronne and the others, and fight the whole settlement? Étienne wanted to do so for a moment. The blood was throbbing in his head, he now looked upon his mates as brutes, he was irritated to see them so unintelligent and barbarous that they wanted to revenge themselves on him for the logic of facts. How stupid it all was! and he felt disgust at his powerlessness to tame them again; and satisfied himself with hastening his steps as though he were deaf to abuse. Soon it became a flight; every house hooted him as he passed, they hastened after his heels, it was a whole nation cursing him with a voice that was becoming like thunder in its overwhelming hatred. It was he, the exploiter, the murderer, who was the sole cause of their misfortune. He rushed out of the settlement, pale and terrified, with this yelling crowd behind his back. When he at last reached the main road most of them left him; but a few persisted, until at the bottom of the slope before the Avantage he met another group coming from the Voreux.

Old Mouque and Chaval were there. Since the death of his daughter Mouquette, and of his son Mouquet, the old man had continued to act as groom without a word of regret or complaint.

Suddenly, when he saw Étienne, he was shaken by fury, tears broke out from his eyes, and a flood of coarse words burst from his mouth, black and bleeding from his habit of chewing tobacco.

'You devil! you bloody swine! you filthy snout! Wait, you 've got to pay me for my poor children; you 'll have to come to it!'

He picked up a brick, broke it, and threw both pieces.

'Yes! yes! clear him off!' shouted Chaval, who was grinning in excitement, delighted at this vengeance. 'Every one gets his turn; now you 're stuck to the wall, you dirty hound!'

And he also attacked Étienne with stones. A savage clamour arose; they all took up bricks, broke them, and threw them, to rip him open, as they would like to have done to the soldiers. He was dazed and could not flee; he faced them, trying to calm them with phrases. His old speeches, once so warmly received, came back to his lips. He repeated the words with which he had intoxicated them at the time when he could keep them in hand like a faithful flock; but this power was dead, and only stones replied to him. He had just been struck on the left arm, and was drawing back, in great peril, when he found himself hemmed in against the front of the Avantage.

For the last few moments Rasseneur had been at his door.

'Come in,' he said simply.

Étienne hesitated; it choked him to take refuge there.

'Come in; then I 'll speak to them.'

He resigned himself, and took refuge at the other end of the parlour, while the innkeeper filled up the doorway with his broad shoulders.

'Look here, my friends, just be reasonable. You know very well that I 've never deceived you. I 've always been in favour of quietness, and if you had listened to me, you certainly wouldn't be where you are now.'

Rolling his shoulders and belly, he went on at length, allowing his facile eloquence to flow with the lulling gentleness of warm water. And all his old success came back; he regained his popularity, naturally and without an effort, as if he had never been hooted and called a coward a month before. Voices arose in approval: 'Very good! we are with you! that is the way to put it!' Thundering applause broke out.

Étienne, in the background, grew faint, and there was bitterness at his heart. He recalled Rasseneur's prediction in the forest, threatening him with the ingratitude of crowds. What imbecile brutality! What an abominable forgetfulness of old services! It was a blind force which constantly devoured itself. And beneath his anger at seeing these brutes spoil their own cause, there was despair at his own fall and the tragic end of his ambition. What! was it already done for! He remembered hearing beneath the beeches three thousand hearts beating to the echo of his own. On that day he had held his popularity in both hands. Those people belonged to him; he felt that he was their master. Mad dreams had then intoxicated him. Montsou at his feet, Paris beyond, becoming a deputy perhaps, crushing the middle class in a speech, the first speech ever pronounced by a workman in a parliament. And it was all over! He awakened, miserable and detested; his people were dismissing him by flinging bricks.

Rasseneur's voice rose higher:

'Never will violence succeed; the world can't be remade in a day. Those who have promised you to change it all at one stroke are either making fun of you or they are rascals!'

'Bravo! bravo!' shouted the crowd.

Who then was the guilty one? And this question which Étienne put to himself overwhelmed him more than ever. Was it in fact his fault, this misfortune which was making him bleed, the wretchedness of some, the murder of others, these women, these children, lean, and without bread? He had had that lamentable vision one evening before the catastrophe. But then a force was lifting him, he was carried away with his mates. Besides, he had never led them, it was they who led him, who obliged him to do things which he would never have done if it were not for the shock of that crowd pushing behind him. At each new violence he had been stupefied by the course of events, for he had neither foreseen nor desired any of them. Could he anticipate, for instance, that his followers in the settlement would one day stone him? These infuriated people lied when they accused him of having promised them an existence all fodder and laziness. And in this justification, in this reasoning, in which he tried to fight against his remorse, was hidden the anxiety that he had not risen to the height of his task; it was the doubts of the half-cultured man still perplexing him. But he felt himself at the end of his courage, he was no longer at heart with his mates; he feared this enormous mass of the people, blind and irresistible, moving like a force of nature, sweeping away everything, outside rules and theories. A certain repugnance was detaching him from them—the discomfort of his new tastes, the slow movement of all his being towards a superior class.

At this moment Rasseneur's voice was lost in the midst of enthusiastic shouts:

'Hurrah for Rasseneur! he 's the fellow! Bravo, bravo!'

The innkeeper shut the door, while the band dispersed; and the two men looked at each other in silence. They both shrugged their shoulders. They finished up by having a drink together.

On the same day there was a great dinner at Piolaine; they were celebrating the betrothal of Négrel and Cécile. Since the previous evening the Grégoires had had the dining-room waxed and the drawing-room dusted. Mélanie reigned in the kitchen, watching over the roasts and stirring the sauces, the odour of which ascended to the attics. It had been decided that Francis, the coachman, should help Honorine to wait. The gardener's wife would wash up, and the gardener would open the gate. Never had the substantial, patriarchal old house been in such a state of gaiety.

Everything went off beautifully. Madame Hennebeau was charming with Cécile, and she smiled at Négrel when the Montsou lawyer gallantly proposed the health of the future household. M. Hennebeau was also very amiable. His smiling face struck the guests. The report circulated that he was rising in favour with the directors, and that he would soon be made an officer of the Legion of Honour, on account of the energetic manner in which he had put down the strike. Nothing was said about recent events; but there was an air of triumph in the general joy, and the dinner became the official celebration of a victory. At last, then, they were saved, and once more they could begin to eat and sleep in peace. A discreet allusion was made to those dead whose blood the Voreux mud had yet scarcely drunk up. It was a necessary lesson; and they were all affected when the Grégoires added that it was now the duty of all to go and heal the wounds in the settlements. They had regained their benevolent placidity, excusing their brave miners, whom they could already see again at the bottom of the mines, giving a good example of everlasting resignation. The Montsou notables, who had now left off trembling, agreed that this question of the wage-system ought to be studied, cautiously. The roasts came on; and the victory became complete when M. Hennebeau read a letter from the bishop announcing Abbé Ranvier's removal. The middle class throughout the province had been roused to anger by the story of this priest who treated the soldiers as murderers. And when the dessert appeared, the lawyer resolutely declared that he was a free-thinker. Deneulin was there with his two daughters. In the midst of the joy, he forced himself to hide the melancholy of his ruin. That very morning he had signed the sale of his Vandame concession to the Montsou Company. With the knife at his throat he had submitted to the directors' demands, at last giving up to them that prey they had been on the watch for so long, scarcely obtaining from them the money necessary to pay off his creditors. He had even accepted, as a lucky chance, at the last moment, their offer to keep him as divisional engineer, thus resigning himself to watch over, as a simple salaried servant, that pit which had swallowed up his fortune. It was the knell of small personal enterprises, the approaching disappearance of the masters, eaten up, one by one, by the ever-hungry ogre of capital, drowned in the rising flood of great companies. He alone paid the expenses of the strike; he understood that they were drinking to his disaster when they drank to M. Hennebeau's rosette. And he only consoled himself a little when he saw the fine courage of Lucie and Jeanne, who looked charming in their done-up toilettes, laughing at the downfall, like happy tomboys disdainful of money.

When they passed into the drawing-room for coffee, M. Grégoire drew his cousin aside and congratulated him on the courage of his decision.

'What would you have? Your real mistake was to risk the million of your Montsou denier over Vandame. You gave yourself a terrible wound, and it has melted away in that dog's labour, while mine, which has not stirred from my drawer, still keeps me comfortably doing nothing, as it will keep my grandchildren's children.'

CHAPTER II

On Sunday Étienne escaped from the settlement at nightfall. A very clear sky, sprinkled with stars, lit up the earth with the blue haze of twilight. He went down towards the canal, and followed the bank slowly, in the direction of Marchiennes. It was his favourite walk, a grass-covered path two leagues long, passing straight beside this geometrical stream, which unrolled itself like an endless ingot of molten silver. He never met any one there. But on this day he was vexed to see a man come up to him. Beneath the pale starlight, the two solitary walkers only recognized each other when they were face to face.

'What! is it you?' said Etienne.

Souvarine nodded his head without replying. For a moment they remained motionless, then side by side they set out towards Marchiennes. Each of them seemed to be continuing his own reflections, as though they were far away from each other.

'Have you seen in the paper about Pluchart's success at Paris?' asked Étienne, at length. 'After that meeting at Belleville, they waited for him on the pavement, and gave him an ovation. Oh! he 's afloat now, in spite of his sore throat. He can do what he likes in the future.'

The engine-man shrugged his shoulders. He felt contempt for fine talkers, fellows who go into politics as one goes to the bar, to get an income out of phrases.

Étienne was now studying Darwin. He had read fragments, summarized and popularized in a five-sou volume; and out of this ill-understood reading he had gained for himself a revolutionary idea of the struggle for existence, the lean eating the fat. the strong people devouring the pallid middle class. But Souvarine furiously attacked the stupidity of the Socialists who accept Darwin, that apostle of scientific inequality, whose famous selection was only good for aristocratic philosophers. His mate persisted, however, wishing to reason out the matter, and expressing his doubts by an hypothesis: supposing the old society were no longer to exist, swept away to the crumbs; well, was it not to be feared that the new world would grow up again, slowly spoilt by the same injustices, some sick and others flourishing, some skilful and intelligent, fattening on everything, and others imbecile and lazy, becoming slaves again? But before this vision of eternal wretchedness, the engine-man shouted out fiercely that if justice was not possible with man, then man must disappear. For every rotten society there must be a massacre, until the last creature was exterminated. And there was silence again.

For a long time, with sunken head, Souvarine walked over the short grass, so absorbed that he kept to the extreme edge, by the water, with the quiet certainty of a sleep-walker on a roof. Then he shuddered causelessly, as though he had stumbled against a shadow. His eyes lifted and his face was very pale; he said softly to his companion:

'Did I ever tell you how she died?'

'Whom do you mean?'

'My wife, over there, in Russia.'

Étienne made a vague gesture, astonished at the tremor in his voice and at the sudden desire for confidence in this lad, who was usually so impassive in his stoical detachment from others and from himself. He only knew that the woman was his mistress, and that she had been hanged at Moscow.

'The affair hadn't gone off,' Souvarine said, with eyes still vacantly following the white stream of the canal between the bluish colonnades of tall trees. 'We had been a fortnight at the bottom of a hole undermining the railway, and it was not the imperial train that was blown up, it was a passenger train. Then they arrested Annutchka. She brought us bread every evening, disguised as a peasant woman. She lit the fuse, too, because a man might have attracted attention. I followed the trial, hidden in the crowd, for six long days.'

His voice became thick, and he coughed as though he were choking.

'Twice I wanted to cry out, and to rush over the people's heads to join her. But what was the good? One man less would be one soldier less; and I could see that she was telling me not to come, when her large eyes met mine.'

He coughed again.

'On the last day in the square I was there. It was raining; they stupidly lost their heads, put out by the falling rain. It took twenty minutes to hang the other four; the cord broke, they could not finish the fourth. Annutchka was standing up waiting. She could not see all, she was looking for me in the crowd. I got on to a post and she saw me, and our eyes never turned from each other. When she was dead she was still looking at me. I waved my hat; I came away.'

There was silence again. The white road of the canal unrolled to the far distance, and they both walked with the same quiet step as though each had fallen back into his isolation. At the horizon, the pale water seemed to open the sky with a little hole of light.

'It was our punishment,' Souvarine went on roughly. 'We were guilty to love each other. Yes, it is well that she is dead; heroes will be born from her blood, and I no longer have any cowardice at my heart. Ah! nothing, neither parents, nor wife, nor friend! Nothing to make my hand tremble on the day when I must take others' lives or give up my own.'

Étienne had stopped, shuddering in the cool night. He discussed no more, he simply said:

'We have gone far; shall we go back?'

They went back towards the Voreux slowly, and he added, after a few paces:

'Have you seen the new placards?'

The Company had that morning put up some more large yellow posters. They were clearer and more conciliatory, and the Company undertook to take back the certificates of those miners who went down on the following day. Everything would be forgotten, and pardon was offered even to those who were most implicated.

'Yes, I 've seen,' replied the engine-man.

'Well, what do you think of it?'

'I think that it 's all up. The flock will go down again. You are all too cowardly.'

Étienne feverishly excused his mates: a man may be brave, a mob which is dying of hunger has no strength. Step by step they were returning to the Voreux; and before the black mass of the pit he continued swearing that he, at least, would never go down; but he could forgive those who did. Then, as the rumour ran that the carpenters had not had time to repair the tubbing, he asked for information. Was it true? Had the weight of the soil against the timber which formed the internal skirt of scaffolding to the shaft so pushed it in that the winding-cages rubbed as they went down for a length of over fifty metres?

Souvarine, who once more became uncommunicative, replied briefly. He had been working the day before, and the cage did, in fact, jar; the engine-men had even had to double the speed to pass that spot. But all the bosses received any observations with the same irritating remark: it was coal they wanted; that could be repaired later on.

'You see that 's smashing up!' Étienne murmured. 'It will be a fine time!'

With eyes vaguely fixed on the pit in the shadow, Souvarine quietly concluded:

'If that does smash up, the mates will know it, since you advise them to go down again.'

Nine o'clock struck at the Montsou steeple; and his companion having said that he was going to bed, he added, without putting out his hand:

'Well, good-bye. I 'm going away.'

'What! you 're going away?'

'Yes, I 've asked for my certificate back. I 'm going elsewhere.'

Étienne, stupefied and affected, looked at him. After walking for two hours he said that to him! And in so calm a voice, while the mere announcement of this sudden separation made his own heart ache. They had got to know each other, they had toiled together; that always makes one sad, the idea of not seeing a person again.

'You 're going away! And where do you go?'

'Over there—I don't know at all.'

'But I shall see you again?'

'No, I think not.'

They were silent and remained for a moment facing each other without finding anything to say.

'Then good-bye.'

'Good-bye.'

While Étienne ascended toward the settlement, Souvarine turned and again went along the canal bank; and there, now alone, he continued to walk, with sunken head, so lost in the darkness that he seemed merely a moving shadow of the night. Now and then he stopped, he counted the hours that struck afar. When he heard midnight strike he left the bank and turned towards the Voreux.

At that time the pit was empty, and he only met a sleepy-eyed captain. It was not until six o'clock that they would begin to get up steam to resume work. First he went to take from a cupboard a jacket which he pretended to have forgotten. Various tools—a drill armed with its screw, a small but very strong saw, a hammer, and a chisel—were rolled up in this jacket. Then he left. But instead of going out through the shed he passed through the narrow corridor which led to the ladder passage. With his jacket under his arm he quietly went down without a lamp, measuring the depth by counting the ladders. He knew that the cage jarred at three hundred and seventy-four metres against the fifth row of the lower tubbing. When he had counted fifty-four ladders he put out his hand and was able to feel the swelling of the planking. It was there. Then with the skill and coolness of a good workman who has been reflecting over his task for a long time, he set to work. He began by sawing a panel in the brattice so as to communicate with the winding-shaft. With the help of matches, quickly lighted and blown out, he was then able to ascertain the condition of the tubbing and of the recent repairs.

Between Calais and Valenciennes the sinking of mine shafts is surrounded by immense difficulties on account of the masses of subterranean water in great sheets at the level of the lowest valleys. Only the construction of tubbings, frameworks jointed like the stays of a barrel, can keep out the springs which flow in and isolate the shafts in the midst of the lakes, which with deep obscure waves beat against the walls. It had been necessary in sinking the Voreux to establish two tubbings: that of the upper level, in the shifting sands and white clays bordering the chalky stratum, and fissured in every part, swollen with water like a sponge; then that of the lower level, immediately above the coal stratum, in a yellow sand as fine as flour, flowing with liquid fluidity; it was here that the Torrent was to be found, that subterranean sea so dreaded in the mine pits of the Nord, a sea with its storms and its shipwrecks, an unknown and unfathomable sea, rolling its dark floods more than three hundred metres beneath the daylight. Usually the tubbings resisted the enormous pressure; the only thing to be dreaded was the piling up of the neighbouring soil, shaken by the constant movement of the old galleries which were filling up. In this descent of the rocks lines of fracture were sometimes produced which slowly extended as far as the scaffolding, at last perforating it and pushing it into the shaft; and there was the great danger of a landslip and a flood filling the pit with an avalanche of earth and a deluge of springs.

Souvarine, sitting astride in the opening he had made, discovered a very serious defect in the fifth row of tubbing. The wood was bellied out from the framework; several planks had even come out of their shoulder-pieces. Abundant filtrations, *pichoux* the miners call them, were jetting out of the joints through the tarred oakum with which they were caulked. The carpenters, pressed for time, had been content to place iron squares at the angles, so carelessly that all the screws were not put in. A considerable movement was evidently going on behind in the sand of the torrent.

Then with his wimble he unscrewed the squares so that another push would tear them all off. It was a foolhardy task, during which he frequently only just escaped from falling headlong down the hundred and eighty metres which separated him from the bottom. He had been obliged to seize the oak guides, the joists along which the cages slid; and suspended over the void he traversed the length of the cross-beams with which they were joined from point to point, slipping along, sitting down, turning over, simply buttressing himself on an elbow or a knee, with tranquil contempt of death. A breath would have sent him over, and three times he caught himself up without a shudder. First he felt with his hand and then worked, only lighting a match when he lost himself in the midst of these slimy beams. After loosening the screws he attacked the wood itself, and the peril became still greater. He had sought for the key, the piece which held the others; he attacked it furiously, making holes in it, sawing it, thinning it so that it lost its resistance; while through the holes and the cracks the water which escaped in small jets blinded him and soaked him in icy rain. Two matches were extinguished. They all became damp and then there was night, the bottomless depth of darkness.

From this moment he was seized by rage. The breath of the invisible intoxicated him, the black horror of this rain-beaten hole urged him to mad destruction. He wreaked his fury at random against the tubbing, striking where he could with his wimble, with his saw, seized by the desire to bring the whole thing at once down on his head. He brought as much ferocity to the task as though he had been digging a knife into the skin of some execrated living creature. He would kill the Voreux at last, that evil beast with ever-open jaws which had swallowed so much human flesh! The bite of his tools could be heard, his spine lengthened, he crawled, climbed down, then up again, holding on by a miracle, in continual movement, the flight of a nocturnal bird amid the scaffolding of a belfry.

But he grew calm, dissatisfied with himself. Why could not things be done coolly? Without haste he took breath, and then went back into the ladder passage, stopping up the hole by replacing the panel which he had sawn. That was enough; he did not wish to raise alarm by excessive damage which would have been repaired immediately. The beast was wounded in the belly; we should see if it was still alive at night. And he had left his mark; the frightened world would know that the beast had not died a natural death. He took his time in methodically rolling up his tools in his jacket, and slowly climbed up the ladders. Then, when he had emerged from the pit without being seen, it did not occur to him to go and change his clothes. Three o'clock struck. He remained standing on the road, waiting.

At the same hour Étienne, who was not asleep, was disturbed by a slight sound in the thick night of the room. He distinguished the low breath of the children, and the snoring of Bonnemort and Maheude; while Jeanlin near him was breathing with a prolonged flute-like whistle. No doubt he had dreamed, and he was turning back when the noise began again. It was the creaking of a palliasse, the stifled effort of someone who is getting up. Then he imagined that Catherine must be ill.

'I say, is it you? What is the matter?' he asked in a low voice.

No one replied, and the snoring of the others continued. For five minutes nothing stirred. Then there was fresh creaking. Feeling certain this time that he was not mistaken, he crossed the room, putting his hands out into the darkness to feel the opposite bed. He was surprised to find the young girl sitting up, holding in her breath, awake and on the watch.

'Well! why don't you reply? What are you doing, then?'

At last she said:

'I 'm getting up.'

'Getting up at this hour?'

'Yes, I 'm going back to work at the pit.'

Étienne felt deeply moved, and sat down on the edge of the palliasse, while Catherine explained her reasons to him. She suffered too much by living thus in idleness, feeling continual looks of reproach weighing on her; she would rather run the risk of being knocked about down there by Chaval. And if her mother refused to take her money when she brought it, well! she was big enough to act for herself and make her own soup.

'Go away; I want to dress. And don't say anything, will you, if you want to be kind?'

But he remained near her; he had put his arms round her waist in a caress of grief and pity. Pressed one against the other in their shirts, they could feel the warmth of each other's naked flesh, at the edge of this bed, still moist with night's sleep. She had at first tried to free herself; then she began to cry quietly, in her turn taking him by the neck to press him against her in a despairing clasp. And they remained, without any further desires, with the past of their unfortunate love, which they had not been able to satisfy. Was it, then, done with for ever? Would they never dare to love each other some day, now that they were free? It only needed a little happiness to dissipate their shame— that awkwardness which prevented them from coming together because of all sorts of ideas which they themselves could not read clearly.

'Go to bed again,' she whispered. 'I don't want to light up, it would wake mother. It is time; leave me.'

He could not hear; he was pressing her wildly, with a heart drowned in immense sadness. The need for peace, an irresistible need for happiness, was carrying him away; and he saw himself married, in a neat little house, with no other ambition than to live and to die there, both of them together. He would be satisfied with bread; and if there were only enough for one, she should have it. What was the good of anything else? Was there anything in life worth more?

But she was unfolding her naked arms.

'Please, leave me.'

Then, in a sudden impulse, he said in her ear:

'Wait, I 'm coming with you.'

And he was himself surprised at what he had said. He had sworn never to go down again; whence then came this sudden decision, arising from his lips without thought of his, without even a moment's discussion? There was now such calm within him, so complete a cure of his doubts, that he persisted like a man saved by chance, who has at last found the only harbour from his torment. So he refused to listen to her when she became alarmed, understanding that he was devoting himself for her and fearing the ill words which would greet him at the pit. He laughed at everything; the placards promised pardon and that was enough.

'I want to work; that 's my idea. Let us dress and make no noise.'

They dressed themselves in the darkness, with a thousand precautions. She had secretly prepared her miner's clothes the evening before; he took a jacket and breeches from the cupboard; and they did not wash themselves for fear of knocking the bowl. All were asleep, but they had to cross the narrow passage where the mother slept. When they started, as ill-luck would have it, they stumbled against a chair. She woke and asked drowsily:

'Eh! what is it?'

Catherine had stopped, trembling, and violently pressing Étienne's hand.

'It 's me; don't trouble yourself,' he said. 'I feel stifled and am going outside to breathe a bit.'

'Very well.'

And Maheude fell asleep again. Catherine dared not stir. At last she went down into the parlour and divided a slice of bread-and-butter which she had reserved from a loaf given by a Montsou lady. Then they softly closed the door and went away.

Souvarine had remained standing near the Avantage, at the corner of the road. For half an hour he had been looking at the colliers who were returning to work in the darkness, passing by with the dull tramp of a herd. He was counting them, as a butcher counts his beasts at the entrance to the slaughter-house, and he was surprised at their number; even his pessimism had not foreseen that the number of cowards would have been so great. The stream continued to pass by, and he grew stiff, very cold, with clenched teeth and bright eyes.

But he started. Among the men passing by, whose faces he could not distinguish, he had just recognized one by his walk. He came forward and stopped him.

'Where are you going to?'

Étienne, in surprise, instead of replying, stammered:

'What! you 've not set out yet!'

Then he confessed he was going back to the pit. No doubt he had sworn; only it could not be called life to wait with crossed arms for things which would perhaps happen in a hundred years; and, besides, reasons of his own had decided him.

Souvarine had listened to him, shuddering. He seized him by the shoulder, and pushed him towards the settlement.

'Go home again; I want you to. Do you understand?'

But Catherine having approached, he recognized her also. Étienne protested, declaring that he allowed no one to judge his conduct. And the engine-man's eyes went from the young girl to her companion, while he stepped back with a sudden, relinquishing movement. When there was a woman in a man's heart, that man was done for; he might die. Perhaps he saw again in a rapid vision his mistress hanging over there at Moscow, that last link cut from his flesh, which had rendered him free of the lives of others and of his own life. He said simply:

'Go.'

Étienne, feeling awkward, was delaying, and trying to find some friendly word, so as not to separate in this manner.

'Then you 're still going?'

'Yes.'

'Well, give me your hand, old chap. A pleasant journey, and no ill-feeling.'

The other stretched out an icy hand. Neither friend nor wife.

'Good-bye for good this time.'

'Yes, good-bye.'

And Souvarine, standing motionless in the darkness, watched Étienne and Catherine entering the Voreux.

CHAPTER III

At four o'clock the descent began. Dansaert, who was personally installed at the marker's office in the lamp cabin, wrote down the name of each worker who presented himself and had a lamp given to him. He took them all, without remark, keeping to the promise of the placards. When, however, he noticed Étienne and Catherine at the wicket, he started and became very red, and was opening his mouth to refuse their names; then, he contented himself with the triumph, and a jeer. Ah! ah! so the strong man was thrown? The Company was, then, in luck since the terrible Montsou wrestler had come back to it to ask for bread? Étienne silently took his lamp and went towards the shaft with the putter.

But it was there, in the receiving-room, that Catherine feared the mates' bad words. At the very entrance she recognized Chaval, in the midst of some twenty miners, waiting till a cage was free. He came furiously towards her, but the sight of Étienne stopped him. Then he affected to sneer with an offensive shrug of the shoulders. Very good! he didn't care a hang, since the other had come to occupy the place that was still warm; good riddance! It only concerned the gentleman if he liked the leavings; and beneath the exhibition of this contempt he was again seized by a tremor of jealousy, and his eyes flamed. For the rest, the mates did not stir, standing silent, with eyes lowered. They contented themselves with casting a sidelong look at the new-comers; then, dejected and without anger, they again stared fixedly at the mouth of the shaft, with their lamps in their hands, shivering beneath their thin jackets in the constant draughts of this large room. At last the cage was wedged on to the keeps, and they were ordered to get in. Catherine and Étienne were squeezed in one tram, already containing Pierron and two pikemen. Beside them, in the other tram, Chaval was loudly saying to Father Mouque that the directors had made a mistake in not taking advantage of the opportunity to free the pits of the blackguards who were corrupting them; but the old groom, who had already fallen back into the dog-like resignation of his existence, no longer grew angry over the death of his children, and simply replied by a gesture of conciliation.

The cage freed itself and slipped down into the darkness. No one spoke. Suddenly, when they were in the middle third of the descent, there was a terrible jarring. The iron creaked, and the men were thrown on to each other.

'By God!' growled Étienne, 'are they going to flatten us? We shall end by being left here for good, with their confounded tubbing. And they talk about having repaired it!'

The cage had, however, cleared the obstacle. It was now descending beneath so violent a rain, like a storm, that the workmen anxiously listened to the pouring. A number of leaks must then have appeared in the caulking of the joints.

Pierron, who had been working for several days, when asked about it did not like to show his fear, which might be considered as an attack on the management, so he only replied:

'Oh, no danger! it 's always like that. No doubt they 've not had time to caulk the leaks.'

The torrent was roaring over their heads, and they at last reached the pit-eye beneath a veritable waterspout. Not one of the captains had thought of climbing up the ladders to investigate the matter. The pump would be enough, the carpenters would examine the joints the following night. The reorganization of work in the galleries gave considerable trouble. Before allowing the pikemen to return to their hewing cells, the engineer had decided that for the first five days all the men should execute certain works of consolidation which were extremely urgent. Landslips were threatening everywhere; the passages had suffered to such an extent that the timbering had to be repaired along a length of several hundred metres. Gangs of ten men were therefore formed below, each beneath the control of a captain. Then they were set to work at the most damaged spots. When the descent was complete, it was found that three hundred and twenty-two miners had gone down, about half of those who worked there when the pit was in full swing.

Chaval belonged to the same gang as Catherine and Étienne. This was not by chance; he had at first hidden behind his mates, and had then forced the captain's hand. This gang went to the end of the north gallery, nearly three kilometres away, to clear out a landslip which was stopping up a gallery in the Dix-Huit-Pouces seam. They attacked the fallen rocks with shovel and pick. Étienne, Chaval, and five others cleared away the rubbish while Catherine, with the trammers, wheeled the earth up to the upbrow. They seldom spoke, and the captain never left them. The putter's two lovers, however, were on the point of coming to blows. While growling that he had had enough of this trollop, Chaval was still thinking of her, and slyly hustling her about, so that Étienne had threatened to settle him if he did not leave her alone. They eyed each other fiercely, and had to be separated.

Towards eight o'clock Dansaert passed to give a glance at the work. He appeared to be in a very bad humour, and was furious with the captain; nothing had gone well, what was the meaning of such work, the planking would everywhere have to be done over again! And he went away declaring that he would come back with the engineer. He had been waiting for Négrel since morning, and could not understand the cause of this delay.

Another hour passed by. The captain had stopped the removal of the rubbish to employ all his people in supporting the roof. Even the putter and the two trammers left off wheeling to prepare and bring pieces of timber. At this end of the gallery the gang formed a sort of advance guard at the very extremity of the mine, now without communication with the other stalls. Three or four times strange noises, distant rushes, made the workers turn their heads to listen. What was it, then? One would have said that the passages were being emptied and the mates already returning at a running pace. But the sound was lost in the deep silence, and they set to wedging their wood again, dazed by the loud blows of the hammer. At last they returned to the rubbish, and the wheeling began once more. Catherine came back from her first journey in terror, saying that no one was to be found at the upbrow.

'I called, but there was no reply. They 've all cleared out of the place.'

The bewilderment was so great that the ten men threw down their tools to rush away. The idea that they were abandoned, left alone at the bottom of the mine, so far from the pit-eye, drove them wild. They only kept their lamps and ran in single file—the men, the boys, the putter; the captain himself lost his head and shouted out appeals, more and more frightened at the silence in this endless desert of galleries. What then had happened that they did not meet a soul? What accident could thus have driven away their mates? Their terror was increased by the uncertainty of the danger, this threat which they felt there without knowing what it was.

When they at last came near the pit-eye, a torrent barred their road. They were at once in water to the knees, and were no longer able to run, laboriously fording the flood with the thought that one minute's delay might mean death.

'By God! it 's the tubbing that 's given way,' cried Étienne. 'I said we should be left here for good.'

Since the descent Pierron had anxiously observed the increase of the deluge which fell from the shaft. As with two others he loaded the trams he raised his head, his face covered with large drops, and his ears ringing with the roar of the tempest above. But he trembled especially when he noticed that the sump beneath him, that pit ten metres deep, was filling; the water was already spurting through the floor and covering the metal plates. This showed that the pump was no longer sufficient to fight against the leaks. He heard it panting with the groan of fatigue. Then he warned Dansaert, who swore angrily, replying that they must wait for the engineer. Twice he returned to the charge without extracting anything else but exasperated shrugs of the shoulder. Well! the water was rising; what could he do?

Mouque appeared with Bataille, whom he was leading to work, and he had to hold him with both hands, for the sleepy old horse had suddenly reared up, and, with a shrill neigh, was stretching his head towards the shaft.

'Well, philosopher, what troubles you? Ah! it 's because it rains. Come along, that doesn't concern you.'

But the beast quivered all over his skin, and Mouque forcibly drew him to the haulage gallery.

Almost at the same moment as Mouque and Bataille were disappearing at the end of a gallery, there was a crackling in the air, followed by the prolonged noise of a fall. It was a piece of tubbing which had got loose and was falling a hundred and eighty metres down, rebounding against the walls. Pierron and the other porters were able to get out of the way, and the oak plank only smashed an empty tram. At the same time, a mass of water, the leaping flood of a broken dyke, rushed down. Dansaert proposed to go up and examine; but, while he was still speaking, another piece rolled down. And in terror before the threatening catastrophe, he no longer hesitated, but gave the order to go up, sending captains to warn the men in their stalls.

Then a terrible hustling began. From every gallery rows of workers came rushing up, trying to take the cages by assault. They crushed madly against each other in order to be taken up at once. Some who had thought of trying the ladder passage came down again shouting that it was already stopped up. That was the terror they all felt each time that the cage rose; this time it was able to pass, but who knew if it would be able to pass again in the midst of the obstacles obstructing the shaft? The downfall must be continuing above, for a series of low detonations was heard, the planks were splitting and bursting amid the continuous and increasing roar of a storm. One cage soon became useless, broken in and no longer sliding between the guides, which were doubtless broken. The other jarred to such a degree that the cable would certainly break soon. And there remained a hundred men to be taken up, all panting, clinging to one another, bleeding and half-drowned. Two were killed by falls of planking. A third, who had seized the cage, fell back fifty metres up and disappeared in the sump.

Dansaert, however, was trying to arrange matters in an orderly manner. Armed with a pick he threatened to open the skull of the first man who refused to obey; and he tried to arrange them in file, shouting that the porters were to go up last after having sent up their mates. He was not listened to, and he had to prevent the pale and cowardly Pierron from entering among the first. At each departure he pushed him aside with a blow. But his own teeth were chattering, a minute more and he would be swallowed up; everything was smashing up there, a flood had broken loose, a murderous rain of scaffolding. A few men were still running up when, mad with fear, he jumped into a tram, allowing Pierron to jump in behind him. The cage rose.

At this moment the gang to which Étienne and Chaval belonged had just reached the pit-eye. They saw the cage disappear and rushed forward, but they had to draw back from the final downfall of the tubbing; the shaft was stopped up and the cage would not come down again. Catherine was sobbing, and Chaval was choked with shouting oaths. There were twenty of them; were those bloody bosses going to abandon them thus? Father Mouque, who had brought back Bataille without hurrying, was still holding him by the bridle, both of them stupefied, the man and the beast, in the face of this rapid flow of the inundation. The water was already rising to their thighs. Étienne in silence, with clenched teeth, supported

Catherine between his arms. And the twenty yelled with their faces turned up, obstinately gazing at the shaft like imbeciles, that shifting hole which was belching out a flood and from which no help could henceforth come to them.

At the surface, Dansaert, on arriving, perceived Négrel running up. By some fatality, Madame Hennebeau had that morning delayed him on rising, turning over the leaves of catalogues for the purchase of wedding presents. It was ten o'clock.

'Well! what 's happening, then?' he shouted from afar.

'The pit is ruined,' replied the head captain.

And he described the catastrophe in a few stammered words, while the engineer incredulously shrugged his shoulders. What! could tubbing be demolished like that? They were exaggerating; he would make an examination.

'I suppose no one has been left at the bottom?'

Dansaert was confused. No, no one; at least, so he hoped. But some of the men might have been delayed.

'But,' said Négrel, 'what in the name of creation have you come up for, then? You can't leave your men!'

He immediately gave orders to count the lamps. In the morning three hundred and twenty-two had been distributed, and now only two hundred and fifty-five could be found; but several men acknowledged that in the hustling and panic they had dropped theirs and left them behind. An attempt was made to call over the men, but it was impossible to establish the exact number. Some of the miners had gone away, others did not hear their names. No one was agreed as to the number of the missing mates. It might be twenty, perhaps forty. And the engineer could only make out one thing with certainty: there were men down below, for their yells could be distinguished through the sound of the water and the fallen scaffolding, on leaning over the mouth of the shaft.

Négrel's first care was to send for M. Hennebeau, and to try to close the pit; but it was already too late. The colliers who had rushed to the Deux-Cent-Quarante settlement, as though pursued by the cracking tubbing, had frightened the families; and bands of women, old men, and little ones came up running, shaken by cries and sobs. They had to be pushed back, and a line of overseers was formed to keep them off, for they would have interfered with the operations. Many of the men who had come up from the shaft remained there stupidly without thinking of changing their clothes, riveted by fear before this terrible hole in which they had nearly remained for ever. The women, rushing wildly around them, implored them for names. Was So-and-so among them? and that one? and this other? They did not know, they stammered; they shuddered terribly, and made gestures like madmen, gestures which seemed to be pushing away some abominable vision which was always present to them. The crowd rapidly increased, and lamentations arose from the roads. And up there on the pit-bank, in Bonnemort's cabin, on the ground was seated a man, Souvarine, who had not gone away, who was looking on.

'The names! the names!' cried the women, with voices choked by tears.

Négrel appeared for a moment, and said hurriedly:

'As soon as we know the names they shall be given out, but nothing is lost so far: every one will be saved. I am going down.'

Then, silent with anguish, the crowd waited. The engineer, in fact, with quiet courage was preparing to go down. He had had the cage unfastened, giving orders to replace it at the end of the cable by a tub; and as he feared that the water would extinguish his lamp, he had another fastened beneath the tub, which would protect it.

Several captains, trembling and with white, disturbed faces, assisted in these preparations.

'You will come with me, Dansaert,' said Négrel, abruptly.

Then, when he saw them all without courage, and that the head captain was tottering, giddy with terror, he pushed him aside with a movement of contempt.

'No, you will be in my way. I would rather go alone.'

He was already in the narrow bucket, which swayed at the end of the cable; and holding his lamp in one hand and the signal-cord in the other, he shouted to the engine-man:

'Gently!'

The engine set the drums in movement, and Négrel disappeared in the gulf, from which the yells of the wretches below still arose.

At the upper part nothing had moved. He found that the tubbing here was in good condition. Balanced in the middle of the shaft he lighted up the walls as he turned round; the leaks between the joints were so slight that his lamp did not suffer. But at three hundred metres, when he reached the lower tubbing, the lamp was extinguished, as he expected, for a jet had filled

N 897

the tub. After that he was only able to see by the hanging lamp which preceded him in the darkness, and, in spite of his courage, he shuddered and turned pale in the face of the horror of the disaster. A few pieces of timber alone remained; the others had fallen in with their frames. Behind, enormous cavities had been hollowed out, and the yellow sand, as fine as flour, was flowing in considerable masses; while the waters of the Torrent, that subterranean sea with its unknown tempests and shipwrecks, were discharging in a flow like a weir. He went down lower, lost in the midst of these chasms which continued to multiply, beaten and turned round by the waterspout of the springs, so badly lighted by the red star of the lamp moving on below, that he seemed to distinguish the roads and squares of some destroyed town far away in the play of the great moving shadows. No human work was any longer possible. His only remaining hope was to attempt to save the men in peril. As he sank down he heard the cries becoming louder, and he was obliged to stop; an impassable obstacle barred the shaft—a mass of scaffolding, the broken joists of the guides, the split brattices entangled with the metal-work torn from the pump. As he looked on for a long time with aching heart, the yelling suddenly ceased. No doubt, the rapid rise of the water had forced the wretches to flee into the galleries, if, indeed, the flood had not already filled their mouths.

Négrel resigned himself to pulling the signal-cord as a sign to draw up. Then he had himself stopped again. He could not conceive the cause of this sudden accident. He wished to investigate it, and examined those pieces of the tubbing which were still in place. At a distance the tears and cuts in the wood had surprised him. His lamp, drowned in dampness, was going out, and, touching with his fingers, he clearly recognized the marks of the saw and of the wimble—the whole abominable labour of destruction. Evidently this catastrophe had been intentionally produced. He was stupefied, and the pieces of timber, cracking and falling down with their frames in a last slide, nearly carried him with them. His courage fled. The thought of the man who had done that made his hair stand on end, and froze him with a supernatural fear of evil, as though, mixed with the darkness, the man were still there paying for his immeasurable crime. He shouted and shook the cord furiously; and it was, indeed, time, for he perceived that the upper tubbing, a hundred metres higher, was in its turn beginning to move. The joints were opening, losing their oakum caulking, and streams were rushing through. It was now only a question of hours before the tubbing would all fall down.

At the surface M. Hennebeau was anxiously waiting for Négrel.

'Well, what?' he asked.

But the engineer was choked, and could not speak; he felt faint.

'It is not possible; such a thing was never seen. Have you examined?'

He nodded with a cautious look. He refused to talk in the presence of some captains who were listening, and led his uncle ten metres away, and not thinking this far enough, drew still farther back; then, in a low whisper, he at last told of the outrage, the torn and sawn planks, the pit bleeding at the neck and groaning. Turning pale, the manager also lowered his voice, with that instinctive need of silence in face of the monstrosity of great orgies and great crimes. It was useless to look as though they were trembling before the ten thousand Montsou men; later on they would see. And they both continued whispering, overcome at the thought that a man had had the courage to go down, to hang in the midst of space, to risk his life twenty times over in this terrible task. They could not even understand this mad courage in destruction; they refused to believe, in spite of the evidence, just as we doubt those stories of celebrated escapes of prisoners who fly through windows thirty metres above the ground.

When M. Hennebeau came back to the captains a nervous spasm was drawing his face. He made a gesture of despair, and gave orders that the mine should be evacuated at once. It was a kind of funeral procession, in silent abandonment, with glances thrown back at those great masses of bricks, empty and still standing, but which nothing henceforth could save.

And as the manager and the engineer came down last from the receiving-room, the crowd met them with its clamour, repeating obstinately:

'The names! the names! Tell us the names!'

Maheude was now there, among the women. She recollected the noise in the night; her daughter and the lodger must have gone away together, and they were certainly down at the bottom. And after having cried that it was a good thing, that they deserved to stay there, the heartless cowards, she had run up, and was standing in the first row, trembling with anguish. Besides, she no longer dared to doubt; the discussion going on around her informed her as to the names of those who were down. Yes, yes, Catherine was among them, Étienne also—a mate had seen them. But there was not always agreement with regard to the others. No, not this one; on the contrary, that one, perhaps Chaval, with whom, however, a trammer declared that he had ascended. The Levaque and Pierronne, although none of their people were in danger, cried out and lamented as loudly as the others. Zacharie, who had come up among the first, in spite of his inclination to make fun of everything had weepingly kissed his wife and mother, and remained near the latter, quivering, and showing an unexpected degree of affection for his sister, refusing to believe that she was below so long as the bosses made no authoritative statement.

'The names! the names! For pity's sake, the names!'

Négrel, who was exhausted, shouted to the overseers:

'Can't you make them be still? It 's enough to kill one with vexation! We don't know the names!'

Two hours passed away in this manner. In the first terror no one had thought of the other shaft at the old Réquillart mine. M. Hennebeau was about to announce that the rescue would be attempted from that side, when a rumour ran round: five men had just escaped the inundation by climbing up the rotten ladders of the old unused passage, and Father Mouque was named. This caused surprise, for no one knew he was below. But the narrative of the five who had escaped increased the weeping; fifteen mates had not been able to follow them, having gone astray, and been walled up by falls. And it was no longer possible to assist them, for there were already ten metres of water in Réquillart. All the names were known, and the air was filled with the groans of a slaughtered multitude.

'Will you make them be still?' Négrel repeated furiously. 'Make them draw back! Yes, yes, to a hundred metres! There is danger; push them back, push them back!'

It was necessary to struggle against these poor people. They were imagining all sorts of misfortunes, and they had to be driven away so that the deaths might be concealed; the captains explained to them that the shaft would destroy the whole mine. This idea rendered them mute with terror, and they at last allowed themselves to be driven back step by step; the guards, however, who kept them back had to be doubled, for they were fascinated by the spot and continually returned. Thousands of people were hustling each other along the road; they were running up from all the settlements, and even from Montsou.

And the man above, on the pit-bank, the fair man with the girlish face, smoked cigarettes to occupy himself, keeping his clear eyes fixed on the pit.

Then the wait began. It was midday; no one had eaten, but no one moved away. In the misty sky, of a dirty grey colour, rusty clouds were slowly passing by. A big dog, behind Rasseneur's hedge, was barking furiously without cessation, irritated by the living breath of the crowd. And the crowd had gradually spread over the neighbouring ground, forming a circle at a hundred metres round the pit. The Voreux arose in the centre of the great space. There was not a soul there, not a sound; it was a desert. The windows and the doors, left open, showed the abandonment within; a forgotten red cat, divining the peril in this solitude, jumped from a staircase and disappeared. No doubt the stoves of the boilers were scarcely extinguished, for the tall brick chimney gave out a light smoke beneath the dark clouds; while the weathercock on the steeple creaked in the wind with a short, shrill cry, the only melancholy voice of these vast buildings which were about to die.

At two o'clock nothing had moved. M. Hennebeau, Négrel, and other engineers who had hastened up, formed a group in black coats and hats standing in front of the crowd; and they, too, did not move away, though their legs were aching with fatigue, and they were feverish and ill at their impotence in the face of such a disaster, only whispering occasional words as though at a dying person's bedside. The upper tubbing must nearly all have fallen in, for sudden echoing sounds could be heard as of deep broken falls, succeeded by silence. The wound was constantly enlarging; the landslip which had begun below was rising and approaching the surface. Négrel was seized by nervous impatience; he wanted to see, and he was already advancing alone into this awful void when he was seized by the shoulders. What was the good? he could prevent nothing. An old miner, however, circumventing the overseers, rushed into the shed; but he quietly reappeared, he had gone for his sabots.

Three o'clock struck. Still nothing. A falling shower had soaked the crowd, but they had not withdrawn a step. Rasseneur's dog had begun to bark again. And it was at twenty minutes past three only that the first shock was felt. The Voreux trembled, but continued solid and upright. Then a second shock followed immediately, and a long cry came from open mouths; the tarred screening-shed, after having tottered twice, had fallen down with a terrible crash. Beneath the enormous pressure the structures broke and jarred each other so powerfully that sparks leapt out. From this moment the earth continued to tremble, the shocks succeeded one another, subterranean downfalls, the rumbling of a volcano in eruption. Afar the dog was no longer barking, but he howled plaintively as though announcing the oscillations which he felt coming; and the women, the children, all these people who were looking on, could not keep back a clamour of distress at each of these blows which shook them. In less than ten minutes the slate roof of the steeple fell in, the receiving-room and the engine-room were split open, leaving a considerable breach. Then the sounds ceased, the downfall stopped, and there was again deep silence.

For an hour the Voreux remained thus, broken into, as though bombarded by an army of barbarians. There was no more crying out; the enlarged circle of spectators merely looked on. Beneath the piled-up beams of the sifting-shed, fractured tipping-cradles could be made out with broken and twisted hoppers. But the rubbish had especially accumulated at the receiving-room, where there had been a rain of bricks, and large portions of wall and masses of plaster had fallen in. The iron scaffold which bore the pulleys had bent, half buried in the pit; a cage was still suspended, a torn cable-end was hanging; then there was a hash of trams, metal plates, and ladders. By some chance the lamp cabin remained standing, exhibiting on the left its bright rows of little lamps. And at the end of its disembowelled chamber, the engine could be seen seated squarely on its massive foundation of masonry; its copper was shining, and its huge steel limbs seemed to possess indestructible muscles. The enormous crank, bent in the air, looked like the powerful knee of some giant quietly reposing in his strength.

After this hour of respite, M. Hennebeau's hopes began to rise. The movement of the soil must have come to an end, and there would be some chance of saving the engine and the remainder of the buildings. But he would not yet allow any one to approach, considering another half-hour's patience desirable. This waiting became unbearable; the hope increased the anguish, and all hearts were beating quickly. A dark cloud, growing large at the horizon, hastened the twilight, a sinister dayfall over this wreck of earth's tempests. Since seven o'clock they had been there without moving or eating.

And suddenly, as the engineers were cautiously advancing,

a supreme convulsion of the soil put them to flight. Subterranean detonations broke out; a whole monstrous artillery was cannonading in the gulf. At the surface, the last buildings were tipped over and crushed. At first a sort of whirlwind carried away the rubbish from the sifting-shed and the receiving-room. Next, the boiler buildings burst and disappeared. Then it was the low square tower, where the pumping-engine was groaning, which fell on its face like a man mown down by a bullet. And then a terrible thing was seen; the engine, dislocated from its massive foundation, with broken limbs was struggling against death; it moved, it straightened its crank, its giant's knee, as though to rise; but, crushed and swallowed up, it was dying. The chimney alone, thirty metres high, still remained standing, though shaken, like a mast in the tempest. It was thought that it would be crushed to fragments and fly to powder, when suddenly it sank in one block, drunk down by the earth, melted like a colossal candle; and nothing was left, not even the point of the lightning conductor. It was done for; the evil beast crouching in this hole, gorged with human flesh, was no longer breathing with its thick, long respiration. The Voreux had been swallowed whole by the abyss.

The crowd rushed away yelling. The women hid their eyes as they ran. Terror drove the men along like a pile of dry leaves. They wished not to shout and they shouted, with swollen breasts, and arms in the air, before the immense hole which had been hollowed out. This crater, as of an extinct volcano, fifteen metres deep, extended from the road to the canal for a space of at least forty metres. The whole square of the mine had followed the buildings, the gigantic platforms, the footbridges with their rails, a complete train of trams, three wagons; without counting the wood supply, a forest of cut timber, gulped down like straw. At the bottom it was only possible to distinguish a confused mass of beams, bricks, iron, plaster, frightful remains, piled up, entangled, soiled in the fury of the catastrophe. And the hole became larger, cracks started from the edges, reaching afar, across the fields. A fissure ascended as far as Rasseneur's bar, and his front wall had cracked. Would the settlement itself pass into it? How far ought they to flee to reach shelter at the end of this abominable day, beneath this leaden cloud which also seemed about to crush the earth?

A cry of pain escaped Négrel. M. Hennebeau, who had drawn back, was in tears. The disaster was not complete; one bank of the canal gave way, and the canal emptied itself like one bubbling sheet through one of the cracks. It disappeared there, falling like a cataract down a deep valley. The mine drank down this river; the galleries would now be submerged for years. Soon the crater was filled and a lake of muddy water occupied the place where once stood the Voreux, like one of those lakes beneath which sleep accursed towns. There was a terrified silence, and nothing now could be heard but the fall of this water rumbling in the bowels of the earth.

Then on the shaken pit-bank Souvarine rose up. He had recognized Maheude and Zacharie sobbing before this downfall, the weight of which was so heavy on the heads of the wretches who were in agony beneath. And he threw down his last cigarette; he went away, without looking back, into the now dark night. Afar his shadow diminished and mingled with the darkness. He was going over there, to the unknown. He was going tranquilly to extermination, wherever there may be dynamite to blow up towns and men. He will be there, without doubt, when the middle class in agony shall hear the pavement of the streets bursting up beneath their feet.

CHAPTER IV

On the night that followed the collapse of the Voreux M. Hennebeau started for Paris, wishing to inform the directors in person before the newspapers published the news. And when he returned on the following day he appeared to be quite calm, with his usual correct administrative air. He had evidently freed himself from responsibility; he did not appear to have decreased in favour. On the contrary, the decree appointing him officer of the Legion of Honour was signed twenty-four hours afterwards.

But if the manager remained safe, the Company was tottering beneath the terrible blow. It was not the few million francs that had been lost, it was the wound in the flank, the deep incessant fear of the morrow in face of this massacre of a mine. The Company was so impressed that once more it felt the need of silence. What was the good of stirring up this abomination? If the villain were discovered, why make a martyr of him in order that his awful heroism might turn other heads, and give birth to a long line of incendiaries and murderers? Besides, the real culprit was not suspected. The Company came to think that there was an army of accomplices, not being able to believe that a single man could have had courage and strength for such a task; and it was precisely this thought which weighed on them, this thought of an ever-increasing threat to the existence of their mines. The manager had received orders to organize a vast system of espionage, and then to dismiss quietly, one by one, the dangerous men who were suspected of having had a hand in the crime. They contented themselves with this method of purification—a prudent and politic method.

There was only one immediate dismissal, that of Dansaert, the head captain. Ever since the scandal at Pierronne's house he had become impossible. A pretext was made of his attitude in danger, the cowardice of a captain abandoning his men. This was also a prudent sop thrown to the miners, who hated him.

Among the public, however, many rumours had circulated, and the directors had to send a letter of correction to one newspaper, contradicting a story in which mention was made of a barrel of powder lighted by the strikers. After a rapid inquiry the Government inspector had concluded that there had been a natural rupture of the tubbing, occasioned by the piling up of the soil; and the Company had preferred to be silent, and to accept the blame of a lack of superintendence. In the Paris press, after the third day, the catastrophe had served to increase the stock of general news; nothing was talked of but the men perishing at the bottom of the mine, and the telegrams published every morning were eagerly read. At Montsou people grew pale and speechless at the very name of the Voreux, and a legend had formed which made the boldest tremble as they whispered it. The whole country showed great pity for the victims; visits were organized to the destroyed pit, and whole families hastened up to shudder at the ruins which lay so heavily over the heads of the buried wretches.

Deneulin, who had been appointed divisional engineer, came upon the midst of the disaster on beginning his duties; and his first care was to turn the canal back into its bed, for this torrent increased the damage every hour. Extensive works were necessary, and he at once set a hundred men to construct a dyke. Twice over the impetuosity of the stream carried away the first dams. Now pumps were set up and a furious struggle was going on; step by step the vanished soil was being violently reconquered.

But the rescue of the engulfed miners was a still more absorbing work. Négrel was appointed to attempt a supreme effort, and arms were not lacking to help him; all the colliers rushed to offer themselves in an outburst of brotherhood. They forgot the strike, they did not trouble themselves at all about payment; they might get nothing, they only asked to risk their lives as soon as there were mates in danger of death. They were all there with their tools, quivering as they waited to know where they ought to strike. Many of them, sick with fright after the accident, shaken by nervous tremors, soaked in cold sweats, and the prey of continual nightmares, got up in spite of everything, and were as eager as any in their desire to fight against the earth, as though they had a revenge to take on it. Unfortunately, the difficulty began when the question arose, What could be done? how could they go down? from what side could they attack the rocks?

Négrel's opinion was that not one of the unfortunate people was alive; the fifteen had surely perished, drowned or suffocated. But in these mine catastrophes the rule is always to assume that buried men are alive, and he acted on this supposition. The first problem which he proposed to himself was to decide where they could have taken refuge. The captains and old miners whom he consulted were agreed on one point: in the face of the rising water the men had certainly come up from gallery to gallery to the highest cuttings, so that they were, without doubt, driven to the end of the upper passages. This agreed with Father Mouque's information, and his confused narrative even gave reason to suppose that in the wild flight the band had separated into smaller groups, leaving fugitives on the road at every level. But the captains were not unanimous when the discussion of possible attempts at rescue arose. As the passages nearest to the surface were a hundred and fifty metres down, there could be no question of sinking a shaft. Réquillart remained the one means of access, the only point by which they could approach. The worst was that the old pit, now also inundated, no longer communicated with the Voreux; and above the level of the water only a few ends of galleries belonging to the first level were left free. The pumping process would require years, and the best plan would be to visit these galleries and ascertain if any of them approached the submerged passages at the end of which the distressed miners were suspected to be. Before logically arriving at this point, much discussion had been necessary to dispose of a crowd of impracticable plans.

Négrel now began to stir up the dust of the archives; he discovered the old plans of the two pits, studied them, and decided on the points at which their investigations ought to be carried on. Gradually this hunt excited him; he was, in his turn, seized by a fever of devotion, in spite of his ironical indifference to men and things. The first difficulty was in going down at Réquillart; it was necessary to clear out the rubbish from the mouth of the shaft, to cut down the mountain ash, and raze the sloes and the hawthorns; they had also to repair the ladders. Then they began to feel around. The engineer, having gone down with ten workmen, made them strike the iron of their tools against certain parts of the seam which he pointed out to them; and in deep silence they each placed an ear to the coal, listening for any distant blows to reply. But they went in vain through every practicable gallery; no echo returned to them. Their embarrassment increased. At what spot should they cut into the bed? Towards whom should they go, since no one appeared to be there? They persisted in seeking, however, notwithstanding the exhaustion produced by their growing anxiety.

On the first day, Maheude came in the morning to Réquillart. She sat down on a beam in front of the shaft, and did not stir from it till the evening. When a man came up, she rose and questioned him with her eyes: Nothing? No, nothing! And she sat down again, and waited still, without a word, with hard, fixed face. Jeanlin also, seeing that his den was invaded, prowled around with the frightened air of a beast of prey whose burrow will betray his booty. He thought of the little soldier lying beneath the rocks, fearing lest they should trouble his sound sleep; but that side of the mine was beneath the water, and, besides, their investigations were directed more to the left, in the west gallery. At first, Philomène had also come, accompanying Zacharie, who was one of the gang; then she became wearied at catching cold, without need or result, and went back to the settlement, dragging through her days, a limp, indifferent woman, occupied from morning to night in coughing. Zacharie, on the contrary, lived for nothing else; he would have devoured the soil to get back his sister. At night he shouted out that he saw her, he heard her, very lean from hunger, her chest sore with calling for help. Since that he had tried to dig without orders, saying that it was there, that he was sure of it.

The engineer would not let him come down any more, and he would not go away from the pit, from which he was driven off; he could not even sit down and wait near his mother, he was so deeply stirred by the need to act, which drove him constantly on.

It was the third day. Négrel, in despair, had resolved to abandon the attempt in the evening. At midday, after lunch, when he came back with his men to make one last effort, he was surprised to see Zacharie, red and gesticulating, come out of the mine shouting:

'She 's there! She 's replied to me! Come along, quickly!'

He had slid down the ladders, in spite of the watchman, and was declaring that he had heard hammering over there, in the first passage of the Guillaume seam.

'But we have already been twice in that direction,' Négrel observed, sceptically. 'Anyhow, we 'll go and see.'

Maheude had risen, and had to be prevented from going down. She waited, standing at the edge of the shaft, gazing down into the darkness of the hole.

Négrel, down below, himself struck three blows, at long intervals. He then applied his ear to the coal, cautioning the workers to be very silent. Not a sound reached him, and he shook his head; evidently the poor lad was dreaming. In a fury, Zacharie struck in his turn, and listened anew with bright eyes and limbs trembling with joy. Then the other workmen tried the experiment, one after the other, and all grew animated, hearing the distant reply, very far away. The engineer was astonished; he again applied his ear, and was at last able to catch a sound of aerial softness, a rhythmical roll scarcely to be distinguished, the well-known cadence beaten by the miners when they are fighting against the coal in the midst of danger. The coal transmits the sound with crystalline limpidity for a very great distance. A captain who was there estimated that the thickness of the block which separated them from their mates could not be less than fifty metres. But it seemed as if they could already stretch out a hand to them, and general gladness broke out. Négrel decided to begin at once the work of approach.

When Zacharie, up above, saw Maheude again, they embraced each other.

'It won't do to get excited,' Pierronne, who had come for a visit of inquisitiveness, was cruel enough to say. 'If Catherine isn't there, it would be such a grief afterwards!'

That was true; Catherine might be somewhere else.

'Just leave me alone, will you? Damn it!' cried Zacharie, in a rage. 'She 's there; I know it!'

Maheude sat down again in silence, with motionless face, continuing to wait.

As soon as the story was spread at Montsou, a new crowd arrived. Nothing was to be seen; but they remained there all the same, and had to be kept at a distance. Down below, the work went on day and night. For fear of meeting an obstacle, the engineer had had three descending galleries opened in the seam, converging to the point where the enclosed miners were supposed to be. Only one pikeman could hew at the coal on the narrow face of the tube; he was relieved every two hours, and the coal piled in was passed up, from hand to hand, by a chain of men, increased as the hole was hollowed out. The work at first proceeded very quickly; they did six metres a day.

Zacharie had secured a place among the workers chosen for the hewing. It was a post of honour which was disputed over, and he became furious when they wished to relieve him after his two hours of regulation labour. He robbed his mates of their turn, and refused to let go the pick. His gallery was soon in advance of the others. He fought against the coal so fiercely that his breath could be heard coming from the tube like the roar of a forge within his breast. When he came out, black and muddy, dizzy with fatigue, he fell to the ground and had to be wrapped up in a covering. Then, still tottering, he plunged back again, and the struggle began anew—the low, deep blows, the stifled groans, the victorious fury of massacre. The worst was that the coal now became hard; he twice broke his tool, and was exasperated that he could not get on so fast. He suffered also from the heat, which increased with every metre of advance, and was unbearable at the end of this narrow hole where the air could not circulate. A hand ventilator worked well, but aeration was so inadequate that on three occasions it was necessary to take out fainting hewers who were being asphyxiated.

Négrel lived below with his men. His meals were sent down to him, and he sometimes slept for a couple of hours on a truss of straw, rolled in a cloak. The one thing that kept them up was the supplication of the wretches beyond, the call which was sounded ever more distinctly to hasten on the rescue. It now rang very clearly with a musical sonority, as though struck on the plates of a harmonica. It led them on; they advanced to this crystalline sound as men advance to the sound of cannon in battle. Every time that a pikeman was relieved, Négrel went down and struck, then applying his ear; and every time, so far, the reply had come, rapid and urgent. He had no doubt remaining; they were advancing in the right direction, but with what fatal slowness! They would never arrive soon enough. On the first two days they had indeed hewn through thirteen metres; but on the third day they fell to five, and then on the fourth to three. The coal was becoming closer and harder, to such an extent that they now with difficulty struck through two metres. On the ninth day, after superhuman efforts, they had advanced thirty-two metres, and calculated that some twenty must still be left before them. For the prisoners it was the beginning of the twelfth day; twelve times over had they passed twenty-four hours without bread, without fire, in that icy darkness! This awful idea moistened the eyelids and stiffened the arms of the workers. It seemed impossible that Christians could live longer. The distant blows had become weaker since the previous day, and every moment they trembled lest they should stop.

Maheude came regularly every morning to sit at the mouth of the shaft. In her arms she brought Estelle, who could not remain alone from morning to night. Hour by hour she followed the workers, sharing their hopes and fears. There was feverish expectation among the groups standing around, and even as far as Montsou, with endless discussion. Every heart in the district was beating down there beneath the earth.

On the ninth day, at the breakfast hour, no reply came from Zacharie when he was called for the relay. He was like a madman, working on furiously with oaths. Négrel, who had come up for a moment, was not there to make him obey, and only a captain and three miners were below. No doubt Zacharie, infuriated with the feeble vacillating light, committed the imprudence of opening his lamp, although severe orders had been given, for leakages of fire-damp had taken place, and the gas remained in enormous masses in these narrow, unventilated passages. Suddenly, a roar of thunder was heard, and a spout of fire darted out of the tube as from the mouth of a cannon charged with grapeshot. Everything flamed up and the air caught fire like powder, from one end of the galleries to the other. This torrent of flame carried away the captain and three workers, ascended the pit, and leapt up to the daylight in an eruption which hid the rocks and the ruins around. The inquisitive fled, and Maheude arose, pressing the frightened Estelle to her breast.

When Négrel and the men came back they were seized by a terrible rage. They struck their heels on the earth as on a step-mother who was killing her children at random in the imbecile whims of her cruelty. They were devoting themselves, they were coming to the help of their mates, and still they must lose some of their men! After three long hours of effort and danger they reached the galleries once more, and the melancholy ascent of the victims took place. Neither the captain nor the workers were dead, but they were covered by awful wounds which gave out an odour of grilled flesh; they had drunk of fire, the burns had got into their throats, and they constantly moaned and prayed to be left alone. One of the three miners was the man who had smashed the pump at Gaston-Marie with a final blow of the shovel during the strike; the two others still had scars on their hands, and grazed, torn fingers from the energy with which they had thrown bricks at the soldiers. The pale and shuddering crowd took off their hats when they were carried by.

Maheude stood waiting. Zacharie's body at last appeared. The clothes were burnt, the body was nothing but black charcoal, calcined and unrecognizable. The head had been smashed by the explosion and no longer existed. And when these awful remains were placed on a stretcher, Maheude followed them mechanically, her burning eyelids without a tear. With Estelle drowsily lying in her arms, she went along, a tragic figure, her hair lashed by the wind. At the settlement Philomène seemed stupid; her eyes were turned into fountains and she was quickly relieved. But the mother had already returned with the same step to Réquillart; she had accompanied her son, she was returning to wait for her daughter.

Three more days passed by. The rescue work had been resumed amid incredible difficulties. The galleries of approach had fortunately not fallen after the fire-damp explosion; but the air was so heavy and so vitiated that more ventilators had to be installed. Every twenty minutes the pikemen relieved one another. They were advancing; scarcely two metres separated them from their mates. But now they worked feeling cold at their hearts, striking hard only out of vengeance; for the noises had ceased, and the low, clear cadence of the call no longer sounded. It was the twelfth day of their labours, the fifteenth since the catastrophe; and since the morning there had been a death-like silence.

The new accident increased the curiosity at Montsou, and the inhabitants organized excursions with such spirit that the

Grégoires decided to follow the fashion. They arranged a party, and it was agreed that they should go to the Voreux in their carriage, while Madame Hennebeau took Lucie and Jeanne there in hers. Deneulin would show them over his yards, and then they would return by Réquillart, where Négrel would tell them the exact state of things in the galleries, and if there was still hope. Finally, they would dine together in the evening.

When the Grégoires and their daughter Cécile arrived at the ruined mine, toward three o'clock, they found Madame Hennebeau already there, in a sea-blue dress, protecting herself under her parasol from the pale February sun. The warmth of spring was in the clear sky. M. Hennebeau was there with Deneulin, and she was listening, with listless ear, to the account which the latter gave her of the efforts which had been made to dam up the canal. Jeanne, who always carried a sketch-book with her, began to draw, carried away by the horror of the subject; while Lucie, seated beside her on the remains of a wagon, was crying out with pleasure, and finding it awfully jolly. The incomplete dam allowed numerous leaks, and frothy streams fell in a cascade down the enormous hole of the engulfed mine. The crater was being emptied, however, and the water, drunk by the earth, was sinking, and revealing the fearful ruin at the bottom. Beneath the tender azure of this beautiful day there lay a sewer, the ruins of a town drowned and melted in mud.

'And people come out of their way to see that!' exclaimed M. Grégoire, disillusioned.

Cécile, rosy with health and glad to breathe so pure an air, was cheerfully joking, while Madame Hennebeau made a little grimace of repugnance as she murmured:

'The fact is, this is not pretty at all.'

The two engineers laughed. They tried to interest the visitors, taking them round and explaining to them the working of the pumps and the manipulation of the stamper which drove in the piles. But the ladies became anxious. They shuddered when they knew that the pumps would have to work for six or seven years before the shaft was reconstructed and all the water exhausted from the mine. No, they would rather think of something else; this destruction was only good to give bad dreams.

'Let us go,' said Madame Hennebeau, turning towards her carriage.

Lucie and Jeanne protested. What! so soon! and the drawing which was not finished! They wanted to remain; their father would bring them to dinner in the evening. M. Hennebeau alone took his place with his wife in the carriage, for he wished to question Négrel.

'Very well! go on before,' said M. Grégoire. 'We will follow you; we have a little visit of five minutes to make over there at the settlement. Go on, go on! we shall be at Réquillart as soon as you.'

He got up behind Madame Grégoire and Cécile, and while the other carriage went along by the canal, theirs gently ascended the slope.

Their excursion was to be completed by a visit of charity. Zacharie's death had filled them with pity for this tragical Maheu family, about whom the whole country was talking. They had no pity for the father, that brigand, that slayer of soldiers, who had to be struck down like a wolf. But the mother touched them, that poor woman who had just lost her son after having lost her husband, and whose daughter was perhaps a corpse beneath the earth; to say nothing of an invalid grandfather, a child who was lame as the result of a landslip, and a little girl who died of starvation during the strike. So that, though this family had in part deserved its misfortunes by the detestable spirit it had shown, they had resolved to assert the breadth of their charity, their desire for forgetfulness and conciliation, by themselves bringing an alms. Two parcels, carefully wrapped up, had been placed beneath a seat of the carriage.

An old woman pointed out to the coachman Maheude's house, No. 16 in the second block. But when the Grégoires alighted with the parcels, they knocked in vain; at last they struck their fists against the door, still without reply; the house echoed mournfully, like a house emptied by grief, frozen and dark, long since abandoned.

'There 's no one there,' said Cécile, disappointed. 'What a nuisance! What shall we do with all this?'

Suddenly the door at the side opened, and the Levaque woman appeared.

'Oh, sir! I beg pardon, ma'am. Excuse me, miss. It 's the neighbour that you want? She 's not there; she 's at Réquillart.'

With a flow of words she told them the story, repeating to them that people must help one another, and that she was keeping Lénore and Henri in her house to allow the mother to go and wait over there. Her eyes had fallen on the parcels, and she began to talk about her poor daughter, who had become a widow, displaying her own wretchedness, while her eyes shone with covetousness. Then, in a hesitating way, she muttered:

'I 've got the key. If the lady and gentleman would really like—— The grandfather is there.'

The Grégoires looked at her in stupefaction. What! The grandfather was there! But no one had replied. He was sleeping, then? And when the Levaque made up her mind to open the door, what they saw stopped them on the threshold. Bonnemort was there alone, with large fixed eyes, nailed to his chair in front of the cold fireplace. Around him the room appeared larger without the clock or the polished deal furniture which formerly animated it; there only remained against the green crudity of the walls the portraits of the emperor and empress, whose rosy lips were smiling with official benevolence. The old man did not stir nor wink his eyelids beneath the sudden light from the door; he seemed imbecile, as though he had not seen all these people come in. At his feet lay his plate, garnished with ashes, such as is placed for cats for ordure.

'Don't mind if he 's not very polite,' said the Levaque woman, obligingly. 'Seems he 's broken something in his brain. It 's a fortnight since he left off speaking.'

But Bonnemort was shaken by some agitation, a deep scraping which seemed to arise from his belly, and he expectorated into the plate a thick black expectoration. The ashes were soaked into a coaly mud, all the coal of the mine which he drew from his chest. He had already resumed his immobility. He stirred no more, except at intervals, to spit.

Uneasy, and with stomachs turned, the Grégoires endeavoured to utter a few friendly and encouraging words.

'Well, my good man,' said the father, 'you have a cold, then?'

The old man, with his eyes to the wall, did not turn his head. And a heavy silence fell once more.

'They ought to make you a little gruel,' added the mother.

He preserved his mute stiffness.

'I say, papa,' murmured Cécile, 'they certainly told us he was an invalid; only we did not think of it afterwards——'

She interrupted herself, much embarrassed. After having placed on the table a *pot-au-feu* and two bottles of wine, she undid the second parcel and drew from it a pair of enormous boots. It was the present intended for the grandfather, and she held one boot in each hand, in confusion, contemplating the poor man's swollen feet, which would never walk more.

'Eh! they come a little late, don't they, my worthy fellow?' said M. Grégoire again, to enliven the situation. 'It doesn't matter, they 're always useful.'

Bonnemort neither heard nor replied, with his terrible face as cold and as hard as a stone.

Then Cécile furtively placed the boots against the wall. But in spite of her precautions the nails clanked; and those enormous boots stood oppressively in the room.

'He won't say thank you,' said the Levaque woman, who had cast a look of deep envy on the boots. 'Might as well give a pair of spectacles to a duck, asking your pardon.'

She went on; she was trying to draw the Grégoires into her own house, where she hoped to gain their pity. At last she thought of a pretext; she praised Henri and Lénore, who were so good, so gentle, and so intelligent, answering like angels the questions that they were asked. They would tell the lady and gentleman all that they wished to know.

'Will you come for a moment, my child?' asked the father, glad to get away.

'Yes, I 'll follow you,' she replied.

Cécile remained alone with Bonnemort. What kept her there trembling and fascinated, was the thought that she seemed to recognize this old man: where then had she met this square livid face, tattooed with coal? Suddenly she remembered; she saw again a mob of shouting people who surrounded her, and she felt cold hands pressing her neck. It was he; she saw the man again; she looked at his hands placed on his knees, the hands of an invalid workman whose whole strength is in his wrists, still firm in spite of age. Gradually Bonnemort seemed to awake, he perceived her and examined her in his turn. A flame mounted to his cheeks, a nervous spasm drew his mouth, from which flowed a thin streak of black saliva. Fascinated, they remained opposite each other—she flourishing, plump, and fresh from the long idleness and sated comfort of her race; he swollen with water, with the pitiful ugliness of a foundered beast, destroyed from father to son by a century of work and hunger.

At the end of ten minutes, when the Grégoires, surprised at not seeing Cécile, came back into the Maheus' house, they uttered a terrible cry. Their daughter was lying on the ground, with livid face, strangled. At her neck fingers had left the red imprint of a giant's hand. Bonnemort, tottering on his dead legs, had fallen beside her without power to rise. His hands were still hooked, and he looked round with his imbecile air and large open eyes. In his fall he had broken his plate, the ashes were spread round, the mud of the black expectoration had stained the floor; while the great pair of boots, safe and sound, stood side by side against the wall.

It was never possible to establish the exact facts. Why had Cécile come near? How could Bonnemort, nailed to his chair, have been able to seize her throat? Evidently, when he held her, he must have become furious, constantly pressing, overthrown with her, and stifling her cries to the last groan. Not a sound, not a moan had traversed the thin partition to the neighbouring house. It seemed to be an outbreak of sudden madness, a longing to murder before this white young neck. Such savagery was stupefying in an old invalid, who had lived like a worthy man, an obedient brute, opposed to new ideas. What rancour, unknown to himself, by some slow process of poisoning, had risen from his bowels to his brain? The horror of it led to the conclusion that he was unconscious, that it was the crime of an idiot.

The Grégoires, meanwhile, on their knees, were sobbing, choked with grief. Their idolized daughter, that daughter desired so long, on whom they had lavished all their goods, whom they used to watch sleeping, on tiptoe, whom they never thought sufficiently well nourished, never sufficiently plump! It was the downfall of their very life; what was the good of living, now that they would have to live without her?

The Levaque woman in distraction cried:

'Ah, the old beggar! what 's he done there? Who would have expected such a thing? And Maheude, who won't come back till evening! Shall I go and fetch her?'

The father and mother were crushed, and did not reply.

'Eh? It will be better. I 'll go.'

But, before going, the Levaque woman looked at the boots. The whole settlement was excited, and a crowd was already hustling around. Perhaps they would get stolen. And then the Maheus had no man, now, to put them on. She quietly carried them away. They would just fit Bouteloup's feet.

At Réquillart the Hennebeaus, with Négrel, waited a long time for the Grégoires. Négrel, who had come up from the pit, gave details. They hoped to communicate that very evening with the prisoners, but they would certainly find nothing but corpses, for the death-like silence continued. Behind the engineer, Maheude, seated on the beam, was listening with white face, when the Levaque woman came up and told her the old man's strange deed. And she only made a sweeping gesture of impatience and irritation. She followed her, however.

Madame Hennebeau was much affected. What an abomination! That poor Cécile, so merry that very day, so full of life an hour before! M. Hennebeau had to lead his wife for a moment into old Mouque's hovel. With his awkward hands he unfastened her dress, troubled by the odour of musk which her open bodice exhaled. And as with streaming tears she clasped Négrel, terrified at this death which cut short the marriage, the husband watched them lamenting together, and was delivered from one anxiety. This misfortune would arrange everything; he preferred to keep his nephew for fear of his coachman.

CHAPTER V

At the bottom of the shaft the abandoned wretches were yelling with terror. The water now came up to their hips. The noise of the torrent dazed them, the final falling in of the tubbing sounded like the last crack of doom; and their bewilderment was completed by the neighing of the horses shut up in the stable, the terrible, unforgettable death-cry of an animal that is being slaughtered.

Mouque had let go Bataille. The old horse was there, trembling, with its dilated eye fixed on this water which was constantly rising. The pit-eye was rapidly filling; the greenish flood slowly enlarged under the red gleam of the three lamps which were still burning under the roof. And suddenly, when he felt this ice soaking his coat, he set out in a furious gallop, and was engulfed and lost at the end of one of the haulage galleries.

Then there was a general rush, the men following the beast.

'Nothing more to be done in this damned hole!' shouted Mouque. 'We must try at Réquillart.'

The idea that they might get out by the old neighbouring pit if they arrived before the passage was cut off, now carried them away. The twenty hustled one another as they went in single file, holding their lamps in the air so that the water should not extinguish them. Fortunately, the gallery rose with an imperceptible slope, and they proceeded for two hundred metres, struggling against the flood, which was not now gaining on them. Sleeping beliefs reawakened in these distracted souls; they invoked the earth, for it was the earth that was avenging herself, discharging the blood from the vein because they had cut one of her arteries. An old man stammered forgotten prayers, bending his thumbs backwards to appease the evil spirits of the mine.

But at the first turning disagreement broke out; the groom proposed turning to the left, others declared that they could make a short cut by going to the right. A minute was lost.

'Well, die there! what the devil does it matter to me?' Chaval brutally exclaimed. 'I go this way.'

He turned to the right, and two mates followed him. The others continued to rush behind Father Mouque, who had grown up at the bottom of Réquillart. He himself hesitated, however, not knowing where to turn. They lost their heads; even the old men could no longer recognize the passages, which lay like a tangled skein before them. At every bifurcation they were pulled up short by uncertainty, and yet they had to decide.

Étienne was running last, delayed by Catherine, who was paralysed by fatigue and fear. He would have gone to the right with Chaval, for he thought that the better road; but he had not, preferring to part from Chaval. The rush continued, however; some of the mates had gone from their side, and only seven were left behind old Mouque.

'Hang on to my neck and I will carry you,' said Étienne to the young girl, seeing her grow weak.

'No, let me be,' she murmured. 'I can't do more; I would rather die at once.'

They delayed and were left fifty metres behind; he was lifting her, in spite of her resistance, when the gallery was suddenly stopped up; an enormous block fell in and separated them from the others. The inundation was already soaking the soil, which was shifting on every side. They had to retrace their steps; then they no longer knew in what direction they were going. There was an end of all hope of escaping by Réquillart. Their only remaining hope was to gain the upper workings, from which they might perhaps be delivered if the water sank.

Étienne at last recognized the Guillaume seam.

'Good!' he exclaimed. 'Now I know where we are. By God! we were in the right road; but we may go to the devil now! Here, let us go straight on; we will climb up the passage.'

The flood was beating against their breasts, and they walked very slowly. As long as they had light they did not despair, and they blew out one of the lamps to economize the oil, meaning to empty it into the other lamp. They had reached the chimney passage, when a noise behind made them turn. Was it some mates, then, who had also found the road barred and were returning? A roaring sound came from afar; they could not understand this tempest which approached them, spattering foam. And they cried out when they saw a gigantic whitish mass coming out of the shadow and trying to rejoin them between the narrow timbering in which it was being crushed.

It was Bataille. On leaving the pit-eye he had wildly galloped along the dark galleries. He seemed to know his road in this subterranean town which he had inhabited for eleven years, and his eyes saw clearly in the depths of the eternal night in which he had lived. He galloped on and on, bending his head, drawing up his feet, passing through these narrow tubes in the earth, filled by his great body. Road succeeded to road, and the forked turnings were passed without any hesitation. Where was he going? Over there, perhaps, towards that vision of his youth, to the mill where he had been born on the bank of the Scarpe, to the confused recollection of the sun burning in the air like a great lamp. He desired to live, his beast's memory awoke; the longing to breathe once more the air of the plains drove him straight onwards to the discovery of that hole, the exit beneath the warm sun into light. Rebellion carried away his ancient resignation; this pit was murdering him after having blinded him. The water which pursued him was lashing him on the flanks and biting him on the crupper. But as he went deeper in, the galleries became narrower, the roofs lower, and the walls protruded. He galloped on in spite of everything, grazing himself, leaving fragments of his limbs on the timber. From every side the mine seemed to be pressing on to him to take him and to stifle him.

Then Étienne and Catherine, as he came near them, perceived that he was strangling between the rocks. He had stumbled and broken his two front legs. With a last effort, he dragged himself a few metres, but his flanks could not pass; he remained hemmed in and garrotted by the earth. With his bleeding head stretched out, he still sought for some crack with his great troubled eyes. The water was rapidly covering him; he began to neigh with that terrible prolonged death-rattle with which the other horses had already died in the stable. It was a sight of fearful agony, this old beast fractured and motionless, struggling at this depth, far from the daylight. The flood was drowning his mane, and his cry of distress never ceased; he uttered it more hoarsely, with his large open mouth stretched out. There was a last rumble, the hollow sound of a cask which is being filled; then deep silence fell.

'Oh, my God! take me away!' Catherine sobbed. 'Ah, my God! I 'm afraid; I don't want to die. Take me away take me away!'

She had seen death. The fallen shaft, the inundated mine, nothing had seized her with such terror as this clamour of Bataille in agony. And she constantly heard it; her ears were ringing with it; all her flesh was shuddering with it.

'Take me away! take me away!'

Étienne had seized her and lifted her; it was, indeed, time. They ascended the chimney passage, soaked to the shoulders. He was obliged to help her, for she had no strength to cling to the timber. Three times over he thought that she was slipping from him and falling back into that deep sea of which the tide was roaring beneath them. However, they were able to breathe for a few minutes when they reached the first gallery, which was still free. The water reappeared, and they had to hoist themselves up again. And for hours this ascent continued, the flood chasing them from passage to passage, and constantly forcing them to ascend. At the sixth level a respite rendered them feverish with hope, and it seemed that the waters were becoming stationary. But a more rapid rise took place, and they had to climb to the seventh and then to the eighth level. Only one remained, and when they had reached it they anxiously watched each centimetre by which the water gained on them. If it did not stop they would then die like the old horse, crushed against the roof, and their chests filled by the flood.

Landslips echoed every moment. The whole mine was shaken, and its distended bowels burst with the enormous flood which gorged them. At the end of the galleries the air, driven back, pressed together and crushed, exploded terribly amid split rocks and overthrown soil. It was a terrifying uproar of interior cataclysms, a remnant of the ancient battle when deluges overthrew the earth, burying the mountains beneath the plains.

And Catherine, shaken and dazed by this continuous downfall, joined her hands, stammering the same words without cessation:

'I don't want to die! I don't want to die!'

To reassure her, Étienne declared that the water was not now moving. Their flight had lasted for fully six hours, and they would soon be rescued. He said six hours without knowing, for they had lost all count of time. In reality, a whole day had already passed in their climb up through the Guillaume seam.

Drenched and shivering, they settled themselves down. She undressed herself without shame and wrung her clothes; then she put on again the jacket and breeches, and let them finish drying on her. As her feet were bare, he made her take his own sabots. They could wait patiently now; they had lowered the wick of the lamp, leaving only the feeble gleam of a night-light. But their stomachs were torn by cramp, and they both realized that they were dying of hunger. Up till now they had not felt that they were living. The catastrophe had occurred before breakfast, and now they found their bread-and-butter, swollen by the water and changed into sop. She had to become angry before he would accept his share. As soon as she had eaten she fell asleep from weariness, on the cold earth. He was devoured by insomnia, and watched over her with fixed eyes and forehead between his hands.

How many hours passed by thus? He would have been unable to say. All that he knew was that before him, through the hole they had ascended, he had seen the flood reappear, black and moving, the beast whose back was ceaselessly swelling out to reach them. At first it was only a thin line, a supple serpent stretching itself out; then it enlarged into a crawling, crouching flank; and soon it reached them, and the sleeping girl's feet were touched by it. In his anxiety he yet hesitated to wake her. Was it not cruel to snatch her from this repose of unconscious ignorance, which was, perhaps, lulling her with a dream of the open air and of life beneath the sun? Besides, where could they fly? And he thought and remembered that the upbrow established at this part of the seam communicated end to end with that which served the upper level. That would be a way out. He let her sleep as long as possible, watching the flood gain on them, waiting for it to chase them away. At last he lifted her gently, and a great shudder passed over her.

'Ah, my God! it 's true! it 's beginning again, my God!'

She remembered, she cried out, again finding death so near.

'No! calm yourself,' he whispered. 'We can pass, upon my word!'

To reach the upbrow they had to walk doubled up, again wetted to the shoulders. And the climbing began anew, now more dangerous, through this hole entirely of timber, a hundred metres long. At first they wished to pull the cable so as to fix one of the carts below, for if the other should come down during their ascent, they would be crushed. But nothing moved, some obstacle interfered with the mechanism. They ventured in, not daring to make use of the cable which was in their way, and tearing their nails against the smooth framework. He came behind, supporting her by his head when she slipped with torn hands. Suddenly they came across the splinters of a beam which barred the way. A portion of the soil had fallen down and prevented them from going any higher. Fortunately a door opened here and they passed into a passage. They were stupefied to see the flicker of a lamp in front of them. A man cried wildly to them:

'More clever people as big fools as I am!'

They recognized Chaval, who had found himself blocked by the landslip which filled the upbrow; his two mates who had set out with him had been left on the way with fractured skulls. He was wounded in the elbow, but had had the courage to go back on his knees, take their lamps, and search them to steal their bread-and-butter. As he escaped, a final downfall behind his back had closed the gallery.

He immediately swore that he would not share his victuals with these people who came up out of the earth. He would sooner knock their brains out. Then he, too, recognized them; his anger fell, and he began to laugh with a laugh of evil joy.

'Ah! it 's you, Catherine! you 've broken your nose, and you want to join your man again. Well, well! we 'll play out the game together.'

He pretended not to see Étienne. The latter, overwhelmed by this encounter, made a gesture as though to protect the putter, who was pressing herself against him. He must, however, accept the situation. Speaking as though they had left each other good friends an hour before, he simply asked:

'Have you looked down below? We can't pass through the cuttings, then?'

Chaval still grinned.

'Ah, bosh! the cuttings! They 've fallen in too; we are between two walls, a real mousetrap. But you can go back by the brow if you are a good diver.'

The water, in fact, was rising; they could hear it rippling. Their retreat was already cut off. And he was right; it was a mousetrap, a gallery-end obstructed before and behind by considerable falls of earth. There was not one issue; all three were walled up.

'Then you 'll stay?' Chaval added, jeeringly. 'Well, it 's the best you can do, and if you 'll just leave me alone, I shan't even speak to you. There 's still room here for two men. We shall soon see which will die first, provided they don't come to us, which seems a tough job.'

The young man said:

'If we were to hammer, they would, perhaps, hear us.'

'I 'm tired of hammering. Here, try yourself with this stone.'

Étienne picked up the fragment of sandstone which the other had already broken off, and against the seam at the end he struck the miner's call, the prolonged roll by which workmen in peril signal their presence. Then he placed his ear to listen. Twenty times over he persisted; no sound replied.

During this time Chaval affected to be coolly attending to his little household. First he arranged the three lamps against the wall; only one was burning, the others could be used later on. Afterwards, he placed on a piece of timber the two slices of bread-and-butter which were still left. That was the side-board; he could last quite two days with that, if he were careful. He turned round saying:

'You know, Catherine, there will be half for you when you are famished.'

The young girl was silent. It completed her unhappiness to find herself again between these two men.

And their awful life began. Neither Chaval nor Étienne opened their mouths, seated on the earth a few paces from each other. At a hint from the former the latter extinguished his lamp, a piece of useless luxury; then they sank back into silence. Catherine was lying down near Étienne, restless under the glances of her former lover. The hours passed by; they heard the low murmur of the water for ever rising; while from time to time deep shocks and distant echoes announced the final settling down of the mine. When the lamp was empty and they had to open another to light it, they were, for a moment, disturbed by the fear of fire-damp; but they would rather have been blown up at once than live on in darkness. Nothing exploded, however; there was no fire-damp. They stretched themselves out again, and the hours continued to pass by.

A noise aroused Étienne and Catherine, and they raised their heads. Chaval had decided to eat; he had cut off half a slice of bread-and-butter, and was chewing it slowly, to avoid the temptation of swallowing it all. They gazed at him, tortured by hunger.

'Well, do you refuse?' he said to the putter, in his provoking way. 'You 're wrong.'

She had lowered her eyes, fearing to yield; her stomach was torn by such cramps that tears were swelling beneath her eyelids. But she understood what he was asking; in the morning he had breathed over her neck; he was seized again by one of his old furies of desire on seeing her near the other man. The glances with which he called her had a flame in them which she knew well, the flame of his crisis of jealousy when he would fall on her with his fists, accusing her of committing abominations with her mother's lodger. And she was not willing; she trembled lest, by returning to him, she should throw these two men on to each other in this narrow cave, where they were all in agony together. Good God! why could they not end together in comradeship!

Etienne would have died of inanition rather than beg a mouthful of bread from Chaval. The silence became heavy; an eternity seemed to be prolonging itself with the slowness of monotonous minutes which passed by, one by one, without hope. They had now been shut up together for a day. The second lamp was growing pale, and they lighted the third.

Chaval started on his second slice of bread-and-butter, and growled:

'Come then, stupid!'

Catherine shivered. Étienne had turned away in order to leave her free. Then, as she did not stir, he said to her in a low voice:

'Go, my child.'

The tears which she was stifling then rushed forth. She wept for a long time, without even strength to rise, no longer knowing if she was hungry, suffering with pain which she felt all over her body. He was standing up, going backwards and forwards, vainly beating the miner's call, enraged at this remainder of life which he was obliged to live here tied to a rival whom he detested. Not even enough space to die away from each other! As soon as he had gone ten paces he must come back and knock up against this man. And she, this sorrowful girl whom they were disputing over even in the earth! She would belong to the one who lived longest; that man would steal her from him should he go first. There was no end to it; the hours followed the hours; the revolting promiscuity became worse, with the poison of their breaths and the ordure of their necessities satisfied in common. Twice he rushed against the rocks as though to open them with his fists.

Another day was done, and Chaval had seated himself near Catherine, sharing with her his last half-slice. She was chewing the mouthfuls painfully; he made her pay for each with a caress, in his jealous obstinacy not willing to die until he had had her again in the other man's presence. She abandoned herself in exhaustion. But when he tried to take her she complained.

'Oh, leave me! you 're breaking my bones.'

Étienne, with a shudder, had placed his forehead against the timber so as not to see. He came back with a wild leap.

'Leave her, by God!'

'Does it concern you?' said Chaval. 'She 's my wife; I suppose she belongs to me!'

And he took her again and pressed her, out of bravado, crushing his red moustaches over her mouth, and continuing:

'Will you leave us alone, eh? Will you be good enough to look over there if we are at it?'

But Étienne, with white lips, shouted:

'If you don't let her go, I 'll do for you!'

The other quickly stood up, for he had understood by the hiss of the voice that his mate was in earnest. Death seemed to them too slow; it was necessary that one of them should immediately yield his place. It was the old battle beginning over again, down in the earth where they would soon sleep side by side; and they had so little room that they could not swing their fists without grazing them.

'Look out!' growled Chaval. 'This time I 'll have you.'

From that moment Étienne became mad. His eyes seemed drowned in red vapour, his chest was congested by the flow of blood. The need to kill seized him irresistibly, a physical need, like the irritation of mucus which causes a violent spasm of coughing. It rose and broke out beyond his will, beneath the pressure of the hereditary disease. He had seized a sheet of slate in the wall and he shook it and tore it out, a very large, heavy piece. Then with both hands and with tenfold strength he brought it down on Chaval's skull.

The latter had no time to jump backwards. He fell, his face crushed, his skull broken. The brains had bespattered the roof of the gallery, and a purple jet flowed from the wound, like the continuous jet of a spring. Immediately there was a pool, which reflected the smoky star of the lamp. Darkness was invading the walled-up cave, and this body, lying on the earth, looked like the black boss of a mass of rough coal.

Leaning over, with wide eyes, Étienne looked at him. It was done, then; he had killed. All his struggles came back to his memory confusedly, that useless fight against the poison which slept in his muscles, the slowly accumulated alcohol of his race. He was, however, only intoxicated by hunger; the remote intoxication of his parents had been enough. His hair stood up before the horror of this murder; and yet, in spite of the revolt which came from his education, a certain gladness made his heart beat, the animal joy of an appetite at length satisfied. He felt pride, too, the pride of the stronger man. The little soldier appeared before him, with his throat opened by a knife, killed by a child. Now he, too, had killed.

But Catherine, standing erect, uttered a loud cry:

'My God! he is dead!'

'Are you sorry?' asked Étienne, fiercely.

She was choking, she stammered. Then, tottering, she threw herself into his arms.

'Ah, kill me too! Ah, let us both die!'

She clasped him, hanging to his shoulders, and he clasped her; and they hoped that they would die. But Death was in no hurry, and they unlocked their arms. Then, while she hid her eyes, he dragged away the wretch, and threw him down the upbrow, to remove him from the narrow space in which they still had to live. Life would no longer have been possible with that corpse beneath their feet. And they were terrified when they heard it plunge into the midst of the foam which leapt up. The water had already filled that hole, then? They saw it; it was entering the gallery.

Then there was a new struggle. They had lighted the last lamp; it was becoming exhausted in illuminating this floor, with its regular, obstinate rise which never ceased. At first the water came up to their ankles; then it wetted their knees. The passage sloped up, and they took refuge at the end. This gave them a respite for some hours. But the flood caught them up, and bathed them to the waist. Standing up, brought to bay, with their spines close against the rock, they watched it ever and ever increasing. When it should reach their mouths, all would be over. The lamp, which they had fastened up, threw a yellow light on the rapid surge of the little waves. It was becoming pale; they could distinguish no more than a constantly diminishing semicircle, as though eaten away by the darkness which seemed to grow with the flood; and suddenly the darkness enveloped them. The lamp had gone out, after having spit forth its last drop of oil. There was now complete and absolute night, that night of the earth which they would have to sleep through without ever again opening their eyes to the brightness of the sun.

'By God!' Étienne swore, in a low voice.

Catherine, as though she had felt the darkness seize her, sheltered herself against him. She repeated, in a whisper, the miner's saying:

'Death is blowing out the lamp.'

Yet in the face of this threat their instincts struggled, the fever of living animated them. He violently set himself to hollow out the slate with the hook of the lamp, while she helped him with her nails. They formed a sort of elevated bench, and when they had both hoisted themselves up to it, they found themselves seated with hanging legs and bent backs, for the vault forced them to lower their heads. They now only felt the icy water at their heels; but before long the cold was at their ankles, their calves, their knees, with its invincible, truceless movement. The bench, not properly smoothed, was soaked in moisture, and so slippery that they had to hold themselves on vigorously to avoid slipping off. It was the end; what could they expect, reduced to this niche where they dared not move, exhausted, starving, having neither bread nor light? and they suffered especially from the darkness, which would not allow them to see the coming of death. There was deep silence; the mine, being gorged with water, no longer stirred. They had nothing beneath them now but the sensation of that sea, swelling out its silent tide from the depths of the galleries.

The hours succeeded one another, all equally black; but they were not able to measure their exact duration, becoming more and more vague in their calculation of time. Their tortures, which might have been expected to lengthen the minutes, rapidly bore them away. They thought that they had only been shut up for two days and a night, when in reality the third day had already come to an end. All hope of help had gone; no one knew they were there, no one could come down to them. And hunger would finish them off if the inundation spared them. For one last time it occurred to them to beat the call, but the stone was lying beneath the water. Besides, who would hear them?

Catherine was leaning her aching head against the seam, when she sat up with a start.

'Listen!' she said.

At first Étienne thought she was speaking of the low noise of the ever-rising water. He lied in order to quiet her.

'It 's me you hear; I 'm moving my legs.'

'No, no; not that! Over there, listen!'

And she placed her ear to the coal. He understood, and did likewise. They waited for some seconds, with stifled breath. Then, very far away and very weak, they heard three blows at long intervals. But they still doubted; their ears were ringing; perhaps it was the cracking of the soil. And they knew not what to strike with in answer.

Étienne had an idea.

'You have the sabots. Take them off and strike with the heels.'

She struck, beating the miner's call; and they listened and again distinguished the three blows far off. Twenty times over they did it, and twenty times the blows replied. They wept and embraced each other, at the risk of losing their balance. At last the mates were there, they were coming. An overflowing joy and love carried away the torments of expectation and the rage of their vain appeals, as though their rescuers had only to split the rock with a finger to deliver them.

'Eh!' she cried merrily; 'wasn't it lucky that I leant my head?'

'Oh, you 've got an ear!' he said in his turn. 'Now, *I* heard nothing.'

From that moment they relieved each other, one of them always listening, ready to answer at the least signal. They soon caught the sounds of the pick; the work of approaching them was beginning, a gallery was being opened. But their joy sank. In vain they laughed to deceive each other; despair was gradually seizing them. At first they entered into long explanations; evidently they were being approached from Réquillart. The gallery descended in the bed; perhaps several were being opened, for there were always three men hewing. Then they talked less, and were at last silent when they came to calculate the enormous mass which separated them from their mates. They continued their reflections in silence, counting the days and days that a workman would take to penetrate such a block. They would never be reached soon enough; they would have time to die twenty times over. And no longer venturing to exchange a word in this redoubled anguish, they gloomily replied to the appeals by a roll of the sabots, without hope, only retaining the mechanical need to tell the others that they were still alive.

Thus passed a day, two days. They had been at the bottom six days. The water had stopped at their knees, neither rising nor falling, and their legs seemed to be melting away in this icy bath. They could certainly keep them out for an hour or so, but their position then became so uncomfortable that they were twisted by horrible cramps, and were obliged to let their feet fall in again. Every ten minutes they hoisted themselves back by a jerk on the slippery rock. The fractures of the coal struck into their spines, and they felt at the back of their necks a fixed intense pain, through having to keep constantly bent in order to avoid striking their heads. And their suffocation increased; the air, driven back by the water, was compressed into a sort of bell, in which they were shut up. Their voices were muffled, and seemed to come from afar. Their ears began to buzz, they heard the peals of a furious tocsin, the tramp of a flock beneath a storm of hail, going on unceasingly.

At first Catherine suffered horribly from hunger. She pressed her poor shrivelled hands against her breasts, her breathing was deep and hollow, a continuous tearing moan, as though tongs were tearing her stomach.

Étienne, choked by the same torture, was feeling feverishly round him in the darkness, when his fingers came upon a half-rotten piece of timber, which his nails could crumble. He gave a handful of it to the putter, who swallowed it greedily. For two days they lived on this worm-eaten wood, devouring it all, in despair when it was finished, grazing their hands in the effort to crush the other planks which were still solid with resisting fibres. Their torture increased, and they were enraged that they could not chew the cloth of their clothes. A leather girdle, which he wore round the waist, relieved them a little. He bit small pieces from it with his teeth, and she chewed them, and endeavoured to swallow them. This occupied their jaws, and gave them the illusion of eating. Then, when the girdle was finished, they went back to their clothes, sucking them for hours.

But soon these violent crises subsided; hunger became only a low deep ache with the slow progressive languor of their strength. No doubt they would have succumbed if they had not had as much water as they desired. They merely bent

O 897

down and drank from the hollow of the hand, and that very frequently, parched by a thirst which all this water could not quench.

On the seventh day Catherine was bending down to drink when her hand struck some floating body before her.

'I say, look! What 's this?'

Étienne felt in the darkness.

'I can't make out; it seems like the cover of the ventilation door.'

She drank, but as she was drawing up a second mouthful the body came back, striking her hand. And she uttered a terrible cry.

'My God! it 's he!'

'Whom do you mean?'

'Him! You know well enough. I felt his moustache.'

It was Chaval's corpse, risen from the upbrow and pushed on to them by the flow. Étienne stretched out his arm; he, too, felt the moustache and the crushed nose, and shuddered with disgust and fear. Seized by horrible nausea, Catherine had spat out the water which was still in her mouth. It seemed to her that she had been drinking blood, and that all the deep water before her was now that man's blood.

'Wait!' stammered Étienne. 'I 'll push him off!'

He kicked the corpse, which moved off. But soon they felt it again striking against their legs.

'By God! Get off!'

And the third time Étienne had to leave it. Some current always brought it back. Chaval would not go; he desired to be with them, against them. It was an awful companion, at last poisoning the air. All that day they never drank, struggling, preferring to die. It was not until the next day that their suffering decided them: they pushed away the body at each mouthful and drank in spite of it. It had not been worth while to knock his brains out, for he came back between him and her, obstinate in his jealousy. To the very end he would be there, even though he was dead, preventing them from coming together.

A day passed, and again another day. At every shiver of the water Étienne received a slight blow from the man he had killed, the simple elbowing of a neighbour who is reminding you of his presence. And every time it came he shuddered. He continually saw it there, swollen, greenish, with the red moustache and the crushed face. Then he no longer remembered; he had not killed him; the other man was swimming and trying to bite him.

Catherine was now shaken by long endless fits of crying, after which she was completely prostrated. She fell at last into a condition of irresistible drowsiness. He would arouse her, but she stammered a few words and at once fell asleep again without even raising her eyelids; and fearing lest she should be drowned, he put his arm round her waist. It was he now who replied to the mates. The blows of the pick were now approaching, he could hear them behind his back. But his strength, too, was diminishing; he had lost all courage to strike. They were known to be there; why weary oneself more? It no longer interested him whether they came or not. In the stupefaction of waiting he would forget for hours at a time what he was waiting for.

One relief comforted them a little: the water sank, and Chaval's body moved off. For nine days the work of their deliverance had been going on, and they were for the first time taking a few steps in the gallery when a fearful commotion threw them to the ground. They felt for each other and remained in each other's arms like mad people, not understanding, thinking the catastrophe was beginning over again. Nothing more stirred, the sound of the picks had ceased.

In the corner where they were seated holding each other, side by side, a low laugh came from Catherine.

'It must be good outside. Come, let 's go out of here.'

Étienne at first struggled against this madness. But the contagion was shaking his stronger head, and he lost the exact sensation of reality. All their senses seemed to go astray, especially Catherine's. She was shaken by fever, tormented now by the need to talk and move. The ringing in her ears had become the murmur of flowing water, the song of birds; she smelled the strong odour of crushed grass, and could see clearly great yellow patches floating before her eyes, so large that she thought she was out of doors, near the canal, in the meadows on a fine summer day.

'Eh? how warm it is! Take me, then; let us keep together. Oh, always, always!'

He pressed her, and she rubbed herself against him for a long time, continuing to chatter like a happy girl:

'How silly we have been to wait so long! I would have liked you at once, and you did not understand; you sulked. Then, do you remember, at our house at night, when we could not sleep, with our faces out listening to each other's breathing, with such a longing to come together?'

He was won by her gaiety, and joked over the recollection of their silent tenderness.

'You struck me once. Yes, yes, blows on both cheeks!'

'It was because I loved you,' she murmured. 'You see, I prevented myself from thinking of you. I said to myself that it was quite done with, and all the time I knew that one day or another we should get together. It only wanted an opportunity—some lucky chance. Wasn't it so?'

A shudder froze him. He tried to shake off this dream; then he repeated slowly:

'Nothing is ever done with; a little happiness is enough to make everything begin again.'

'Then you 'll keep me, and it will be all right this time?'

And she slipped down fainting. She was so weak that her low voice died out. In terror he kept her against his heart.

'Are you in pain?'

She sat up surprised.

'No, not at all. Why?'

But this question aroused her from her dream. She gazed at the darkness with distraction, wringing her hands in another fit of sobbing.

'My God, my God, how black it is!'

It was no longer the meadows, the odour of the grass, the song of larks, the great yellow sun; it was the fallen, inundated mine, the stinking gloom, the melancholy dripping of this cellar where they had been groaning for so many days. Her perverted senses now increased the horror of it; her childish superstitions came back to her; she saw the Black Man, the old dead miner who returns to the pit to twist naughty girls' necks.

'Listen! did you hear?'

'No, nothing; I heard nothing.'

'Yes, the Man—you know? Look! he is there. The earth has let all the blood out of the vein to revenge itself for being cut into; and he is there—you can see him—look! blacker than night. Oh, I 'm so afraid, I 'm so afraid!'

She became silent, shivering. Then in a very low voice she whispered:

'No, it 's always the other one.'

'What other one?'

'Him who is with us; who is not alive.'

The image of Chaval haunted her, she talked of him confusedly, she described the dog's life she led with him, the only day when he had been kind to her at Jean-Bart, the other days of follies and blows, when he would kill her with caresses after having covered her with kicks.

'I tell you that he 's coming, that he will still keep us from being together! His jealousy is coming on him again. Oh, push him off! Oh, keep me, keep me close!'

With a sudden impulse she hung on to him, seeking his mouth and pressing her own passionately to it. The darkness lighted up, she saw the sun again, and she laughed a quiet laugh of love. He shuddered to feel her thus against his flesh, half naked beneath the tattered jacket and trousers, and he seized her with a reawakening of his virility. It was at length their wedding night, at the bottom of this tomb, on this bed of mud, the longing not to die before they had had their happiness, the obstinate longing to live and make live one last time. They loved each other in despair of everything, in death.

After that there was nothing more. Étienne was seated on the ground, always in the same corner, and Catherine was lying motionless on his knees. Hours and hours passed by. For a long time he thought she was sleeping; then he touched her; she was very cold, she was dead. He did not move her, however, for fear of arousing her. The idea that he was the first who had possessed her as a woman, and that she might be pregnant, filled him with tenderness. Other ideas, the desire to go away with her, joy at what they would both do later on, came to him at moments, but so vaguely that it seemed only as though his forehead had been touched by a breath of sleep. He grew weaker, he only had strength to make a little gesture, a slow movement of the hand, to assure himself that she was certainly there, like a sleeping child in her frozen stiffness. Everything was being annihilated; the night itself had disappeared, and he was nowhere, out of space, out of time. Something was certainly striking beside his head, violent blows were approaching him; but he had been too lazy to reply, benumbed by immense fatigue; and now he knew nothing, he only dreamed that she was walking before him, and that he heard the slight clank of her sabots. Two days passed; she had not stirred; he touched her with his mechanical gesture, reassured to find her so quiet.

Étienne felt a shock. Voices were sounding, rocks were rolling to his feet. When he perceived a lamp he wept. His blinking eyes followed the light, he was never tired of looking at it, enraptured by this reddish point which scarcely stained the darkness. But some mates carried him away, and he allowed them to introduce some spoonfuls of soup between his clenched teeth. It was only in the Réquillart gallery that he recognized someone standing before him, the engineer, Negrel; and these two men, who felt contempt for each other—the rebellious workman and the sceptical master—threw themselves on each other's necks, sobbing loudly in the deep upheaval of all the humanity within them. It was an immense sadness, the misery of generations, the extremity of grief to which life can fall.

At the surface, Maheude, stricken down near dead Catherine, uttered a cry, then another, then another—very long, deep, incessant moans. Several corpses had already been brought up, and placed in a row on the ground: Chaval, who was thought to have been crushed beneath a landslip, a trammer, and two hewers, also crushed, with brainless skulls and bellies swollen with water. Women in the crowd went out of their minds, tearing their skirts and scratching their faces. When Étienne was at last taken out, after having been accustomed to the lamps and fed a little, he appeared fleshless, and his hair was quite white. People turned away and shuddered at this old man. Maheude left off crying to stare at him stupidly with her large fixed eyes.

CHAPTER VI

It was four o'clock in the morning, and the fresh April night was growing warm at the approach of day. In the limpid sky the stars were twinkling out, while the east grew purple with dawn. And a slight shudder passed over the drowsy black country, the vague rumour which precedes awakening.

Étienne, with long strides, was following the Vandame road. He had just passed six weeks at Montsou, in bed at the hospital. Though very thin and yellow, he felt strength to go, and he went. The Company, still trembling for its pits, was constantly sending men away, and had given him notice that he could not be kept on. He was offered the sum of one hundred francs, with the paternal advice to leave off working in mines, as it would now be too severe for him. But he refused the hundred francs. He had already received a letter from Pluchart, calling him to

Paris, and enclosing money for the journey. His old dream would be realized. The night before, on leaving the hospital, he had slept at the Bon-Joyeux, Widow Désir's. And he rose early; only one desire was left, to bid his mates farewell before taking the eight o'clock train at Marchiennes.

For a moment Étienne stopped on the road, which was now becoming rose-coloured. It was good to breathe that pure air of the precocious spring. It would turn out a superb day. The sun was slowly rising, and the life of the earth was rising with it. And he set out walking again, vigorously striking with his brier stick, watching the plain afar, as it rose from the vapours of the night. He had seen no one; Maheude had come once to the hospital, and, probably, had not been able to come again. But he knew that the whole settlement of the Deux-Cent-Quarante was now going down at Jean-Bart, and that she too had taken work there. Little by little the deserted roads were peopled, and colliers constantly passed Étienne with pallid, silent faces. After two and a half months of strike, when they had returned to the pits, conquered by hunger, they had been obliged to accept the timbering tariff, that disguised decrease in wages, now the more hateful because stained with the blood of their mates. They were being robbed of an hour's work, they were being made false to their oath never to submit; and this imposed perjury stuck in their throats like gall. Work was beginning again everywhere, at Mirou, at Madeleine, at Crèvecœur, at the Victoire. Everywhere, in the morning haze, along the roads lost in darkness, the flock was tramping on, rows of men trotting with faces bent towards the earth, like cattle led to the slaughter-house. They shivered beneath their thin garments, crossing their arms, rolling their hips, expanding their backs with the humps formed by the brick between the shirt and the jacket. And in this wholesale return to work, in these mute shadows, all black, without a laugh, without a look aside, one felt the teeth clenched with rage, the hearts swollen with hatred, a simple resignation to the necessity of the belly.

The nearer Étienne approached the pit the more their number increased. They nearly all walked alone; those who came in groups were in single file, already exhausted, tired of one another and of themselves. He noticed one who was very old, with eyes that shone like hot coals beneath his livid forehead. Another, a young man, was panting with the restrained fury of a storm. Many had their sabots in their hands; one could scarcely hear the soft sound of their coarse woollen stockings on the ground. It was an endless rustling, a general downfall, the forced march of a beaten army, moving on with lowered heads, sullenly absorbed in the desire to renew the struggle and achieve revenge.

When Étienne arrived, Jean-Bart was emerging from the shade; the lanterns, hooked on to the platform, were still burning in the growing dawn. Above the obscure buildings a trail of steam arose like a white plume delicately tinted with carmine. He passed up the sifting-staircase to go to the receiving-room.

The descent was beginning, and the men were coming from the shed. For a moment he stood by, motionless amid the noise and movement. The rolling of the trams shook the metal floor, the drums were turning, unrolling the cables in the midst of cries from the trumpet, the ringing of bells, blows of the mallet on the signal block; he found the monster again swallowing his daily ration of human flesh, the cages rising and plunging, engulfing their burden of men, without ceasing, with the facile gulp of a voracious giant. Since his accident he had a nervous horror of the mine. The cages, as they sank down, tore his bowels. He had to turn away his head; the pit exasperated him.

But in the vast and still sombre hall, feebly lighted up by the exhausted lanterns, he could perceive no friendly face. The miners, who were waiting there with bare feet and their lamps in their hands, looked at him with large restless eyes, and then lowered their faces, drawing back with an air of shame. No doubt they knew him and no longer had any spite against him; they seemed, on the contrary, to fear him, blushing at the thought that he would reproach them with cowardice. This attitude made his heart swell; he forgot that these wretches had stoned him, he again began to dream of changing them into heroes, of directing a whole people, this force of Nature which was devouring itself. A cage was embarking its men, and the batch disappeared; as others arrived he saw at last one of his lieutenants in the strike, a worthy fellow who had sworn to die.

'You too!' he murmured, with aching heart.

The other turned pale and his lips trembled; then, with a movement of excuse:

'What would you have? I've got a wife.'

Now in the new crowd coming from the shed he recognized them all.

'You too!—you too!—you too!'

And all shrank back, stammering in choked voices:

'I have a mother.'—'I have children.'—'One must get bread.'

The cage did not reappear; they waited for it mournfully, with such sorrow at their defeat that they avoided meeting each other's eyes, obstinately gazing at the shaft.

'And Maheude?' Étienne asked.

They made no reply. One made a sign that she was coming. Others raised their arms, trembling with pity. Ah, poor woman! what wretchedness! The silence continued, and when Étienne stretched out his hand to bid them farewell, they all pressed it vigorously, putting into that mute squeeze their rage at having yielded, their feverish hope of revenge. The cage was there; they got into it and sank, devoured by the gulf.

Pierron had appeared with his naked captain's lamp fixed into the leather of his cap. For the past week he had been chief of the gang at the pit-eye, and the men moved away, for promotion had rendered him bossy. The sight of Étienne annoyed him; he came up, however, and was at last reassured when the young man announced his departure. They talked. His wife now kept the Estaminet du Progrès, thanks to the support of all those gentlemen, who had been so good to her. But he interrupted himself and turned furiously on to Father Mouque, whom he accused of not sending up the dung-heap from his stable at the regulation hour. The old man listened with bent shoulders. Then, before going down, suffering from this reprimand, he, too, gave his hand to Étienne, with the same long pressure as the others, warm with restrained anger and quivering with future rebellion. And this old hand which trembled in his, this old man who was forgiving him for the loss of his dead children, affected Étienne to such a degree that he watched him disappear without saying a word.

'Then Maheude is not coming this morning?' he asked Pierron after a time.

At first the latter pretended not to understand, for there was ill-luck even in speaking of her. Then, as he moved away, under the pretext of giving an order, he said at last:

'Eh! Maheude? There she is.'

In fact, Maheude had reached the shed with her lamp in her hand, dressed in trousers and jacket, with her head confined in the cap. It was by a charitable exception that the Company, pitying the fate of this unhappy woman, so cruelly afflicted, had allowed her to go down again at the age of forty; and as it seemed difficult to set her again at haulage work, she was employed to manipulate a small ventilator which had been installed in the north gallery, in those infernal regions beneath Tartaret, where there was no movement of air. For ten hours, with aching back, she turned her wheel at the bottom of a burning tube, baked by forty degrees of heat. She earned thirty sous.

When Étienne saw her, a pitiful sight in her male garments —her breast and belly seeming to be swollen by the dampness of the cuttings—he stammered with surprise, trying to find words to explain that he was going away and that he wished to say good-bye to her.

She looked at him without listening, and said at last, speaking familiarly:

'Eh? it surprises you to see me. It 's true enough that I threatened to wring the neck of the first of my children who went down again; and now that I 'm going down I ought to wring my own, ought I not? Ah, well! I should have done it by now if it hadn't been for the old man and the little ones at the house.'

And she went on in her low, fatigued voice. She did not excuse herself, she simply narrated things—that they had been nearly starved, and that she had made up her mind to it, so that they might not be sent away from the settlement.

'How is the old man?' asked Étienne.

'He is always very gentle and very clean. But he is quite off his nut. He was not brought up for that affair, you know. There was talk of shutting him up with the madmen, but I was not willing; they would have done for him in his soup. His story has, all the same, been very bad for us, for he 'll never get his pension; one of those gentlemen told me that it would be immoral to give him one.'

'Is Jeanlin working?'

'Yes, those gentlemen found something for him to do at the top. He gets twenty sous. Oh! I don't complain; the bosses have been very good, as they told me themselves. The brat's twenty sous and my thirty, that makes fifty. If there were not six of us we should get enough to eat. Estelle devours now, and the worst is that it will be four or five years before Lénore and Henri are old enough to come to the pit.'

Étienne could not restrain a movement of pain.

'They, too!'

Maheude's pale cheeks turned red, and her eyes flamed. But her shoulders sank as if beneath the weight of destiny.

'What would you have? They after the others. They have all been done for there; now it 's their turn.'

She was silent; some landers, who were rolling trams, disturbed them. Through the large dusty windows the early sun was entering, drowning the lanterns in grey light; and the engine moved every three minutes, the cables unrolled, the cages continued to swallow down men.

'Come along, you loungers, look sharp!' shouted Pierron. 'Get in; we shall never have done with it to-day.'

Maheude, whom he was looking at, did not stir. She had already allowed three cages to pass, and she said, as though arousing herself and remembering Étienne's first words:

'Then you 're going away?'

'Yes, this morning.'

'You 're right; better be somewhere else if one can. And I 'm glad to have seen you, because you can know now, anyhow, that I 've nothing on my mind against you. For a moment I could have killed you, after all that slaughter. But one thinks, doesn't one? One sees that when all 's reckoned up it 's nobody's fault. No, no! it 's not your fault; it 's the fault of everybody.'

Now she talked with tranquillity of her dead, of her man, of Zacharie, of Catherine; and tears only came into her eyes when she uttered Alzire's name. She had resumed her calm reasonableness, and judged things sensibly. It would bring no luck to the middle class to have killed so many poor people. Sure enough, they would be punished for it one day, for everything has to be paid for. There would even be no need to interfere; the whole thing would explode by itself. The soldiers would fire on the masters just as they had fired on the men. And in her everlasting resignation, in that hereditary discipline under which she was again bowing, a conviction had established itself, the certainty that injustice could not last longer; and that, if there were no good God left, another would spring up to avenge the wretched.

She spoke in a low voice, with suspicious glances round. Then, as Pierron was coming up, she added, aloud:

'Well, if you 're going, you must take your things from our house. There are still two shirts, three handkerchiefs, and an old pair of trousers.'

Étienne, with a gesture, refused these few things saved from the dealers.

'No, it 's not worth while; they can be for the children. At Paris I can arrange for myself.'

Two cages had once more gone down, and Pierron decided to speak straight to Maheude.

'I say now, over there, they are waiting for you! Is that little chat nearly done?'

But she turned her back. Why should he be so zealous, this man who had sold himself? The descent didn't concern him. His men hated him enough already on his level. And she persisted, with her lamp in her hand, frozen amid the draughts in spite of the mildness of the season. Neither Étienne nor she found anything more to say. They remained facing each other with hearts so full that they would like to speak once more.

At last she spoke for the sake of speaking.

'The Levaque is in the family-way. Levaque is still in prison; Bouteloup is taking his place meanwhile.'

'Ah, yes! Bouteloup.'

'And, listen! did I tell you? Philomène has gone away.'

'What! gone away?'

'Yes, gone away with a Pas-de-Calais miner. I was afraid she would leave the two brats on me. But no, she took them with her. Eh? A woman who spits blood and looks as if she were always swallowing her tongue!'

She mused for a moment, and then went on in a slow voice:

'There 's been talk on my account. You remember they said I slept with you. Lord! After my man's death that might very well have happened if I had been younger. But now I 'm glad it wasn't so, for we should have regretted it, sure enough.'

'Yes, we should have regretted it,' Étienne repeated, simply.

That was all; they spoke no more. A cage was waiting for her; she was being called angrily, threatened with a fine. Then she made up her mind, and pressed his hand. Deeply moved, he still looked at her, so worn and worked out, with her livid face, her discoloured hair escaping from the blue cap, her body as of a good over-fruitful beast, deformed beneath the jacket and trousers. And in this last pressure of the hands he felt again the long, silent pressure of his mates, giving him a rendezvous for the day when they would begin again. He understood perfectly. There was a tranquil faith in the depths of her eyes. It would be soon, and that time it would be the final blow.

'A damned shammer!' exclaimed Pierron.

Pushed and hustled, Maheude squeezed into a tram with four others. The signal-cord was drawn to strike for meat, the cage was unhooked and fell into the night, and there was nothing more but the rapid flight of the cable.

Then Étienne left the pit. Below, beneath the screening-shed, he noticed a creature seated on the earth, with legs stretched out, in the midst of a thick pile of coal. It was Jeanlin, who was employed there to clean the large coal. He held a block of coal between his thighs, and freed it with a hammer from the fragments of slate. A fine powder drowned him in such a flood of soot that the young man would never have recognized him if the child had not lifted his ape-like face, with the protruding ears and small greenish eyes. He laughed, with a joking air, and, giving a final blow to the block, disappeared in the black dust which arose.

Outside, Étienne followed the road for a while, absorbed in his thoughts. All sorts of ideas were buzzing in his head. But he felt the open air, the free sky, and he breathed deeply. The sun was appearing in glory at the horizon, there was a reawakening of gladness over the whole country. A flood of gold rolled from the east to the west in the immense plain. This heat of life was expanding and extending in a tremor of youth, in which vibrated the sighs of the earth, the song of birds, all the murmuring sounds of the waters and the woods. It was good to live, and the old world wanted to live through one more spring.

And penetrated by that hope, Étienne slackened his walk, his eyes wandering to right and to left amid the gaiety of the new season. He thought about himself, he felt himself strong, seasoned by his hard experiences at the bottom of the mine. His education was complete, he was going away armed, a rational soldier of the revolution, having declared war against society as he saw it and as he condemned it. The joy of rejoining Pluchart and of being, like Pluchart, a leader who was listened to, inspired him with speeches, and he began to arrange the phrases. He was meditating an enlarged programme; that middle-class refinement, which had raised him above his class, had deepened his hatred of the middle class. He felt the need of glorifying these workers, whose odour of wretchedness was now unpleasant to him; he would show that they alone were great and stainless, the only nobility and the only strength in which humanity could be dipped afresh. He already saw himself in the tribune, triumphing with the people, if the people had not devoured him.

The loud song of a lark made him look up towards the sky. Little red clouds, the last vapours of the night, were melting in the limpid blue; and the vague faces of Souvarine and Rasseneur came to his memory. Decidedly, all was spoilt when each man tried to get power for himself. Thus that famous International which was to have renewed the world had impotently miscarried, and its formidable army had been cut up and crumbled away from internal dissensions. Was Darwin right, then, and the world only a battlefield, where the strong ate the weak for the sake of the beauty and continuance of the race? This question troubled him, although he settled it like a man who is satisfied with his knowledge. But one idea dissipated his doubts and enchanted him—that of taking up his old explanation of the theory the first time that he should speak. If any class must be devoured, would not the people, still new and full of life, devour the middle class, exhausted by enjoyment? The new society would arise from new blood. And in this expectation of an invasion of barbarians, regenerating the old decayed nations, reappeared his absolute faith in an approaching revolution, the real one—that of the workers—the fire of which would inflame this century's end with that purple of the rising sun which he saw like blood on the sky. He still walked, dreaming, striking his brier stick against the flints on the road, and when he glanced around him he recognized the various places. Just there, at the Fourche-aux-Bœufs, he remembered that he had taken command of the band that morning when the pits were sacked. To-day the brutish, deathly, ill-paid work was beginning over again. Beneath the earth, down there at seven hundred metres, it seemed to him he heard low, regular, continuous blows; it was the men he had just seen go down, the black workers, who were hammering in their silent rage. No doubt they were beaten. They had left their dead and their money on the field; but Paris would not forget the volleys fired at the Voreux, and the blood of the Empire, too, would flow from that incurable wound. And if the industrial crisis was drawing to an end, if the workshops were opening again one by one, a state of war was not less declared, and peace was henceforth impossible. The colliers had reckoned up their men; they had tried their strength, with their cry for justice arousing the workers all over France. Their defeat, therefore, reassured no one. The Montsou bourgeois, in their victory, felt the vague uneasiness that arises on the morrow of a strike, looking behind them to see if their end did not lie inevitably over there, in spite of all, beyond that great silence. They understood that the revolution would be born again unceasingly, perhaps to-morrow, with a general strike—the common understanding of all workers having general funds, and so able to hold out for months, eating their own bread. This time a push only had been given to a ruinous society, but they had heard the rumbling beneath their feet, and they felt more shocks arising, and still more, until the old edifice would be crushed, fallen in and swallowed, going down like the Voreux to the abyss.

Étienne took the Joiselle road, to the left. He remembered that he had prevented the band from rushing on to Gaston-Marie. Afar, in the clear sky he saw the steeples of several pits —Mirou to the right, Madeleine and Crèvecœur side by side. Work was going on everywhere; he seemed to be able to catch the blows of the pick at the bottom of the earth, striking now from one end of the plain to the other, one blow, and another blow, and yet more blows, beneath the fields and roads and villages which were laughing in the light, all the obscure labour of the underground prison, so crushed by the enormous mass of the rocks that one had to know it was underneath there to distinguish its great painful sigh. And he now thought that, perhaps, violence would not hasten things. Cutting cables, tearing up rails, breaking lamps, what a useless task it was! It was not worth while for three thousand men to rush about in a devastating band doing that. He vaguely divined that lawful methods might one day be more terrible. His reason was ripening, he had sown the wild oats of his spite. Yes, Maheude had well said, with her good sense, that that would be the great blow—to organize quietly, to know one another, to unite in associations when the laws would permit it; then, on the morning when they felt their strength, and millions of workers would be face to face with a few thousand idlers, to take the power into their own hands and become the masters. Ah! what a reawakening of truth and justice! The sated and crouching god would at once get his deathblow, the monstrous idol hidden in the depths of his sanctuary, in that unknown distance where poor wretches fed him with their flesh without ever having seen him.

But Étienne, leaving the Vandame road, now came on to the paved street. On the right he saw Montsou, which was lost in the valley. Opposite were the ruins of the Voreux, the accursed hole where three pumps worked unceasingly. Then there were the other pits at the horizon, the Victoire, Saint-Thomas, Feutry-Cantel; while, towards the north, the tall chimneys of the blast furnaces, and the batteries of coke ovens, were smoking in the transparent morning air. If he was not to lose the eight o'clock train he must hasten, for he had still six kilometres before him.

And beneath his feet, the deep blows, those obstinate blows of the pick, continued. The mates were all there; he heard them following him at every stride. Was not that Maheude beneath the beetroots, with bent back and hoarse respiration accompanying the rumble of the ventilator? To left, to right, farther on, he seemed to recognize others beneath the wheatfields, the hedges, the young trees. Now the April sun, in the open sky, was shining in his glory, and warming the pregnant earth. From its fertile flanks life was leaping out, buds were bursting into green leaves, and the fields were quivering with the growth of the grass. On every side seeds were swelling, stretching out, cracking the plain, filled by the need of heat and light. An overflow of sap was mixed with whispering voices, the sound of the germs expanding in a great kiss. Again and again, more and more distinctly, as though they were approaching the soil, the mates were hammering. In the heated rays of the sun on this youthful morning the country seemed full of that sound. Men were springing forth, a black avenging army, germinating slowly in the furrows, growing towards the harvests of the next century, and this germination would soon overturn the earth.

Made at the Temple Press Letchworth in Great Britain

Part Three
SYNTHESIS AND APPLICATION

CHAPTER I

ATAVISM AND EPILEPSY IN CRIME AND IN PUNISHMENT

§ 199.

ALL that I have set forth in the present book and in those which preceded it (Vol. I and II of the "Homme Criminel") proves clearly the insecurity of the ancient criminological scaffolding. Have I succeeded in substituting a more solid edifice? If pride in a long and painful task has not blinded me, I think that I can answer in the affirmative. The fundamental proposition undoubtedly is that we ought to study not so much the abstract crime as the criminal.

§ 200. Atavism

The born criminal shows in a proportion reaching 33% numerous specific characteristics that are almost always atavistic. Those who have followed us thus far have seen that many of the characteristics presented by savage races are very often found among born criminals. Such, for example, are: the slight development of the pilar system; low cranial capacity; retreating forehead; highly developed frontal sinuses; great frequency of Wormian bones; early closing of the cranial sutures; the simplicity of the sutures; the thickness of the bones of the skull; enormous development of the maxillaries and the zygomata; prognathism; obliquity of the orbits; greater pigmentation of the skin; tufted and crispy hair; and large ears. To these we may add the lemurine appendix; anomalies of the ear; dental diastemata; great agility; relative insensibility to pain; dullness of the sense of touch; great visual acuteness; ability to recover quickly from wounds; blunted affections; precocity as to sensual pleasures;[1] greater resemblance between the sexes; greater incorrigibility of the woman (Spencer); laziness; absence of remorse; impulsiveness; physiopsychic excitability; and especially improvidence, which sometimes appears as courage and again as recklessness changing to cowardice. Besides these there is great vanity; a passion for gambling and alcoholic drinks; violent but fleeting passions; superstition; extraordinary sensitiveness with regard to one's own personality; and a special conception of God and morality. Unexpected analogies are met even in small details, as, for example, the improvised rules of criminal gangs; the entirely personal influence of the chiefs;[2] the custom of tattooing; the not uncommon cruelty of their games; the excessive use of gestures; the onomatopoetic language with personification of inanimate things; and a special literature recalling that of heroic times, when crimes were celebrated and the thought tended to clothe itself in rhythmic form.

This atavism explains the diffusion of certain crimes, such as the pederasty and infanticide, whose extension to whole companies we could not explain if we did not recall the Romans, the Greeks, the Chinese, and the Tahitians, who not only did not regard them as crimes, but sometimes even practiced them as a national custom. Garofalo has admirably summed up the psychical characteristics of the born criminal as being the absence of the feelings of shame, honor, and pity, which are those that are lacking in the savage also.[3] We may add to these the lack of industry and self-control.

To those who, like Reclus and Krapotkin, object that there are savage peoples who are honorable and chaste, we must reply that a certain degree of density of population and of association among men is necessary for crimes to develop. It is not possible for example, to steal when property does not exist, or to swindle when there is no trade. But the proof that these tendencies exist in germ in the savage, is that when they begin to pass from their stage of savagery and take on a little civilization they always develop the characteristics of criminality

1 "Homme Criminel," Vol. I, pp. 136 to 579.
2 Tacitus, "Germ.," VII.
3 "Criminologie," 2d ed., 1895.

in an exaggerated form. As Ferrero has pointed out to us, even when honor, chastity, and pity are found among savages, impulsiveness and laziness are never wanting. Savages have a horror of continuous work, so that for them the passage to active and methodical labor lies by the road of selection or of slavery only. Thus, according to the testimony of Tacitus, the impulsiveness of the ancient Germans frequently resulted in the murder of slaves, committed in a fit of anger, an act which was not regarded as culpable. Tacitus notes also their lack of capacity for work.

"They have," he says, "large bodies, effective for sudden effort, but they lack the patience necessary for regular work. When they are not at war they do nothing . . . they sleep and eat. The strongest and most warlike live in idleness, leaving the care of the house and the field to the women, the old men, and the weak, becoming themselves brutalized in sloth."

At times, on the other hand, impulsiveness, rather than sluggishness, seems to ally itself with a ceaseless need of movement, which asserts itself in savage peoples in a life of incessant vagabondage. Thus the Andaman Islanders, as Hovelacque tells us, have so restless a disposition that they remain not more than two or three days in the same place, and their wanderings have no other reason than the need of movement. This attitude seems to be the result of a passage between physiopsychic inertia and an intermittent need of violent and unrestrained physical and moral excitation, which always goes with inertia and impulsiveness. Thus it is that those peoples who are normally most lazy and indolent have the most unrestrained and noisy dances, which they carry on until they get into a kind of delirium, and fall down utterly exhausted. "When the Spaniards," writes Robertson, "first saw the American Indians, they were astonished at their mad passion for dancing, and at the dizzy activity which this people, almost always cold and passive, displayed when they gave themselves up to this amusement." "The negroes of Africa," writes Du Chaillu, "dance madly when they hear the sound of the tom-tom, and lose all command of themselves." "It is," says Letourneau, "a real dancing madness, which makes them forget their troubles, public or private."

We may add that the atavism of the criminal, when he lacks absolutely every trace of shame and pity, may go back far beyond the savage, even to the brutes themselves. Pathological anatomy helps prove our position by showing in the case of the criminal a greater development of the cerebellum, a rarer union of the calcarine fissure with the parieto-occipital, the absence of folds in the passage of Gratiolet, the gutter-like shape of the nasal incisure, the frequency of the olecranial foramen, extra ribs and vertebræ, and especially the histological anomalies discovered by Roncoroni in the cortex of the cerebrum of criminals, that is to say, the frequent absence of granular layers, and the presence of nerve cells in the white matter, and immense pyramidal cells. In seeking for analogies beyond our own race we come upon the explanation of the union of the atlas with the occipital bone, the prominence of the canine teeth, the flattening of the palate, and the median occipital fossa, occurring among criminals as with the lemurs and rodents;[1] as also the prehensile foot, the simplicity of the lines of the palm, motor and sensory left-handedness. We recall also the tendency to cannibalism even without desire for vengeance, and still more that form of sanguinary ferocity, mingled with lubricity, of which examples are furnished us by Gille, Verzeni, Legier, Bertrand, Artusio, the Marquis of Sade, and others, with whom atavism was accompanied by epilepsy, idiocy, or general paralysis, but who always recall the pairing of animals, preceded by ferocious and sanguinary contests to overcome the reticence of the female or to conquer rivals.[2]

These facts prove clearly that the most horrible crimes have their origin in those animal instincts of which childhood gives us a pale reflection. Repressed in civilized man by education, environment, and the fear of punishment, they suddenly break out in the born criminal without apparent cause, or under the influence of certain circumstances, such as sickness, atmospheric influences, sexual excitement, or mob influence. We know that certain morbid conditions, such as injuries to the head, meningitis, and chronic intoxication, or certain physiological

[1] "Homme Criminel," Vol. I, pp. 160, 217, 176, 182.

[2] "Homme Criminel," Vol. I, pp. 449, 513; Vol. II, pp. 95, 96, 123, 139, 144, 147.

conditions like pregnancy and senility, produce derangements in the nutrition of the nervous centers, and in consequence atavistic retrogressions. We can see, then, how they may facilitate the tendency to crime, and when we take into account the short distance that separates the criminal from the savage, we come to understand why convicts so easily adopt savage customs, including cannibalism, as was observed in Australia and Guiana.[1] When we note, further, how children, until they are educated, are ignorant of the difference between vice and virtue, and steal, strike, and lie without the least compunction, we easily understand the great precocity in crime, and see why it is that the majority of abandoned children and orphans end by becoming criminals.[2] Further, atavism shows us the inefficacy of punishment for born criminals and why it is that they inevitably have periodic relapses into crime, so that the greatest variation shown by the number of crimes against persons is not more than $\frac{1}{25}$, and by those against property not more than $\frac{1}{50}$.[3]

We see, as Maury very truly remarks, that we are governed by silent laws, which never fall into desuetude and rule society much more surely than the laws inscribed in the codes.

§ 201. Epilepsy

The same phenomena which we observe in the case of born criminals appear again in the rare cases of moral insanity,[4] but may be studied minutely, and on a large scale, in epileptics, criminal or not,[5] as the table given below will prove. There we shall see that not one of the atavistic phenomena shown by criminals is lacking in epilepsy; though epileptics show also certain purely morbid phenomena, such as cephalea, atheroma, delirium, and hallucination. In born criminals also we find, besides the atavistic characteristics, certain others that appear to be entirely pathological, or which at first sight seem more nearly allied to disease than to atavism. Such are, for example, in the anatomical field, excessive asymmetry, cranial capacity

1 Bouvier, "Voyage à la Guyane," 1866.
2 "Homme Criminel," Vol. I, pp. 92 to 108.
3 Maury, "Mouvemente Moral de la Société," Paris, 1860.
4 "Homme Criminel," Vol. II, pp. 2–13.
5 "Homme Criminel," Vol. II, pp. 50–201.

and face too large or too small, sclerosis, traces of meningitis, hydrocephalous forehead, oxycephaly, acrocephaly, cranial depressions, numerous osteophytes, early closing of the cranial sutures, thoracic asymmetry, late grayness of hair, late baldness, and abnormal and early wrinkles; in the biological field, alterations of the reflexes and pupillary inequalities. To these we may add peripheral scotomata of the visual field, which one never finds in savages, with whom, on the contrary, the field of vision is remarkably wide and regular, as we see in the case of the Dinkas. There is also to be added the alteration of hearing, taste, and smell, the predilection for animals, precocity in sexual pleasures, amnesia, vertigo, and maniac and paranoiac complications. These abnormalities, which are found in greater proportion among idiots, cretins, and degenerates in general, are to be explained by the fact that in these cases alcoholic intoxication is added to the effect of atavism, and still more to that of epilepsy.

However, the participation of epilepsy in producing the effect does not exclude atavism, since they equally involve characteristics at once atavistic and pathological, like macrocephaly, cranial sclerosis, Wormian bones, rarity of beard; and in the biological field, left-handedness, analgesis, obtuseness of all senses except that of sight, impulsiveness, pederasty, obscenity, sluggishness, superstition, frequent cannibalism, choleric and impetuous disposition, tendency to reproduce the cries and actions of animals; and especially the histological anomalies of the cortex, which we have noted among criminals, and which reproduce the conditions of the lower animals; and finally anomalies of the teeth. These latter might appear to have no connection with the brain, but are, on the contrary, intimately connected with it, since the teeth proceed from the same embryonic membrane as the brain does.[1]

We may recall here that Gowers, having often noted in epileptics acts peculiar to animals, such as biting, barking, and mewing, concludes from this "that these are manifestations of that instinctive animalism which we possess in the latent state." [2]

[1] "Homme Criminel," Vol. I, p. 232, n.
[2] "Epilepsy," London, 1880.

		Criminals	Epileptics	Atavism	Phenomena of Arrested Development	Morbid Phenomena	Atypical Phenomena
Cranium	Volume too great	+	+	...	...	...	+
	Volume too small	+	+	+	+	+	+
	Sclerosis	+	+	+	+	+	...
	Exostosis	+	+	...	...	+	...
	Asymmetry	+	+	...	...	+	...
	Median occipital fossa	+	+	...	+	...	...
	Cranial index too great	+	+	...	...	+	...
	Strongly arched brows	+	+	...	+	...	...
	Low, retreating forehead	+	+	...	+	...	...
	Hydrocephalous forehead	+	+	...	...	+	...
	Cranial osteophytes	+	+	...	...	+	...
	Numerous Wormian bones	+	+	+	+	...	...
	Frontal suture	+	+	+	+	...	...
	Early synostosis	+	+	...	...	+	...
	Oblique orbits	+	+	+	...	...	...
Face	Lemurine appendix	+	+	+	...	...	...
	Maxillaries too large	+	+	+	...	...	...
	Large and prominent zygomata	+	+	+	...	...	...
	Large, outstanding ears	+	+	+	...	+	...
	Facial asymmetry	+	+	...	...	+	...
	Strabismus	+	+	...	...	+	...
	Masculine face in women	+	+	+	...	...	...
	Dental diastemata	+	+	+	...	...	...
	Anomalies of bones of nose	+	+	...	...	...	...
	Anomalies of teeth	+	+	+	...	...	...
	Bones of face too large	+	+	+	...	...	...
Brain	Anomalies of fissures	+	+	+	+	...	...
	Small weight	+	+	+	+	...	...
	Hypertrophy of cerebellum	+	+	+	...	...	...
	Histological changes of cortex	+	+	+	+	...	...
	Traces of meningitis	+	+	...	...	...	...
Body	Asymmetry of thorax	+	+	...	...	+	...
	Prehensile foot	+	+	+	...	...	...
	Left-handedness	+	+	...	...	...	...
	Hernia	+	+	+	+	+	...
	Simplicity of lines of palm	+	+	+	...	...	+
	Visceral lesions	+	+	...	...	...	+

If fully developed epileptic fits are often lacking in the case of the born criminal, this is because they remain latent, and only show themselves later under the influence of the causes assigned (anger, alcoholism), which bring them to the surface. With both criminals and epileptics there is to be noted an insufficient development of the higher centers. This manifests

		Criminals	Epileptics	Atavism	Phenomena of Arrested Development	Morbid Phenomena	Atypical Phenomena
Skin	Abnormal wrinkles	+	+	+	...	...	...
	Sparse beard	+	+	+	...	+	...
	Yellowish tint	+	+	+	...	...	...
	Tattooing	+	..	+	...	...	...
	Crispy hair	+	+	+	...	...	...
Motor Anomalies	Left-handedness and ambidextry	+	+	+	...	...	...
	Abnormalities of reflexes	+	+	..	...	+	...
	Unequal pupils	+	+	..	...	+	...
	Abnormal agility	+	+	+	...	...	...
Sensorial Anomalies	Obtuseness of sense of touch	+	+	+	...	+	...
	Relative insensibility to pain	+	+	+	...	+	...
	Great visual acuteness	+	+	+	...	...	...
	Obtuseness of hearing, taste, and smell	+	+	+	...	+	...
	Sensorial left-handedness	+	+	+	...	+	...
	Peripheral scotomata of the field of vision	+	+	..	...	+	...
Psychic Anomalies	Limited intelligence	+	+	+	+	...	...
	Superstition	+	+	+	...	...	...
	Emotional obtuseness	+	+	+	...	+	...
	Lack of moral sensibility	+	+	+	...	+	...
	Absence of remorse	+	+	+	...	...	...
	Cannibalism, ferocity, lack of self-control	+	+	+	+	+	...
	Pederasty, onanism, obscenity	+	+	+	...	+	...
	Exaggerated religious beliefs	+	+	+	...	...	...
	Vagrancy	+	+	+	...	...	...
	Sexual precocity	+	+	+	...	+	...
	Vanity	+	+	+	...	...	...
	Simulation	+	+	..	...	+	...
	Laziness, inertia	+	+	+	...	+	...
	Improvidence	+	+	+	...	...	...
	Cowardice	+	+	+	...	...	...
	Passion for gambling	+	+	+	...	...	...
	Mania, paranoia, delirium	+	+	..	...	+	...
	Vertigo	+	+	..	...	+	...
Causes	Heredity (alcoholism, insanity, epilepsy, old age of parents)	+	+	..	...	...	...
	Alcoholism, etc.	+	+	..	...	...	...

itself in a deterioration in the moral and emotional sensibilities, in sluggishness, physiopsychic hyperexcitability, and especially in a lack of balance in the mental faculties, which, even when distinguished by genius and altruism, nevertheless always show gaps, contrasts, and intermittent action.

§ 202. Combination of Morbid Anomalies with Atavism

Very often, moreover, certain common characteristics of criminals and epileptics have been classed as abnormal or morbid and not as atavistic, entirely because of the insufficiency of our embryological and phylogenetic knowledge. Many of the characteristics given in the preceding table (which, however, is only schematic) are atavistic and morbid at the same time, such as microcephaly, cranial sclerosis, etc. Facial asymmetry would also appear to be atavistic when we recall, for example, the flat-fishes (Penta); so likewise the abnormally wrinkled face, taking us back to the Hottentots and the apes. Hernia, also, as Féré rightly remarks, recalls conditions that are normal in the lower vertebrates and in the embryo.

Very often morbidity and atavism go back to a common cause, as Wagner [1] observes in a magnificent dissertation.

"The idea," he writes, "that the atavism of criminals is associated with some specific disease of the fœtus has been completely confirmed by the discoveries of Ettinghausen. If, for example, we freeze the roots of an oak so as partly to kill it, the following year it will put out leaves that are not like the leaves of the modern oak, but like those of the oak of the tertiary period. This fact explains the reappearance of intermediate and indistinct fossil forms. We see very clearly, then, that influences capable of producing a disease can bring about atavistic morphological retrogressions."

The epileptic background upon which the clinical and anatomical picture of the moral lunatic and the born criminal is drawn (a picture that would otherwise be lost in vague semi-juridical, semi-psychiatric hypotheses) explains the instantaneousness, periodicity, and paradoxical character of their symptoms, which are doubtless their most marked characteristics. Note, for example, in this class, the coexistence and interchange of kindness and ferocity, of cowardice and the maddest recklessness, and of genius and complete stupidity.

§ 203. The Criminaloid

Criminaloids, while quite separable from born criminals, do not lack some connection with epilepsy and atavism. Thus

[1] Wagner von Jauregg, "Antrittsvorlesung an der Psychiatrischen Klinik," Vienna, 1895.

there are more epileptics among them (10% among pickpockets) than among normal men, and a greater proportion of criminal types (17%), but there are also certain specific anomalies, such as left-handedness, common among swindlers.[1]

In the biology of the criminaloid we observe a smaller number of anomalies in touch, sensibility to pain, psychometry, and especially less early baldness and grayness, and less tattooing. But, on the other hand, we meet with a larger number of strictly morbid anomalies, depending upon the abuse of alcoholic drinks, such as atheromata, paresis, and scars. Psychic anomalies are especially less frequent with the criminaloid, who has not the cynicism of the born criminal nor the passion for doing evil for its own sake; he confesses his fault more easily and with more sincerity, and repents more often. But he is more lascivious, and more often given to alcoholism; and the criminaloid women are more susceptible to suggestion. The criminaloid is more precocious and relapses oftener, — at least this is the case with pickpockets and simple thieves. They are often drawn into crime by a greater opportunity, although the lack of self-control which makes the epileptic commit crime without reason is sometimes found in the criminaloid also. We may recall how Casenova confessed that when he committed a fraud he never premeditated it, but "seemed to yield to a superior will." A pickpocket said to me, "When the inspiration comes to us we cannot resist." Dostojevsky depicts smugglers of the prison as carrying on their occupation almost without returns, notwithstanding the grave risks they run and in spite of repeated promises not to relapse. Mendel and Benedict describe the impulsive nature of the vagabond, which keeps him moving without object and without rest.

Criminaloids, then, differ from born criminals in degree, not in kind. This is so true that the greater number of them, having become habitual criminals, thanks to a long sojourn in prison, can no longer be distinguished from born criminals except by the slighter character of their physical marks of criminality.

Still less different from born criminals are those latent crimi-

[1] "Homme Criminel," Vol. II, pp. 216, 514, 518.

nals, high in power, whom society venerates as its chiefs. They bear the marks of congenital criminality, but their high position generally prevents their criminal character from being recognized. Their families, of which they are the scourges, may discover it; or their depraved nature may be revealed all too late at the expense of the whole country, at the head of which their own shamelessness, seconded by the ignorance and cowardice of the majority, has caused them to be placed. Even this strange species of criminal monomaniac, who seems to differ from the epileptic in the motive of his crime and the manner of carrying it out,[1] shows nevertheless the epileptic and atavistic origin of his criminality by obsessions, interrupted periods of ideation, lack of self-control, exaggerated importance given to certain details, exhaustion after his criminal crises, fondness for symbolism, excessive and intermittent activity, and finally by hereditary stigmata.

§ 204. Criminal Insane

Even among the true insane criminals those forms predominate which we may call the hypertrophy of crime, the exaggeration of the born criminal, not only in bodily and functional characteristics but also in the manner of committing the crime and in conduct afterward.[2] These serve to explain to us the extent of the impulsive, obscene, and cruel tendencies of the criminal insane, who are almost always obscure epileptics or born criminals upon whom melancholia and monomania have grafted themselves, according to the natural tendency of different forms of psychic disorders to take root together upon the corrupted soil of degeneracy. We have seen, likewise, how hysterical persons, alcoholics, dipsomaniacs, pyromaniacs, kleptomaniacs, the temporarily insane, reproduce many of the characteristics of the epileptic. Even the mattoid, who on account of his habitual calm and the absence of signs of degeneracy and heredity, seems far removed from epilepsy, yet shows at times this epileptic form, which we have seen to be the kernel of crime.[3]

1 "Homme Criminel," Vol. II, pp. 94, 97, 418.
2 "Homme Criminel," Vol. I, pp. 34 to 228; Vol. II, p. 213.
3 "Homme Criminel," Vol. II, p. 646.

§ 205. Criminals by Passion

Criminals of this class form a species apart, and are in complete contrast with the born criminal, both in the harmonious lines of the body, the beauty of the soul, and great nervous and emotional sensitiveness, as well as in the motives of their crimes, always noble and powerful, such as love or politics. Nevertheless they show some points of resemblance with epileptics, such as their tendency to excesses, impulsiveness, suddenness in their outbreaks, and frequent amnesia.[1]

§ 206. Occasional Criminals

Occasional criminals, or better, pseudo-criminals, are those who do not seek the occasion for the crime but are almost drawn into it, or fall into the meshes of the code for very insignificant reasons. These are the only ones who escape all connection with atavism and epilepsy; but, as Garafalo observes, these ought not, properly speaking, to be called criminals.

§ 207. Causes

The study of the causes of crime does not lessen the fatal influence to be assigned to the organic factor, which certainly amounts to 35% and possibly even 40%; the so-called causes of crime being often only the last determinants and the great strength of congenital impulsiveness the principal cause. This we have proved in some cases by the continual relapses occasioned by very small causes, or even without causes, when not only the economic environment has been changed, but when all the circumstances that might encourage crime have been removed; and we have proved it especially by the increasing recidivism in London, notwithstanding the great efforts made by Great Britain to suppress the causes which produce crime. Finally, we have seen that certain circumstances have so strong an action upon criminaloids that they are equivalent to organic causes, and we may even say that they become organic. Among these circumstances should be noted the effect of excessive

[1] "Homme Criminel," Vol. II, p. 226.

heat upon rapes, assaults, assassinations, and revolts, and the effect of alcohol and heredity upon the whole gamut of crime; and to these must be added the effect of race, which in Italy through the Semitic race, and in France through the Ligurian race, increases the crimes of blood.

A fact of the greatest importance is that the same causes which diminish certain crimes increase others, making it difficult for the statesman to devise a remedy. Thus we have seen that education and wealth cause a decrease in certain brutal crimes, especially homicides and assassinations, but at the same time increase others, or even create new crimes, such as bankruptcy and swindling. And if, for example, too great a density is the cause of many crimes, such as frauds and thefts, a sparse population, in its turn, favors brigandage and crimes of blood. Scarcity favors thefts from the forests, forgeries, insurrections, and incendiary fires, while cheapness of grain multiplies the rapes, homicides, and crimes against persons generally.

Alcohol, which next to heat is the most powerful crime-producer, increases, when it is cheap, all the crimes against persons and against the public administration; and if it is dear, all the crimes against property. Yet it presents this strange contradiction, that the more serious crimes are least numerous where alcohol is most abused, doubtless because this abuse takes place in just those localities where there is a higher degree of civilization, and this, by favoring inhibition, decreases the more barbarous crimes.

The school, likewise is a cause of crime, but where education is most general it diminishes the number and seriousness of the crimes.

§ 208. Necessity of Crime

Statistics as well as anthropological investigations show us crime, then, as a natural phenomonon, — a phenomenon (some philosophers would say) as necessary as birth, death, or conception.

This idea of the necessity of crime, however bold it may appear, is nevertheless not so new nor so heterodox as one might believe at first sight. Centuries ago Casaubon expressed the same truth when he said, "Man does not sin, but he is coerced in various degrees"; and St. Bernard likewise said, "Which one of us, however experienced he may be, can distinguish among his own wishes the influence of the *morsus serpentis* from that of the *morbus mentis?*" And further: "The sin is less in our heart, and we do not know whether we ought to ascribe it to ourselves or to the enemy: it is hard to know what the heart does and what it is obliged to do." St. Augustine is still more explicit when he says: "Not even the angels can make the man who wills evil will the good." The boldest and most ardent defender of this theory is a fervent Catholic and a priest of the Tyrol, Ruf.[1]

The defenders of theories quite opposed to our own also affirm it indirectly by the contradictions into which they fall in their definitions. If we compare the different attempts at criminal codes we see how difficult it is for the legal expert to fix the theory of irresponsibility and to find an exact definition for it. "The whole world knows what a good or a bad action is, but it is difficult, even impossible, to tell whether the depraved act has been committed with a full, or only an incomplete, knowledge of the evil," says Mittermayer. Way[2] writes: "We have not yet any scientific knowledge of responsibility." And Mahring says:[3] "Irresponsibility is a matter which criminal justice cannot decide with certainty in any special case." In fact, there are men who are afflicted with incipient insanity, or are so profoundly predisposed to it that the slightest cause may make them fall into it. Others are driven by heredity to eccentricity or to immoral excesses. "Knowledge of the act," says Delbrück, "with an examination of the body and the mind before and after it, is not enough to clear up the question of responsibility; it is necessary to know the life of the criminal from the cradle to the dissecting table."[4] Now as long as the criminal is living it is hardly possible to dissect him. Carrara presumes "absolute responsibility where both intellect and will

[1] G. Ruf, "Die Criminaljustiz, ihre Widersprüche und Zukunft," Innsbruck, 1870.

[2] "Die strafrechtliche Zurechnung," 1851.

[3] "Die Zukunft der peinlichen Rechtspflege," p. 188.

[4] "Zeitschrift für Psychiatrie," 1864, p. 72.

combine in the accomplishment of a criminal action," but he adds immediately afterward, "upon the condition that the action of the will has not been lessened by physical, intellectual, or moral causes." Now we have seen that there is no crime in which these causes are lacking.

§ 209. The Right to Punish

Some one replies to us: "But if you deny responsibility, what right have you to punish? You proclaim that a man is not answerable for his conduct, and yet you exact a penalty. How inconsistent, and how harsh!" I shall never forget how a venerable thinker shook his head when he read these pages, and said to me: "Where will you arrive, with such premises? Must we let ourselves be pillaged and murdered by brigands upon the pretext that we cannot decide whether they know they are doing wrong?" I answer: nothing is less logical than to try to be too logical; nothing is more imprudent than to try to maintain theories, even those which are apparently the soundest, if they are going to upset the order of society. If a physician at the bedside of a patient, when there is grave danger, must proceed cautiously even with the best established system of medicine, the sociologist must observe still greater circumspection, for if he puts into operation innovations of an upsetting nature he will simply succeed in demonstrating the uselessness and inefficiency of his science.

Scientific knowledge, however, is happily not at war but in alliance with social order and practice. If crime is a necessary thing, so also is society's resistance to crime, and, consequently, the punishment of crime, which must be measured by the amount of apprehension with which it inspires the individual. Punishment thus becomes less hateful, but also less contradictory and certainly more efficacious.

I do not believe that any theory of punishment has a sound basis, except that of natural necessity and the right of self-defense. This is the old theory of Beccaria and of Romagnosi,[1]

[1] "Society has the right to make punishment follow upon crime as a necessary means for the preservation of its members." ("Genesi del Diritto Penale.") "Penalties which go beyond the necessity of preserving the public weal are unjust." (Beccaria, "Dei Delitti e delle Pene.")

of Carmignani, and, in part, of Rosmini, Mancini, and Ellero, and it has now valiant defenders in Ferri, Garofalo, and, above all, Poletti. In Germany we see this theory put forward by Hommel, Feuerbach, Grollmann, and Hottzendorff; in England by Hobbes and Bentham; and in France by Ortolan and Tissot. Tissot declares that it is impossible to find any moral relationship between crime and punishment.[1] In France a state prosecuting attorney has said:

"Man has no intrinsic right to punish; in order to have this right he would have to have the knowledge of absolute justice. If it were not in the name of the most absolute necessity, how could a man arrogate to himself the right of judging his fellow man? From the fact that man cannot defend himself without inflicting punishment, the conclusion has been drawn that he has the right to punish; but that he really does not have it may be seen from the fact that when this pretended right is taken by itself without reference to the concrete need it ceases to be valid."

Rondeau, governor under Joseph II, in his "Essai physique sur la peine de mort,"[2] denied the freedom of the will, repudiated the universally accepted notions of good and evil, merit and demerit, and in speaking of repressive justice he declared:

"Crime does not exist in nature; it is the law alone that imposes this unjust designation upon acts that are necessary and inevitable. The innumerable and diverse causes which produce the pretended criminality are all material and all independent of our will, like the miasma that produces fever. Anger is a passing fever, jealousy a momentary delirium, the rapacity of the thief and swindler an aberration of disease, and the depraved passions that drive men to sins against nature are organic imperfections. All moral evil is the result of physical evil. The murderer himself is a sick man like all other criminals. Why, and in the name of what principle, could they be punished, unless it is because they disturb the regular course of the social life and impede the normal and legitimate development of the species? On this ground society, or, better, the government,

"The reason for the state's calling a criminal to account is not to exact vengeance for the crime, but to bring it about that crime shall not be committed in the future." (Carmignani.)

[1] "Introduction Philosophique à l'Étude du Droit Pénal," 1874, p. 375.

[2] Frasati, "La Nuova Scuolo di Diritto Penale in Italia ed all' Estero," Turin, 1891.

had the right to place an obstacle in the way of the fatal consequences of their acts, just as a landowner has a right to build a dike against the flood which threatens to inundate his fields. The social power can, then, without scruple and without hesitation, deprive malefactors of their liberty; but the moment that all crime is recognized as the natural product and logical consequence of some disease, punishment must become only a medical treatment. We shall cure the thief and the vagrant by teaching them the joys of honest work. If by an exception, which is unhappily too frequent, they show themselves insensible to medical cure, they must be separated from their fellow citizens."

We see here that our boldest conclusions are already more than a century old.

One might question whether it is from wickedness or from the effect of their own organism that wild beasts devour man; but notwithstanding this doubt, no one would abstain from killing them and tamely allow himself to be devoured by them. Nor would any one, because of a belief in the right of domestic animals to life and liberty, refrain from harnessing them up for work, or slaughtering them for food. And what right have we to confine the insane, if it is not for self-defense? By what other right do we deprive the conscript soldier of his most holy and noble right of forming his own home and family, and send him, many times in spite of himself, to death?

It is just because the principle of punishment is based upon the necessity of defense that it is really not open to objection.

Formerly, punishment, which was made to correspond to the crime and like it had an atavistic origin, did not attempt to conceal the fact that it was either an equivalent [1] or an act of vengeance. The judges were not ashamed to carry out the sentence themselves, as the members of the holy *Vehme* did. Crime was considered not only as an evil, but as the worst of evils, which only death could pay for. If the guilty did not confess, torture was used. When torture was dispensed with,

[1] ποινή, poena, compensation. In the Iliad, Achilles killed twelve Trojans in return for the death of Patroclus. The compensation for the death of a Frank was 200 sous, and thefts also could be paid for. Slaves lost their lives for the same crimes which cost a free man only 45 sous. (Del Giudice, "La Vendetta nel Diritto Longobardo," 1876.)

witnesses sufficed. Later mere presumptions were sufficient, — and such presumptions! Not only did the judges kill the criminal, but they wanted him to taste death slowly. This cruelty did not diminish crime, but it was logical, nevertheless. The theory does not contradict the practice. The conception was that the criminal never improves, and that he begets children like himself. The death of the criminal alone prevented recidivism. Men of that day obeyed the instinct that impelled them to punish one offence by committing another; but they did not conceal this view. But our logic, our sincerity in penal matters, where is it?

We still have this primitive instinct. When we are trying a criminal, we have always a tendency to measure his punishment by the degree of repugnance and horror with which his crime inspires us and to be filled with indignation against the man who has confessed it. So we not infrequently see representatives of the law forgetting their abstract theories and demanding in loud tones that the vengeance of society be visited upon the offender. Yet the same men, when inditing a book upon criminal law or sitting to legislate on the same subject, would repudiate such an attitude with horror. And what logic is there in the theory, which is being brought into vogue again by Roeder, Garelli, Pessina, that punishment is for the purpose of reform, when we know very well that the reform of the guilty is always or nearly always an exception, while the prison not only does not improve him but even makes him worse. Besides, how, with such a theory, could one justify the punishments inflicted for political crimes, or crimes committed through excitement or passion, followed as they almost always are by spontaneous and complete repentance? Oppenheim, after having written that every crime should be followed by a proportionate penalty and that the penalty should not only be an evil but should appear as such, goes on to say (with Mohl and Thur): "Punishment should have for its only aim the reformation and employment of the criminal." But is not this an obvious contradiction? How can you reconcile the theory which has the criminal dishonored with that which pretends to improve him? How can you brand him upon the brow with iron, and say to him, "Make yourself better"? What are the theories of Herbert, Kant, Altomid, and Hegel, but the ancient ideas of vengeance and the *lex talionis* disguised in modern dress?

And with all this the State does not think of the morrow. It shuts the prisoner up, and when he has served the term of his sentence it sets him at liberty again, thus increasing the danger of society, for the criminal always becomes more depraved in the promiscuity of the prison, and goes out more irritated and better armed against society. With this theory it is not possible to justify the increase of the penalty in the case of recidivism nor the adoption of preventive measures.

Some legislators maintain that a criminal ought to be made to expiate his crime. But the conception of expiation is ecclesiastical, and how can we say that a criminal expiates his crime, when it is by force that we take away his life or his liberty?

The theory of intimidation in its turn offers numerous contradictions. Our predecessors cut off nose and ears, quartered, boiled in water and in oil, and poured melted lead down the throat. But they succeeded only in multiplying crimes and making them more horrible, for the frequency and ferocity of the punishments hardened men; in the time of Robespierre even the children played at guillotining.[1] But what do men expect to accomplish by intimidation nowadays, when penalties have been made so much milder and the prisons are almost like comfortable hotels? And then, what sort of justice is that which punishes a man, less for the crime he has committed than to serve as an example to others?

Further, the right to punish, based upon the nature of the deed itself, has nothing absolute in it, since we see the penalty varying according to the temper and habits of the particular judge. Breton affirms that a judge accustomed to deal with great crimes will inflict punishments relatively more severe

[1] The death penalty was visited in France as late as 1100 upon 116 kinds of crimes; thieves were broken on the wheel, murderers were hanged; later all were broken on the wheel. Between 1770 and 1780 a certain L. was broken on the wheel for stealing linen, and another thief for having stolen cheese. In 1666 in Auvergne there were 276 individuals hanged, 44 beheaded, 32 broken on the wheel, 3 burned, and 28 sent to the galleys. In a single province there were more persons executed than are now convicted in all France.

when he comes to deal with minor offenses; he will give months in prison instead of days. No judges, moreover, even in the same country and when it is a question of identically the same crime, agree exactly upon the sentence. Is it possible to believe in an eternal and absolute principle of justice among men when we see this pretended justice vary so greatly within a brief interval of space or time; when we see bigamy and rape punished so differently in England and in Germany; when we see that not so many years ago a Jew who accosted a Catholic prostitute was condemned to death, as was likewise a Catholic who allowed an involuntary blasphemy to escape him, while infanticide, incest, and rape were tolerated? Do we not even to-day see the right of pardon and the theory of limitations still in force, as if the favor of the king or the lapse of time could change the depraved nature of the criminal or make him less likely to relapse into crime?

CHAPTER I

IS DEGENERATION A FACT?

THOSE who speak of decadence are often called "prophets of gloom," "dismal Jimmies" and so forth. But, as a matter of fact, optimists often do more harm than alarmists, lulling us into a false sense of security, while with their soothing arguments they succeed in convincing a world, predisposed to imagine all is well, that prudent action, while there is yet time, is quite unnecessary.

What we need to know really is *the truth.* We do not want to see things through rose-coloured spectacles, nor do we want to be alarmed where there is no cause for alarm. Unfortunately, the question of racial degeneration cannot be dismissed as easily as some optimists would have it. There are not the necessary statistics to prove that all is well, and it is very difficult to prove conclusively the contrary, for want of certain essential figures. It would be very easy for the Government to place the matter beyond the sphere of controversy by giving us from time to time definite statistics with regard to mental defect and all hereditary diseases. But up to the present this has only been done in a very scrappy and unsatisfactory manner.

I think, however, that some of the figures that are available, though leaving much to be desired, cannot be called ambiguous, and that an unbiassed enquirer, who does not allow the wish to become father to the thought, can eventually come to but one conclusion.

Dr. Tredgold, medical expert to the Royal Commission on the feeble-minded, says:

> "My enquiries have convinced me that mental defect is but one manifestation of disease which is slowly but surely eating its way into the heart of the nation."

He also tells us that the lowest estimate that can be made of the total number of persons suffering from different kinds of mental defect in England and Wales is at present well over 260,000.[1] Mr. Eldon Moore, writing in the *Eugenics Review*, has shewn good reason for estimating defectives, exclusive of those in mad-houses, but inclusive of idiots and imbeciles under care, at 155,833. If we add to these the 133,883 notified insane according to the report of the Board of Control for 1926, we get a grand total of 289,716, which is considerably higher than Dr. Tredgold's estimate. Nineteen years ago the Board of Control estimated the total number at 150,000. I do not mean to imply that there has necessarily been an increase of nearly a hundred per cent. in nineteen years, for the term "mental defect" is elastic, and estimates extremely difficult to arrive at with the figures available; but when we find that

[1] *Mental Deficiency*, 4th edition, pp. 14-15.

all authorities are agreed that defectives breed and multiply much more rapidly than normal people—Dr. Gibbons, gynæcologist to the Grosvenor Hospital for Women, says from two to six times as fast—and that most authorities believe mental defect to be increasing rapidly, we have, to say the least, food for reflection.

Sir W. Arbuthnot Lane says:

> "Some optimists are in the habit of stating that our race is improving physically, that we never were better. That this is absolutely false and that the reverse is true can be demonstrated in the clearest manner possible by a comparison of the published description of the physique and teeth of the soldiers of Wellington and their wives, who were exposed to constant hardship, with our soldiers of the present day, living in comparative luxury."[1]

Major Darwin says:

> "Defective intelligence is undoubtedly associated with great fertility; and it is not therefore surprising to learn that thieves and other petty offenders, whose stupidity is unquestionable, spring from amongst the most fertile stocks in the community."[2]

Again:

> "As regards English criminals generally, except those convicted of fraud, they are generally

[1] Signed article in *Daily Mail*, 9th Nov., 1926.
[2] *The Need for Eugenic Reform*, p. 216.

markedly inferior to the general population in stature and bodily weight."[1]

That is to say, the most fertile stocks in the country are those which give us mental defectives, petty offenders, stupid individuals and persons markedly inferior in stature and bodily weight. Major Darwin, whose methods of enquiry are pre-eminently scientific, has come to these conclusions after weighing the evidence contained in Goring's *The English Convict*, and many other sources. His final answer to the question " Is the race deteriorating ?" is this :

> " Almost every class must on balance be receiving additions from below, and must, in consequence, be suffering from the inferior quality of its recruits. In short, as far as we can judge from the facts at our disposal, the nation as a whole is slowly and steadily deteriorating as regards its average inborn qualities."[2]

Dr. Heron told us in 1906 that " where there is more general pauperism . . . there the birth-rate is highest." And this fact has been confirmed by innumerable authorities and by statistics. The continued high birth-rate of the lowest classes, combined with a low and steadily falling birth-rate among the better, must of necessity lead to deterioration.

[1] Ibid, p. 211.
[2] Ibid, p. 327.

Sir George Newman says that in the elementary schools of England and Wales a million children are so physically or mentally retarded, defective or diseased as to be unable to derive reasonable benefit from the education provided by the State; that even in the healthiest areas 10 per cent. were found, just before the war, to suffer from grave physical disability, 20 per cent. from defective vision and 60 per cent. from bad teeth. During the last stages of the war 40 per cent. of recruits were described as C3. Of insured persons to-day more than 50 per cent. receive medical treatment every year.[1]

From the second general report issued by the People's League of Health we glean the following:

During 1924 not less than 82 per cent. of recruits for the army were rejected as unfit, while only one out of every twenty men examined for the police force was found to be of the physical and mental standard required.

One out of every seven men of working age in this country dies of tuberculosis.

It was estimated that something like 50,000 children of school age are to-day suffering from organic heart disease, while 80 per cent. of children attending our council schools are suffering or have suffered from rickets.

According to the report of the Ministry of National Service, for the years 1917-1919, and the Anthropometrical Committee of the British Association, in 1883, the average height, chest girth and weight of

[1] *Public Opinion in Preventive Medicine*, Newman. Ministry of Health, 1920.

boys of 18 years of age for these years respectively, for the whole country, were as follows :

	1883	1917-1919
Height in inches	66.9	65.1
Chest girth in inches	34.2	32.2
Weight in pounds	131	116.5

As to venereal diseases, whose evil effects are passed on " to the third and fourth generation," it is well known that they have become more common since the war. Before the war, it was estimated (in 1913) from facts supplied by the Royal Commission, that their incidence was something like 20 per cent. of the whole population.

If these facts are not enough to convince the reader that there is little ground for optimism, let him consider what *must* be the effect, in the long run, of the differential birth-rate. It is a well known fact, as has already been stated, that fertility is in inverse ratio to social status.[1] Now, social status is not exactly correlated with eugenic value, but it will be found to be so roughly, because social status depends more upon income than on anything else, and income depends, in the majority of cases, roughly, upon efficiency. Whatever those who are embittered by failure in their own lives or their friends' may be inclined to think, it is a fact that, as a general rule, the more a man earns the more efficient he is likely to be as a worker, whether of hand or brain. Let those who doubt it look not

[1] See *Fertility of Social Classes*, by Dr. Stevenson. Royal Statistical Society, May 1920, p. 410.

only to a few exceptionally unfortunate cases, but make a list at random of the names of a hundred workers known to them, with the earnings of each, and then consider which is on the whole the more handy, intelligent or reliable set of men, the fifty who earn more or the fifty who earn less. Of course, inherited wealth and other facts tend to vitiate the correlation, but, in spite of all, it remains generally true.

Now the result of this differential birth-rate, first noticed about the end of last century, but now shewn to have existed much earlier, is that the more efficient sections of the community must necessarily tend to die out, and the less efficient to increase in numbers, relatively to the former.

From *The Population Problem*, by Carr Saunders, we may learn much about these facts; and the extent of the disparity in the birth-rate among classes of different eugenic values may be gathered from the following figures, given by him, and made up from material furnished to the private Birth-rate Commission by the Registrar General, for the year 1911:

Births per 1,000 *married males under* 55 *years of age*

Upper and middle classes	119
Skilled workmen	153
Unskilled workmen	213

Dean Inge, who quotes the above figures and others in the chapter on "Industrialism" of his book, *England*,[1] also compares, as an extreme case, the Finsbury birth-rate of 41 per thousand of the

[1] p. 210.

population, with that for the city of Westminster, which was 9.

Now the significance of this differential birth-rate becomes evident when we consider the fact that science no longer admits any doubt as to the hereditary nature of moral and intellectual characteristics, as well as physical, so that from inferior stock we cannot possibly reap good. This being so, it is certain that only very drastic measures taken by our rulers could save us from the tendency to degeneration which the differential birth-rate brings with it, and to which every previous civilization has eventually succumbed. For recent investigations have shewn that one of the chief causes of the fall of former empires has been this same tendency, when wealth accumulates and the arts and sciences flourish, for the more intelligent and efficient to become relatively sterile, while the shiftless, decadent classes breed like prolific animals. But if, in place of drastic legislation to combat this tendency, our rulers actually foster it and intensify its evil effects, as will be shewn in the next chapter, what are we to expect ?

As just stated, all ancient civilizations appear to have developed in time this differential birth-rate, while continual warfare, and the slaughter of the superior stocks in various ways, hastened the decay. We know this ; we have seen it happen again and again in the world's history, yet, in our blindness, in spite of having access to sources of knowledge and experience which they had not, we follow up our Great War, which killed off the flower of our race on a scale hitherto unknown in history, by legislation more and more dysgenic, better and better calculated to hasten the catastrophe which we refuse to see approaching.

This we owe to democracy, government by an ignorant majority; for, naturally enough, the ignorant have no care for the future of our race, and if they had, having studied neither history nor science, nor even current statistics, they could not be expected to act with foresight in such matters. Their chosen rulers, knowing this, think more about ways and means of adulating the masses than of warding off distant dangers while there is yet time. Thus the price of silk stockings is a matter of more importance to the politician than the decay of the race, for an election might be decided on the former point, but not on the latter; while to bribe a corrupted majority with money extorted from a still virile minority is their chief care.

During four years, the pick of our manhood went to drain its blood on the fields of Flanders and elsewhere. During that time the old, the cowardly, the feeble and the diseased remained at home to procreate. In the last year of war, Sir George Newman tells us, seven men in every nine among the recruits examined were either on a definitely infirm plane of health and strength, or incapable of undergoing more than a very moderate degree of physical exertion, and could be almost described, in view of their age, as physical wrecks, while one in nine was a chronic invalid. The chief preoccupation of Governmen at that time should have been to prevent these men becoming the fathers of a large proportion of the coming generation, which may again be called upon to face a would-be invader. Instead, however, Mr. Lloyd George gave us the pauper vote, in 1918 of the truly disastrous effects of which I shall have more to say in a later chapter.

Such are the effects of war, "charity" modern legislation and curative medicine on the physique of twentieth-century manhood, that in nearly every "Christian" country the stature exacted for military service has had to be constantly lowered of late. In Spain, with her eternal colonial warfare, it has been lowered three times in the last forty years.

There is great confusion of thought in the popular supposition that by continually improving environment we can prevent degeneration, and, in another conclusion derived from the same false premises, that when we see children looking healthier than their parents there can be no degeneration: Both these wrong ideas are due to the astounding ignorance prevalent, even among educated people, of the elements of the science of heredity.

Every individual depends for his development upon two factors—heredity and environment. One of these being defective, much improvement may come about through the other being above the average, but it is impossible in such a case to have as sound an individual as if both had been good. The first of these factors a man gets from the germ-plasm of his parents, and the second through all that happens to him from the moment of conception onward.

Thus you may have an individual with a splendid inheritance from his fathers, but who nevertheless turns out a wreck owing to bad environment. And you can also have an individual with a very defective heritage, who nevertheless looks healthier than did his father at the same age on account of greatly improved environment. It will be seen therefore that appearances cannot tell us definitely whether there is degeneration or not, and also that by improving environment you are likely *to go on for a certain time improving appearances*, but there must come a time, if you neglect the hereditary factor, when improvements in environment can no longer prevent a more or less rapid downfall.

It is no use, therefore, pointing either to the falling death rate—which means, amongst other things, that the inefficient (who are in the majority) are living longer and producing more of their kind, —or to any real or fancied improvement in the appearance of a population, due to improved environment; nor even to a reduced incidence of non-hereditary diseases, since none of these things touch the question of degeneration.

The only way to solve the question as to whether or not degeneration is actually taking place, *in spite of appearances*, is to note the incidence of hereditary defect and hereditary diseases. We find, on enquiry, that it is just these which are either stationary or increasing, in spite of improvements in environment.

It should also be borne in mind that the more defectives are segregated the better will be the superficial appearance of the population seen in the streets, for the worst cases are all hidden away, and nobody realizes their number, which it is very difficult to get at from official statistics, since these inform us almost exclusively of the incidence of ordinary infectious diseases.[1]

It should be realized that nothing is known in the way of environment which can beneficially affect the germ-plasm, though alcohol and venereal poison appear to affect it harmfully. Otherwise, with rare exceptions, each individual passes it on to his descendants just as he received it from his ancestors. Degeneration comes about because some children inherit the worse characteristics of their ancestors and become inefficient; also by injurious hereditary "mutations." If the children shewing these characteristics or mutations survive and multiply more than the others decadence must necessarily follow. The only way by which the germ-plasm can be improved is not by environment but by diminishing the fertility of degenerate stocks and increasing that of those above par.

"No horticulturist of experience would propose

[1] If we could see the hundreds of huge buildings now devoted to the care of imbeciles, madmen, and all suffering from diseases with hereditary tendency—if we could see them all together, covering acre after acre, stretching away for miles toward the horizon in every direction—we should be struck dumb with horror. But if, instead, we take a walk through Hyde Park on a bright Sunday afternoon, and see apparently happy and healthy thousands on every side, we say "all this about degeneration is obviously mere nonsense."

to produce from the dwarf pea a permanently taller race by a process of manuring. There is little doubt but that the seeds from the richly manured and those from the normally grown plant would, when raised under similar conditions, each produce a row of plants indistinguishable in point of size."[1]

In the same way it is certain that the breed of men can never be appreciably improved by environment alone, though the appearance of any particular generation may be changed.

Apart from figures, statistics and scientific considerations, what do we find if we study the life of the people of to-day and the trend of popular taste? Generally speaking, novels, films and plays must be full of sex if they are to be successful. Had Sir Walter Scott lived to-day, one wonders which publishing house would have accepted *Waverley*. In sculpture and painting the cult of the hideous is replacing that of the beautiful. The same is true of music and dancing.

The Government controls British broadcasting, but instead of making any attempt to educate public taste, it allows our ears to be filled night after night with the horrible din of "jazz," which to-day usurps the place of music. I have heard many foreigners remark: "We never listen in to London now, because we are tired of jazz, and seldom get good music. You can hear better music from almost any station on the Continent."

The idiotic "Charleston" and "Black Bottom"

[1] R. C. Punnett, *Mendelism*, Macmillan & Bowes, 1907.

are the latest substitutes for the graceful waltz, and the still more graceful minuet of once upon a time.

Crime in this country is said to be diminishing. Not so, however, crimes by and against children, which are both increasing ; especially brutal assaults on children, often followed by murder. Here again we are able to appeal to statistics, to a certain extent. From the Report of the Departmental Committee on Sexual Offences against Young Persons, 1925, we can, by taking the figures relating to "persons tried," form some conclusions as to this sort of degeneration. The average number of persons tried annually from 1909 to 1913 was 884 ; from 1910 to 1914 it was 917 ; from 1919 to 1923, 987, and from 1920 to 1924, . . . 1058. One would like to have figures relating to convictions, especially in cases of extreme violence, murder and mutilations. But these are not furnished. A large proportion of the offenders are shewn to be mental defectives, but this condition, when present, is frequently unrecognised by the courts. Even where mental defect is evident, it seems that it is often found impossible to certify it, on account of the wording of the Mental Deficiency Act of 1913.

It is perhaps fortunate for us that most "Christian" countries are shewing the same tendencies as ourselves, though in none are they, for the most part, so pronounced as with us, a fact evidently due to "humanitarian" legislation, in which we take an unenviable lead. But what will happen to us after other nations—the Germans for instance—have begun to put eugenic principles into practice, and the breed of their men begins to improve as rapidly as ours deteriorates? Says Mr. Curle:

> "I can see Germany, unrepentant and bursting with people, again attacking in fifty years. The Germans of that day will be weeded out, and highly efficient."

I myself think this is very likely, for the Germans are always ready to make the most of science for advancing national aims and aspirations, whether it be by means of forces destructive in war or constructive in eugenics. But I should put the period of waiting and recuperating down at considerably less than fifty years.

Nothing seems to me more certain than that the nation which first begins to *breed for efficiency*—denying the right of the scum to beget millions of their kind, to be a financial burden, a military danger and a source of suffering untold to future generations,—is the nation destined to rule the earth, whether it be Italy, Germany, Great Britain, America or Japan. Says Major Darwin:

> "Woe betide the nation which sees other nations advancing in racial qualities, whilst allowing itself to become so degenerate as to be unable to defend itself when the pinch does come!"

Are we deteriorating? Surely, if there is any doubt upon a subject of such vital importance, it is the first duty of the Government to enquire into the matter, clear up all doubts with figures that cannot be disputed, and, if it be found that we are decaying, take the necessary steps to prevent the process from going on. Those steps would, I believe, necessarily consist first of the repeal of much so-called social reform, and, secondly, of definite measures for preventing the lowest from procreating, and the offer of inducements to the efficient to have large families, instead of mulcting the latter with intolerable taxation, for the benefit of the former.

It is therefore urgent that the Ministry of Health should give us periodically the statistics needed.

A Royal Commission, too, ought to be appointed without delay:

(1) To enquire into the incidence, past and present, of all diseases alleged to have an hereditary tendency of some sort, and especially mental disorders;

(2) To enquire into the tendency of such incidence to increase or to decrease;

(3) To enquire into the supposed hereditary nature of such diseases, disorders or defects;

(4) And to make recommendations for the reduction of such incidence.

PART 2

The nineteenth century: social activism and social control

It is as one of the noblest triumphs of our faith that we record in this little volume, the foundation of the National Association for the Promotion of Social Science. For many years, Christian men and women had been labouring in various fields of usefulness, and had endeavoured to grapple with the social evils that had developed themselves. Distinct and important Societies, embracing educational and reformatory objects, improved dwelling-houses, and other appliances for remedying every possible form of human want and misery, had been brought into action; but it was felt that no single effort could be on a scale commensurate with existing need, and that, to insure such a result, the scattered rays of light must be gathered into one focus, and unity must be given to these isolated labours.

The spirit of free spontaneous association, whose growth we have on another occasion endeavoured to trace,* was evoked by one of England's noblest sons, and at his bidding there sprung into existence an armed Hercules, representing and embodying the talent, energies, sympathies, influence, and wealth of hundreds, nay, thousands, of labourers in one common cause,†—that of Christian philanthropy.

---

* See "Handbook of the British Association."

† At the Liverpool Meeting, upwards of two thousand Members and Associates were present.

B 5

Humboldt, in his "Cosmos," rightly gave to Astronomy—"The Science of the Universe Without"—the first place in his great picture of physical knowledge; and if this science may well be called the highest glory of the human intellect, shall we not call the science of philanthropy which is now, for the first time, taking its place among us, the highest glory of our common humanity?

Well might an illustrious foreigner, when speaking of the power of free spontaneous association as developed in England, in contradistinction to the centralization of his own country, observe, that "so long as this principle is in force, and in honour, England will have nothing to fear: her glory and her virtue will be able to resist the contagion of continental servitude." "*Supported by Voluntary Contributions*," he adds; "such is the proud and noble inscription which one reads throughout England on the front of the greater part of the hospitals, almshouses, and various asylums for human misery...........We understand well that these words, 'Supported by Voluntary Subscription,' imply also this,—'Governed by the authority of the subscribers.' It is always the same principle,—effort, personal and permanent sacrifice; and then the right and the power which spring from the sacrifice and effort."

The question will be asked by those for whom we write, "What is Social Science?" and we answer, that it is the science which has for its primary and sole object, to promote the well-being and happiness of the human race.

It embraces in its progress and full development, a consideration of all the subjects that the most enlightened philanthropy can suggest, and at the same time realizes for itself, by the foundation on which it rests, the dignity of a science.

As, in physical science, the first process of philosophical

induction is to systematize and extend a knowledge of facts, so, an elder Association, (on whose model the National Association was founded,) many years ago, commenced researches in statistical science, which now furnish data for the formation of a science of the highest importance to society.

The Statistical Society of London was an offshoot from the British Association; and those who know the full value of statistics, to the statesman and philanthropist, will not be slow to acknowledge the debt of gratitude which the practical men, connected with the National Association, owe to the scientific men who first recognised the importance of statistics, and gave it its right position as a science.* It has been well said,—

"Whoever wishes to correct a great national abuse, or remedy a crying social evil, must curb impatient enthusiasm, restrain his general ardour, while he seeks in blue books for his axioms and definitions, from which he may deduce incontrovertible propositions."†

Thus, while statistics reveal that crime is not the necessary attendant upon poverty or low wages, they show that it is found most abundant in closely-crowded houses, in ill-drained localities, while the morals of the poor quickly mani-

---

* "Had statistics," says Dr. Guy, "effected nothing more than the substitution of *figures* for *words*, they would have established a strong claim to our approbation. Nothing can be more variable or worse defined than the meanings of some words in general use. What meaning are we to attach to such vague words as 'sometimes,' 'generally,' 'occasionally,' 'in the majority of cases?' These terms have every possible signification, and vary in their meaning with the varying disposition, and more or less sanguine character, of those who use them. The 'sometimes' of the cautious is the 'often' of the sanguine; the 'always' of the empiric, the 'never' of the sceptic. But the numbers 1, 10, 100, 1,000 have but one meaning for all mankind."—*Journal of Statistical Society, Vol. II.*

† "Meliora," No. 3.

feet an improvement when sanitary reform has been carried out in their dwellings.

Statistics make known the startling fact that there is in our crowded cities a vast waste of human life from causes wholly preventible; that, under favourable sanitary conditions, fever, hitherto the great destroyer of life, is unknown.

Statistics have revealed spiritual and physical destitutions sufficient to appal the hearts of any who fully realize the amount of suffering represented by such figures. The religious condition of the people, the education of their children, the wretched sanitary state of crowded neighbourhoods, the connexion of intemperance with crime, have all been tested and proved by statistical science, and the results brought before the consideration of a Christian people. As surely as a Newton (the glory of his species) expounded the great law of gravitation as the result of his deep and patient study, his often-repeated observations and experiments, so surely have the great laws that govern the life, health, and well-being of the human race been evolved from the science of statistics, which has thus taken its place as the foundation of Social Science.

Social Science has then a mighty mission to fulfil, coincident with the progress of a great nation; and we claim for it, in the future, a place in the systematic teaching of our schools and Universities, believing that it is of the highest importance to the prosperity of the State, that an acquaintance with its laws, and the principles on which it is founded, should be widely diffused.

It has been said, and with truth, that we have arrived precisely at that stage in our national career when a recognition of benevolence, as a social law, is imperatively needed to rescue us from imminent evils.

Great as has been our progress during the last thirty

years, we do not hesitate to say, that the indifference that is still found, even in the religious world, to the important subjects embraced in Social Science, can only be ascribed to a lamentable ignorance of the intimate connexion that exists between the moral and religious growth of a people, and their physical condition.

While we thankfully own that English men and women are wakening up to this great truth, and to the responsibility resting upon them, with relation to this momentous subject, we still need strenuous exertion further to enlighten the minds of all classes of the community, before the full extent will be realized of the weight of obligation resting upon us, as Christian members of that community.

Nor can we help reiterating the truth we have so earnestly advocated on another occasion, that the first step to secure this result must be the wide dissemination through every available channel of that knowledge which is pre-eminently and peculiarly the science of the age,—viz., the philosophy of common life, considered in its relation to the every-day welfare of the human race,—the science of health, which involves not only, as has been well observed, "the mitigation of the sufferings of the masses, the increase of their enjoyments, the comfort of their homes, the prolongation of their lives, their personal tranquillity and progress, and their natural safety and usefulness, but tends to the better moral and religious condition of the working-classes, and the elevation of the depressed orders below them."

We shall now say a few words on the most salient points arresting our attention, in the consideration of the evils needing remedy in our social condition, and endeavour to show that these have arisen from the popular ignorance of the importance of Social Science, in its relation to a high state of civilization, and of its bearing upon our national and individual prosperity.

We have endeavoured, in the "Handbook of the British Association," to give some idea of the heavy losses that have been incurred in every branch of the industrial arts by our neglect of science; and we have traced the origin of this low appreciation of science and its cultivators to the deficiencies in education that have prevailed throughout all classes of society. This has been the case, from the Universities and the public schools, down to the primary schools for the children of the working-classes, in which, till within a very recent period, no attempt had been made to popularize scientific truths.*

Hence has arisen an inconceivable amount of prejudice and ignorance, which, added to the opposition of persons interested in perpetuating such ignorance, has proved one of the greatest impediments to national progress, and to the growth of an enlightened popular opinion on many important subjects. Unaware of the scientific processes by which the abstract researches of the scientific man are made to contribute so eminently to our industrial wealth, such pursuits have been looked down upon as purely theoretical, and as beneath the dignity of those who gloried in being, as they called themselves, practical men.

It is a remarkable fact, that a large section of the people, and their rulers, long dreaded progress and change, and obstructed any reform that might be proposed. Every new science, every important invention, has, in its turn, been persecuted. Popular clamour, the power of the press, and the pulpit, and even brute force, have all, in their turn, been brought to bear upon any amendment of laws and institutions,

* See "Handbook of the British Association," Chapters I. and II.

and the pioneers of progress have been called the friends of revolution, anarchy, and ruin.

Blind to the sublime truth, that progression is stamped on all the works of the Creator, and that those best fulfil His will who are His intelligent instruments to carry forward that progress to the highest perfection, there have been those, in all ages, who would have desired to keep the human mind stationary,—who would have restricted the acquisitions of the intellect to the select few, leaving the multitude to grope in darkness and error.

Nor are prejudice and ignorance yet extinct, as is manifested in the slow and cautious reception given by society to the great truths that are the foundation of social reform. Well might Arago remark, that "Prejudices are like hurtful plants,—the slightest effort suffices to eradicate them if they be at once attended to; on the contrary, they grow with time, become inveterate, extend far and near, and their numerous ramifications seize upon everything that comes within their reach."

Perhaps there is scarcely a social evil that the National Association may desire to redress, that will not find its advocates, and its apologists, actuated either by interest, ignorance, or prejudice, exemplified in the conduct of towns, whose inhabitants refuse the introduction of sanitary measures, or the application of the Free Library Act, and other beneficial measures; masters and mistresses, who dread what they call over education for their servants, and numerous classes of individuals, who are criminally indifferent.

The difficulty experienced by the Board of Health in overcoming local prejudices, and gaining credence from parish authorities to the facts which they substantiate with reference to sickness, either in an epidemic form, or otherwise, is one instance among thousands we could place before our readers. We give an illustration from the Report of the Board of Health, in the year 1849, during the cholera visitation, which will show the difficulties with which they have had to contend in these respects. It is one of several of equal inattention which are adduced, and is selected by us as the illustration given by the Board from *London* Guardians :—

"Exemplifications of the impunity with which some Boards of Guardians consider they may disregard orders made under the authority of the Legislature, with the view of carrying out its provisions, have been afforded by the course adopted by the Guardians of the Whitechapel Union by, &c.........

"At a time when deaths had already occurred from cholera, under extremely painful circumstances, and when there was reason to apprehend a serious outbreak of the disease, the Guardians of the Whitechapel Union entered on their Minutes the following Resolution :—

"'That it is the opinion of the Guardians that, at present, the order of the Board of Health, of the 18th of November last, need not be acted on in this Union.'

"It appears that this deliberate act of disregard to the orders of the Board of Health was resolved upon the same evening when the medical officers of the Union presented to the Guardians a written statement to the effect that malignant cholera had broken out in some parts of the district.

"Here is an instance of a Board, composed for the most part of persons engaged in the daily routine of trade, not having, in general, it must be presumed, the means of judging from any large observation or experience of the matter on which they decide, deliberately acting on a medical question against the opinion of their own medical officers, and in direct opposition to regulations framed on the largest experience, not only of this country, but of the whole of Europe—regulations so framed for the purpose of carrying out the express provisions of the Legislature.

"It further appears that, on the 21st of November, the Clerk of this Union having, in conformity with the order of the Board of Health, laid before the Guardians returns from the four medical officers, and from some of the Registrars of Births and Deaths, of those places in the Union where epidemic, endemic, and contagious diseases have of late been prevalent, the Guardians came to the following Resolution :—

"'That the Clerk forward such particulars to the various Local Boards in the Union, but that the medical officers be *not* called upon to visit the places in question.'

"It must be borne in mind that this Resolution was adopted by the Guardians at a time when cholera was not only actually prevailing in the district, but was spreading there under circumstances of so painful a nature, as to attract the attention of one of the Coroners of Middlesex, Mr. Baker, who, on the 24th of November, addressed a letter of expostulation to them, on the neglect of the measures which the circumstances of the time obviously required.

"On the 14th of December, the Coroner again addressed a letter to the Clerk of the Union, in which he says :—

"'My attention as Coroner has this day been called to several cases of sudden death in the parish of St. Mary, Whitechapel, of a most awful and appalling character, and I cannot but feel that a very heavy responsibility rests not only upon my own shoulders, but upon those also of the Board of Guardians of the Whitechapel Union, in reference to these and all such cases, it appearing that there have been no less than sixteen under the care of the medical officer lately.

"'I have this day been engaged in an inquiry into some of the deaths, more particularly alluded to in that letter, in Hairbrain Court, Rosemary Lane, and have myself viewed this evening the dead bodies of no fewer than three persons, but have been witness-

also to the most agonizing and appalling situation of others in a dying state in the same locality (within a few yards of the former), who were found by me to be in a state of distress and misery, which could not but be most afflicting to my mind, being surrounded by the most fœtid and unwholesome vapours from bad drainage, and filthiness, and much overcrowded; and allow me to say, in such a state as I could scarcely have deemed it possible to have existed, after the publication of the documents to which I have above alluded.'

We believe that the lives of several of the persons who thus perished might have been saved if the Guardians had carried into immediate effect the orders and regulations under the Act; and, after a protracted investigation ofthe circumstances connected with the death of several of those persons, the Coroner's inquest brought in a verdict against the Guardians of 'Very great neglect.'"

When ignorance and criminal indifference thus trifle with human life, who would not desire that legislative power should step in and enforce those laws and regulations which a superior intelligence had enacted?

While we disclaim all wish to see any adoption of a system that should savour of Continental despotism, we desire to see the strong hand of authority, founded on the will of a free people, applied to such crying evils in our social state; and we contend that, so far from such wise and righteous administration of law being antagonistic to the liberty of the subject, it would be securing to the poorest among us all that constitutes the most precious heritage of a free man.

We perfectly agree with Stuart Mill, that "the fullest development of individual natures which is compatible with a security against mutual encroachment, is the point at which the true legislator should aim. The development of the individual, however," he says, "is not a part of his business; that is to be left to nature, and what Governments have to do is simply to remove all obstacles to the process being carried on in a normal manner."

Do our readers realize what is implied in this last sentence? We think we do, when we look forward to the time which will arrive,—or our Christianity must lose its Divine power,—when the social evils that now exist shall be amended, when the individuality of the poor man's child shall be equally respected and developed with that of the rich; when the common blessings of God's providence, which are the unalienable birthright of every human being, shall no longer be withheld from our poorest brother. Until that time come, we contend that liberty is but imperfectly developed even in this privileged land.

Mr. Mill remarks very truly, that in practice no principle has yet been recognised, by which the propriety or impropriety of Government interference is tested in particular cases. "Some people," he observes, "whenever they see any good to be done, would willingly instigate the Government to undertake the business; while others prefer to bear almost any amount of social evil, rather than add one to the departments of human interests amenable to Governmental control."

Among the social reforms which are now loudly called for by public opinion, there are some that can only be carried into effect by the Legislature, and others which may be accomplished by private means, or by public associations.

The National Association has thus a wide sphere opened before it by both these modes of action. "For if," as Lord Bacon observes, "the ultimate object which legislators ought to have in view, and to which all their enactments and sanctions ought to be subservient, is, that the citizens may live happily," we are well assured that the representations of the Association, founded on the experience and deliberate judgment of our most eminent social reformers, will have the same weight with Government on all the great subjects embraced in Social Science, that those of the British Association increasingly have on subjects connected with the material interests of our country.

In the advancement of any needed reform, an immense point is gained when the advocates of such change are listened to with attention, and their opinions widely promulgated; and we own we anticipate the most important results to social progress from the free ventilation that certain subjects are now obtaining, and from the certainty that the unity of action, so long a desideratum in philanthropical labour, is now attained.

In endeavouring to place before our readers some of the most striking evils in our social state, we would remark that various circumstances have contributed to bring about a remarkable change in our social relations. The immense increase in our population, the magnitude of commercial enterprise in the present day, and the consequent grouping afresh of the people, cannot fail entirely to change their character, and demand fresh efforts on the part of all true philanthropists to meet such new phases in our social life.

Nor can we ignore the fact, that there has been a woeful misapprehension on the part of rich and influential members of society with relation to the obligation springing from such position to the poorer classes. Instead of recognising the possession of wealth and influence as talents given to be employed for the good of his fellows, the rich man has too often viewed these accidents of birth as giving him a superiority over his less fortunate brethren, which is to be asserted on every occasion.

In an admirable Paper, "On the Teaching of Social Economy," by W. M. Williams, F.C.S., read at Birmingham, the following remarks occur:—"All who have had to deal with working men have encountered similar manifestations of suspiciousness. Why, then, should they be as a class so suspicious? So general a phenomenon must have some generally operating cause or causes. By looking back a little into the facts of our social history, we may detect causes that have been in operation for many generations, and have produced a set of traditional notions concerning the relations between the different classes of society, that are calculated to render the bulk of the poorer classes very suspicious of anything bearing the aspect of political interference that comes or seems to come from those who are socially above them."

"It is still held as an hereditary article of popular faith, that the leading effort of political and social effort on the part of the rich is to keep the people down, and to secure to themselves the perpetual maintenance of their own existing advantages. Kings and Queens are supposed to regard the country and the people as their own personal goods and chattels, to be used for their own private gratification; the aristocracy is looked upon as their confederates, among whom the national spoil is divided; and the Government mainly as a machine for the suppression of rebellion. These traditional notions, like most other traditions, are founded on fact. Though they are not applicable to the existing state of things, they are but a strongly-coloured view of what was formerly our actual national condition, the remembrance of which has been handed down in a thousand forms of nursery and fireside tales, of the tyrannous rich, and suffering poor, a remembrance which lingers still among us, and shapes our earliest notions of social things. The mass of the people do not yet believe in the existence of political honesty; but then it must be remembered that honesty has but recently been imported into the political world. By political honesty I mean that the motive of political action shall be really what it is avowedly, namely, to benefit the whole community, irrespective of sect, rank, or party. One of the most hopeful characteristics of the present day is, that it is now becoming the common matter of course that statesmen shall cast aside the interests of party, ignore all the distinctions of rank and sect, and sacrifice them without hesitation when the welfare of the nation demands them. Conduct of this kind, now becoming common, was a rare heroism but a generation past. The evidences of this change have not yet penetrated the masses of the people; they are still under the influences of the past in this respect. Men that are not yet grey can remember well when it was a common fashion among lords and gentlemen, and the clergy, to talk both publicly and privately of the danger of affording to the working man an education calculated to give him notions above his station in life. Even at the present day we sometimes hear faint reverberations of this old foolishness. Such expressions are seldom let fall without being picked up and carried into workshops, where they are commented and thought upon with bitterness and indignation by the best and most intelligent of the workmen, who justly regard as their worst enemy he who dares to stand between them and their full and free career of mental and social advancement. The remembrance of such attempts to curb and limit or pervert his mental growth still festers in the mind of the working man who is striving to educate himself, and is one of the most fertile sources of the mistrust referred to.

"How, then, can this suspiciousness be removed? Simply by first of all acknowledging its existence, facing it fairly, admitting frankly that there did once exist some grounds for it, and, at the same time, showing that they exist no longer. Conviction will not, of course, be immediately enforced; but many will be led to self-examination, and to look around upon the present state of things, and thus the truth will come to be understood. Everything that tends to bridge over the chasms that divide the different classes of society must aid in removing all such distrust; and the necessary growth of general intelligence will, in its ordinary course, gradually develope the conviction that we have grown out of the state of class antagonism in which such mistrust originated."

It has long been with us a favourite axiom, that nothing is lost by the full and free acknowledgment of the wrong, where there is the desire to redress that wrong, and the truth of this is seen in the present disposition of the working-classes, who fully appreciate on their part every act of kindness and consideration shown them. A working man thus speaks of the present prospects of society:—"It is a pleasing sign of the times, that the attention and sympathies of the wisest and best of men have, at length, become directed to the best means of bettering the temporal condition of their less fortunate brethren, to elevate the masses of the community; and to raise them in the scale of civil and religious society has become a subject of earnest inquiry. The statesman, the philosopher, and the divine are seeking out the best means for the attainment of so desirable an object. The pulpit, the press, and the forum are more or less engaged in the same philanthropic cause. The bard, who so lately sung the 'Lay of the Labourer,' and the 'Song of the Shirt,' became the subject of royal favour, and, advocating the rights and the interests of the working-classes, has engaged the wits of the age."

"Another animating characteristic is, that the labourers

themselves are consulted upon these important matters. They are invited to tell their own tale, to chronicle their own wrongs, and to point out the most practicable means for their future well-being and improvement." "Surely," continues the working man, "these are tokens for good, and are the heralds of a better and a brighter day that is beginning to dawn, and should inspire the minds of the care-worn sons of toil with confidence and hope. Such are the 'signs of the times,' and they demonstrate that some very powerful principle is at work in the mind of the Christian public. And that principle is none other than the love of Christ, which is influencing directly, and very often indirectly, the state of society, and is bringing about a great moral reformation. It has broken the chains from off the captive, and ameliorated the condition of the slave, and will, ere long, as to others, undo the heavy burdens, 'and let the oppressed go free.'" *

Regarding the National Association as an exemplification of this principle, we believe that it will, from its commencement, prove one of the most efficient workers in Social Reform.

Great as are our sympathies with every effort now making to improve the social condition of the masses of our population by education, improved dwelling-houses, &c., we own that in looking out upon the broad field of labour before us, we are forcibly struck with the conviction, that, in the present condition of society, these instrumentalities do not go to the roots of our social evils; but that the first and greatest step towards improving the condition and habits of the working man, and towards laying the foundation for provident forethought, frugality, and self-respect, would be

* Prize Essay, by a Frame-work Knitter "On the Temporal Advantages of the Sabbath."

in the improvement of the real Property Law, and the abrogation of the Law of Settlement.

The action of these laws at present has a most pernicious effect upon the working-classes, and as long as they continue to exist in their present condition, they will baffle, in a great measure, the efforts of philanthropists and educationalists, to improve the habits of the present and future generations.

The poor man is precluded from resorting to land, as a means of investing his savings, though there is no acquisition so desired by him as that kind of property. Savings' Banks, which only yield about an average of two per cent. interest, he often distrusts from receiving no tangible security, and the eagerness with which Building Societies are resorted to proves the avidity with which he would avail himself of this mode of investment, were it oftener presented to him.

Many are the evils which may be traced directly or indirectly to this prolific source. The present sanitary condition of our poor in crowded dwellings, which is a disgrace to the age, and a natural consequence of the working of these laws, the hopeless pauperism into which many of the labouring classes have fallen, the indifference of the parents to the education of their children, all, in a great measure, originate in a state of society and its laws, which precludes the poor man from any hope of bettering his condition.

Educationalists declare themselves entirely at a loss how to create in the parents a proper sense of their duties towards their children, and all point to some necessary radical change, which, we believe, would be found in those we advocate.

"A vast work," says Horace Mann, in his "Census Report," "evidently lies before us in the education of the parents, ere we can expect them to be in earnest for the education of their children. What is wanted is the creation of a more

C

benignant atmosphere. However carefully the tree of knowledge may be planted, and however diligently tended, it can never grow to fruitfulness or beauty in an uncongenial air. Concurrently with all direct attempts to cultivate the general intelligence, there needs to be a vigorous endeavour to alleviate, if not remove, that social wretchedness which blights all educational promise, and to shed around the growing popular mind an affluence of wholesome light, on which the half-developed plant may feed and thrive."

The Rev. C. H. Bromby adds his testimony to the existing need of some instrumentality that shall tend to improve the condition of the parents:—"Enthusiastic as I am in advocating popular education, my enthusiasm only belongs to it as to one of many instruments of improvement which the present social condition of our people requires. The social evil is too gigantic to be assailed by any single weapon. We need new organizations to raise the educational condition of the people; but co-ordinately, and *pari passu*, we need more efficacious instruments for improving their physical condition. The extension of bath-houses, and (improved) dwelling-houses, the legislative control of Savings' Banks and Insurance Societies, the restriction of the sale of moral poisons, as well in publications as in beer and spirits, the parallel encouragement of popular games, concerts, and public parks, and *whatever else encourages the hoarding, instead of the dissipation, of the wages of industry,* will, at least in another generation, do more to secure the co-operation of parents in behalf of their children, than all the theories and tentative remedies put together."

If it be a fact—and we believe it is—that, while England is the wealthiest country in the world, there is yet more reckless improvidence among her working-classes than in nay other, it surely becomes us seriously to ask, whence the evil originates? The peasantry of France are proverbially sober, provident, and independent, and pauperism is scarcely known in Saxony, Switzerland, the Tyrol, and many districts in Rhenish Prussia. "Saxony and Bohemia," says Mr. Kay, "lie side by side. The condition of the peasant is as different as can well be imagined. In Saxony there is very little pauperism, but the moment the frontier is passed, the traveller finds himself surrounded by beggars of the most miserable appearance."

The contrast is here occasioned in the one country by the land being free from entail, restriction offering no obstacle to the poor man becoming a peasant-proprietor, while, in Bohemia, the country is divided into large estates, belonging to an absentee nobility.

An intelligent foreigner thus speaks of a population without property:—"It is these labourers entirely without possessions, whether they live in the towns or country districts, whether they are found in the factories, or behind the plough, who are the really dangerous members of society: the real ground of their dangerous character lies in their freedom from any moral restraint, caused by the hopelessness and poverty of their social position."

But whence does the pauperism, which is now a standing institution of Great Britain, take its rise? If we look back to an early period of English history, we shall find the progenitor of the English peasant in the villains tied to the land, or bonded serfs. Remnants of this system existed as late as 1799, when, in certain parts of Scotland, there were still to be found bond-slaves, who were tied to the coal estate, till freed in that year.

When, in 1662, the Law of Settlement was enacted, it restored, in some respects, the slavish law of villanage, binding the labourer to his parish, but without the compensating

C 2

advantage of his claim to the cultivation of land for his use,—a claim he had forfeited as the price for his freedom. Four millions of people were thus confined to their own parishes, after having been characterized for some ages by a hearty enjoyment of their freedom. Let it be noted that, in 1650, before the passing of the Law of Settlement, during a free circulation of labour, the Poor-rates for the whole kingdom were £188,811, and that they increased, in 1698, to £819,000, while the law expenses for removals during the next century were £80,000 per annum.

High authorities on Poor Laws thus speak of the action of this law. In 1776, Adam Smith writes:—"There is scarce a poor man in England, of forty years of age, I will venture to say, who has not, in some part of his life, felt himself most cruelly oppressed by this ill-contrived Law of Settlement." Dr. Chalmers says, "Every one knows with what tyrannic rigour the Law of Settlement is enforced, by sending to their proper parishes the adventurous persons who had found no employment at home, thus checking the industrious poor from looking for better employment beyond the limits of their native parishes."

Meanwhile, the system of the old Poor Law administration had rapidly assisted to demoralize the population. It was first acted on, with the best intentions, by Berkshire Magistrates, on the principle, that relief should be given to the poor, whether in or out of employment, so that the wages of all should be alike, according to the family of each labourer. This system continued in existence for forty years, during which time the Poor-rates rose, in some counties, to a frightful amount.

The tie occasioned by the Law of Settlement made pauperism a matter of course, and was equally the enemy of the master and the labourer.

"Is the labourer," asks a friend of the poor man, "under the Law of Settlement, a freeman, when he knows he has no other resource but his own parish,—when he knows that the parish, by its prevention of removal, has bound him to its soil, and has deprived him of the free use of his industry in youth and manhood? and he has not such elevated sentiments as to spurn the instalment, in his need, of the wages which he should have had in his strength. Abolish parish settlement, and the labourer, if he cannot find work at fair wages, will remove to where he can do so. Is the master, under the Law of Settlement, a freeman, when he knows, if he refuses the labourer work, he must pay the penalty of maintaining him in the union workhouse? Abolish parish settlement, and the master is then free to employ whom he likes, those he wants, and no more, and the labour-market will have the supply and demand more equally distributed." *

But the extent to which the action of this law affects the comfort of the poor man can scarcely be realized, except by those who have resided in agricultural parishes, and have watched its workings. We have done so, and know that, as Arthur Young long since said, "Cottages are the perpetual objects of a parish's jealousy, to the injury of the morals and social condition of the poor."

A few years since, in the neighbourhood of Norwich, in Norfolk, the destruction of cottage property had so been carried out, that many of the estates, within eight or ten miles of the city, had been cleared of tenantry, and in Norwich there were five hundred labourers who had to walk various distances, some even seven miles, to their daily labour, thus expending the working power which

* Jasper Peirson, Esq.

should have been retained for their agricultural employments. So strongly does the fear of increasing the pauper population of a given union of parishes operate, that cottages for the industrious poor are most difficult to obtain in the places where their work lies; and as to the labourer possessing a little freehold of his own, where he could, with patient industry, bring up his family in habits of forethought and providence, such a blessing to the poor man cannot be expected under existing laws.

Let us, for one moment, glance at the present position of the agricultural labourer, and trace the early influences which do their part in forming the future pauper. As soon as the boy is old enough to engage in rural labour, he is taken by some farmer, and engaged in any employment suited to his years, and this, often, at so early an age, that any little education he may have received at the village school is soon obliterated. He becomes a full and able-bodied man, and soon feels himself entirely independent of the control of his parents. Perhaps he has an idea of bettering his condition, but, at the age when hope should be highest, when virtuous resolution should be strengthened by the stimulus of possible success, when every inducement should be given to provident and saving habits, he finds that, because he is unmarried, though his strength and skill may be equal to that of any married labourer, he obtains a less rate of wages. The unjust practice of paying single men a lower rate of wages, simply because they are single, is one that has largely obtained in agricultural counties, and has a most injurious effect upon the single labourer. No hope being given of amending his condition, and a premium being absolutely offered for marriage, he sees no motive for the exercise of self-denial, or provident habits. He abandons himself to his fate, marries early, to become the father of a numerous family, described in the vigorous language of Poor Law Guardians as a *pauper warren.**

In an admirable lecture delivered by William Edwards, Esq., of Framlingham, on the agricultural labourer in 1854, he thus speaks on this subject:—"If the single man be able and industrious, he should have the full worth of his labour, and have the opportunity, and be encouraged to make some provision for his future marriage, and for being settled with comfort and decency. The single man too frequently marries for the sake of a married man's wages, and, with this motive as his chief inducement, he marries too early, as well as otherwise improvidently, and so keeps up the surplus of labour, and perpetuates the evils from which he finds, too late, that he has not, as he supposed, freed himself. He settles down into a cottage not half furnished with worn out lumber, instead of having, as he ought, some comforts and conveniencies about him, which he strives in vain to procure for his increasing family, which grows up around him, subject to all the inconveniencies which he has before suffered in his own person."

What hope can the agricultural labourer have of rising in the world? "His horizon is bounded by the high red brick walls of the union-house; his virtual marriage-settlement can only point to such a refuge, if troubles arise; his old age may there have to seek its last shelter."†

Can it be that "Once a peasant, always a peasant," is the rule of Merrie England, for those of her sons of toil, who, by no happy accident, are able to improve their condition? While, on the one hand, English law throws every obstacle in the way of the labourer's investing his hard-earned

* See "Meliora," No. 3.
† "Essay on Juvenile Depravity," by Rev. Henry Worsley.

savings in a little freehold, and, in another way, acts so as to prevent the erection of sufficient and suitable dwellings for his hire, he is nevertheless tied by the Law of Settlement to his parish, and is not free to carry his labour to the best market. We have known cases, and, doubtless, thousands such have occurred, where industrious labourers have, with grief, refused situations at a distance, that offered very superior wages, because they would lose their settle ment by removing.

Our space does not permit us to do more than indicate some of the points that arrest our attention in the condition of the labourer; but we would ask, with laws operating in so prejudicial a manner upon his independence and comfort, can we wonder at any amount of pauperism, or that a million and a half of persons should receive annually relief in the shape of eight millions of pounds, expended from the public rates? But we feel sure that these evils will eventually be remedied, and that an amendment of our laws, suited to present circumstances, will be carried out.

Lord Brougham, in his inaugural address at Birmingham, pointed to certain "redundancies and errors of our laws as the growth of time, the great innovator, as Lord Bacon has it, changing though slowly, even imperceptibly, yet surely and without any interruption, the circumstances of society; while," continued his Lordship, "a reluctance grounded on a wholesome dread of rash alteration, but carried oftentimes to excess, hinders us from adapting our laws to the state of things, and thus makes the retention of the old rule occasion the very shock, the very violence which it was wished to avert, as the steersman, too anxious to avoid the stream, may be stranded on its bank."

Lord John Russell also remarked, in his Address on Jurisprudence and Amendment of the Law, at the Birmingham

Meeting of the National Association, that "we must not be merely satisfied with transcribing ancient laws and calling them new—we must not put errors into new Acts of Parliament, under the name of consolidation : when we attempt to consolidate, we must also amend. I say this," continued Lord John, "because a Bill was introduced into Parliament last session which proposed to inflict the punishment of death for offences which no Judge would think of enforcing. When we make new laws, we must take care that they are such as can be executed, and are in consonance with the spirit of the times. If we merely take a jumble of Acts, containing provisions condemned by the country, and whilst consolidating them only repeat their errors, we are like those persons described by Dryden, who had their portraits taken—

> 'To stand recorded at their own request,
> To future times a libel and a jest.

The laws must be made consistent with the spirit of the times; and this opens for consideration a great question, to which many men of influence and talent have devoted themselves. For my own part, I have come to the conclusion that the consolidation of the laws must be preceded by their amendment."

We are happy to know that these are the opinions of our most enlightened men; and, with such at the helm, we are hopeful for the future of our country.

If once opportunity be offered to the working man of freely acquiring land, we shall soon see a change in every way beneficial.

"If we could," writes an able reviewer, "we would greatly speed the movement of law reform in this matter. Why, in these matters, should England lag so long behind other European nations? What a wonderful effect for good have the land sales produced in Ireland! As if by magic, a race

c 5

of energetic proprietors and farmers, large and small, has sprung up, some of them from the peasant class, and Ireland is being regenerated. "It is not to the intelligence alone," says Mr. Mill, "that the situation of a peasant proprietor is full of improving influences; *it is no less propitious to the moral virtues of prudence, temperance, and self-control.* The labourer who possesses property, whether he can read and write or not, has an educated mind; he has forethought, caution, reflection, guiding every action; he knows the value of restraint, and is in the constant and habitual exercise of it. We may hope to see, from the present lazy, apathetic, reckless, lawless, and improvident Ireland, a new Ireland arise, consisting of peasant proprietors with something to lose, and of hired labourers with something to gain,—the former attached to peace and law through the possession of property, the latter through the hope of it."

CHAPTER III.

Origin of the National Association—Importance of Unity of Action in Social Reform—Meeting at Lord Brougham's House—Formation of the Association—First Meeting at Birmingham—Opening Service—Sermon of the Rev. Sydney Turner—Inaugural Address of Lord Brougham as President—Addresses of Presidents of Departments—Lord John Russell on Jurisprudence, and Amendment of the Law—Sir J. Pakington on Education—Lord Stanley on Public Health—Sir B. Brodie on Social Economy—Character of Papers read in each Department—Jurisprudence and Amendment of the Law—Education—Punishment and Reformation—Public Health—Social Economy, Concluding Meetings—Meeting at Liverpool—Commerce of Liverpool—Its Moral and Sanitary Condition—Programme of the Week's Proceedings—Opening Service—Address of Lord John Russell as President—Lord Brougham on Popular Literature—Addresses of Presidents of Departments—Excursion to the Akbar—Conversazione—The Departments—Meeting at Queen's College—The Departments—Meeting of Working Men—Conversazione—The Departments—Paper on De Fellenburg—Banquet in St. George's Hall—Closing Meeting.

"The true type of the English character is essentially practical. Warmly attached to his institutions, and revering the bequests left to him by the industry, wisdom, and piety of his forefathers, the Englishman attaches little value to any theoretical scheme. It will attract his attention only after having been for some time placed before him; it must have been thoroughly investigated and discussed before he will entertain it. Should it be an empty theory, it will fall to the ground during this time of probation; should it survive this trial, it will be on account of the practical qualities contained in it; but its adoption in the end will entirely depend upon its harmonizing with the national feeling, the historic development of the country, and the peculiar nature of its institutions."—*His Royal Highness Prince Albert.*

We believe that the first suggestion with relation to the formation of the National Association originated with G. W. Hastings, Esq., a gentleman whose labours have long been known and appreciated in some important branches of Social Science.

In 1856, he mentioned the subject to Lord Brougham, as the noble patriarch of education, and as one of the foremost pioneers of national progression, suggesting that his Lordship should take the lead in founding an Association that should afford, to those engaged in various efforts for the improvement of the people, an opportunity of conference upon the great social problems of the day.

It was felt that the close connexion between various branches of social inquiry could scarcely be fully realized even by those who were most anxious for social reform, were there not some connecting link, as in the case of the British Association, which should bring together those who chiefly devote themselves to promote the inquiries and the measures connected with social improvement,—the great advantage which must accrue from the mutual help afforded to one another, having been abundantly manifested in the successful united labours of the scientific men connected with the British Association, which had already served as a model in the formation of several important societies.

As, in physical science, it had been found that men who occupied themselves exclusively, as solitary students, with any one branch of science, became so engrossed and insulated, as to form erroneous views of its bearing and relation to general science, so it was evident that the student of any of the great questions connected with social economy would be liable to the same disadvantage, were he not brought into contact with those men who had exercised equal powers of mind upon other branches of the same great subject. The serious mistake of the philanthropist, who refuses such aid, and persists in pursuing his own plans for the reformation of society, unenlightened by the experience and knowledge of others, may well be illustrated by the folly of a traveller, who, declining to take his view of an extended landscape from an elevated position, contentedly remains in the valley below, unable, from his limited horizon, to estimate aright the bearing of one part of the landscape to another. And thus, while the ignorant constantly condemn as useless that of which, from the low position they take in intellectual attainment, they are incapable of forming a right estimate, so the man who is anxious to benefit his fellow-creature, may greatly err in his modes of action, through ignorance of the principles upon which such philanthropical labours should be based, and the very evil he seeks to obviate may be increased by his means, as pauperism has often been by ill-judged charity.

Nor, as has been well observed, can such isolated exertion be identified with the division of labour, for "the very principle of division of labour imports a common object, for which each works in his allotted sphere;"* and it will, doubtless, be conceded, that, while the human intellect has rarely the capacity to master every branch of science, yet it can supplement its experience by the interchange of information and opinion with other labourers in the same field.

Mrs. Somerville has remarked, that there is a wide difference between the knowledge required to make discoveries in science, and that of the person who can appreciate, and, appreciating, profit by such labours and discoveries, and this remark applies in every particular to Social Science.

For these, and similar weighty reasons, the proposition of Mr. Hastings met with the most cordial support from Lord Brougham; and upon it being made known to many of

* Introduction to First Report of National Association, by G. W. Hastings, Esq.

those eminent men who are actively engaged in philanthropic work, and as promoters of all that is good or great, there was a ready recognition of the need for such an Association.

It was at first intended to commence the important work of organization in 1856, but various causes retarding the design, it was not until the 29th of July, 1857, that the first formal step was taken.

A private Meeting, on that day, was held at Lord Brougham's house, in Grafton Street, to consider the best means of uniting together persons interested in social improvement. Forty-three persons were present, among whose names we find those of Lord Brougham, Viscount Ebrington, Mr. Adderley, Rev. Sydney Turner, Mr. Forsyth, G. W. Hastings, Esq., Dr. Farr, Mr. Simon, Mr. Herbert Broom, and Mr. Gassiot. Several ladies were present, among whom were Mrs. Austin, Mrs. Jameson, Mrs. Howitt, and Miss Emily Taylor; and several gentlemen attended as a deputation from Birmingham. Letters from Lord Goderich, Mr. Akroyd, Mr. George Denman, the Rev. Canon Girdlestone, and others were received, regretting their unavoidable absence, and expressing their entire sympathy with the object of the Meeting.

At the request of Lord Brougham, Mr. Hastings explained the plan of the proposed Association, and the Rev. Sydney Turner then moved a Resolution, seconded by Lord Ebrington, affirming "the necessity for a closer union among the supporters of the various efforts now being made for social advancement, and establishing the NATIONAL ASSOCIATION FOR THE PROMOTION OF SOCIAL SCIENCE."

The Mayor and inhabitants of Birmingham having expressed through their deputation their wish that the first Meeting of the Association should be held in that town and their intention to give it all the support in their power, a Committee was appointed of the gentlemen present to carry out the previous Resolution. Lord Brougham was appointed, provisionally, President of the Association, and thus was laid the foundation of an Institution which, we believe, is destined to be one of the greatest blessings to mankind in ameliorating the existing evils of our social condition.

After much deliberation, the Committee divided the Association into five Departments, in order that the important subjects of Jurisprudence and Amendment of the Law, Education, Prevention and Repression of Crime, and Public Health, might each obtain a due place and be treated separately.

Under a fifth head, that of Social Economy, were grouped other questions of social interest relating to Capital, Labour, and Production.

A most important principle was laid down by the Committee at this early period of their labours,—viz., that the object of the Association being to elicit truth, not to propound dogmas, in every Department any argument coming within the limits of the subjects for discussion, temperately and fairly urged, would be listened to with respect. Aid was sought by the Committee from all persons interested in the great subjects embraced by the Association without reference to classes or opinions, and sincere help was welcomed from every quarter. It will be obvious that, upon this freedom of action and opinion, the future well-being of the Institution and its power for doing good must depend in a great measure.

The Committee now proceeded to add to their own number, and to appoint, for the first Meeting, Officers and Committees for the different Departments, in the following order :—

President:
The Lord Brougham.

Vice-Presidents:
The Mayor of Birmingham.
The Recorder of Birmingham.

General Secretary:
G. W. Hastings, Esq.

Local Secretaries:
J. T. Bunce, Esq., Charles Ratcliff, Esq., J. W. Winfield, Esq.

Local Bankers:
Messrs. Lloyd & Co.

First Department.—Jurisprudence and Amendment of the Law. *President:* The Right Hon. Lord John Russell, M.P. *Secretaries:* J. Stuart Glennie, Esq., J. Napier Higgins, Esq., Arthur Ryland, Esq.

Second Department.—Education. *President:* The Right Hon. Sir J. S. Pakington, Bart., M.P. *Secretaries:* James Chance, Esq., Rev. Nash Stephenson.

Third Department.—Punishment and Reformation. *President:* Bishop of London. *Secretaries:* Rev. J. P. Burt, Alfred Hill, Esq., Martin Ware, Jun., Esq.

Fourth Department.—Public Health. *President:* Lord Stanley, M.P. *Secretaries:* Dr. Headlam Greenhow, Wm. P. Marshall, Esq., H. W. Rumsey, Esq., F.R.C.S.

Fifth Department.—Social Economy. *President:* Sir Benjamin Collins Brodie, F.R.S. *Secretaries:* W. C. Aitken, Esq., T. H. Bastard, Esq., J. D. Goodman, Esq., W. M. Williams, Esq.

It will be seen in how many respects the plan now adopted was formed on that of the British Association; and a still further similarity is found in the efforts made by the Committee to obtain reports on various subjects of social interest, either from societies formed for their promotion, or from individuals especially well informed on such subjects.

The adoption of this plan by the elder Association has led to a collection of most valuable reports on every branch of science from the most eminent scientific men of the day, and its example cannot fail to be followed by similar success in the wide field of Social Science.

In consequence of the short time that elapsed before the Meeting at Birmingham, there was not time completely to realize the original design in this respect, though, owing to the great exertions of the local Committees, much more was effected than could have been anticipated.

A large number of Papers, giving the opinions and experience of persons interested in the different Departments, were received by the Secretary; and all arrangements having been completed, the 12th of October was appointed as the day for the first Meeting of the Association.

It is to us most gratifying to have to chronicle the interesting fact, that a work, which we fully believe to be in entire accordance with the designs of Divine Providence, was commenced by a humble acknowledgment of entire dependence upon Him, through whose aid alone such undertakings can prosper.

In the afternoon of Monday, the 12th of October, a special service was held at St. Philip's Church, when a sermon was preached for the occasion by the Rev. Sydney Turner, M.A., Inspector of Reformatories. The text was taken from the 6th chapter of Galatians and the 7th verse: "Be not deceived; whatsoever a man soweth, that shall he also reap."

In his introduction, the Preacher dwelt upon the great truth of man's responsibility to God, and then, in language which we trust found an echo in many a heart, and which we cannot refrain from quoting, thus continued:—

"But my business to-day is to deal with these words in their connexion with the work which has now called us together.

"Many noble, many wise, many able and earnest men—workmen in our great social factory—are about to meet in this great town, one of the chief centres of our social life, to confer with and animate one another for the promotion of five great branches of our social polity—Improved Laws, Effective Education, Sanitary Regulation, Reformatory Agency, and Economic Science.

"And I think we cannot do better than take with us, as both a stimulus and a direction to our efforts, the twofold truth which Paul here urges—that every man must reap some sort of harvest, and that by the law of his own being, as well as by the irresistible appointment of his Maker, the harvest he reaps depends on the character of his sowing.

"Firstly, let me apply the words to the spirit and the motives by which, in our gathering and discussions here, we should be actuated. Is there among us a deep sense of the responsibility which lies upon us as to the forming and correcting our opinions? Are we meeting with the conviction that we are not independent or isolated beings, who have only their own caprice or fancy to consult, or their own aims and interests to advance, but as members of the wide-spread fellowship of man, citizens of the great English commonwealth, above all, fellow-servants in the redeemed household of God?

"Many of us have thought, studied, and laboured much, in connexion with the subjects of our conference. Have we done this as seeking for truth, and, therefore, in dependence on the teaching of the Spirit of Truth?

"We are privileged to ask the guidance and to have the help of the One highest and most wise Intelligence. We are privileged to look to Him who formed man, who appointed the conditions of his being, the bounds of his habitation; who knows all the wants, understands all the interests, enters into all the circumstances of his life; and who, of all the schoolmasters we can have recourse to, is the wisest and the surest Instructor in all the science in which man is concerned. In seeking to lessen the pressure of mistaken or unjust laws—to lighten the darkness of ignorance—to relieve the physical miseries of the neglected myriads of our populous cities—to prevent and restrain crime, and to promote juster views of individual duty and social prosperity, are we looking to Him for aid and direction, who hateth injustice, and executeth judgment for the oppressed, and who teacheth man knowledge—who hears the cry of the destitute, and treasures up the tears of the wretched—who wills not that the sinner should perish—who delights in the prosperity of His people?

"My friends, the philosopher of old could well feel that the Deity Himself geometrizes. Our noblest musician could well pray before he entered on the composition of the glorious harmonies, that now touch the deepest springs of our heart's emotions. And so well may the political inquirers—the champions of improved laws and better training—the labourers in the fields of social elevation—lay the foundation of their work in prayer, that the Lord, who giveth wisdom, would make them wise; that He who dwelleth in light would shed His brightness on their minds and hearts, dispelling every cloud of prejudice, cleansing every stain of selfishness and error. Remember, I pray you, that every man reaps as he sows. He that taketh counsel only of his own spirit must expect to be left to that spirit's poverty and meanness: he that walks by the light of the sparks, his own mere human efforts and resources kindle, must expect to wander and lie down benighted in the murky atmosphere that those sparks only render darker. You are called to be fellow-workers with God; oh, seek also to be fellow-thinkers. Seek, by prayer and by study of His revealed truth, to have His mind, to perceive His purposes, to think His thoughts, to speak His words, and so, indeed, to do His will."

In the evening the Inaugural Address was delivered by Lord Brougham in the Town Hall, which was crowded to excess.

This Address has been pronounced one of the greatest of his Lordship's productions. "Some present," writes Mr. Hastings, "remembered the oration with which, thirty-two years past, the same voice had filled the Senate House, at Glasgow, discoursing on the art he has so often illustrated. A few thought of the words of Sir James Mackintosh, in that Senate House, when he described Henry Brougham as accomplished in all philosophy—a guiding example to those who would win fame through persevering labour."

After alluding to the honour conferred upon him, as President of the first Meeting of the Association, Lord Brougham proceeded to set forth the views with which it had been proposed, and the advantages expected to result from its adoption. His Lordship first noticed the division of human knowledge into two great branches,—the science which treats of abstract truths, and the science which treats of real existences, matter and mind, or natural and moral science. "Of these, the latter and most important relates

E 2

either to man, as an individual, or a member of the community; and this, or the political or the social branch, is most eminently deserving of attentive study." "When," continued the noble speaker, "the paramount importance of moral and political science is regarded, and the facilities for its cultivation, from the nature of the evidence it rests on, are duly considered, it has struck observers, whose efforts are directed towards human improvement, as somewhat extraordinary that the course should not have been taken here, which has proved so advantageous for assisting the progress of inquiry in the other great department of knowledge—Natural Philosophy; and they have strongly recommended the plan of assembling together occasionally, perhaps periodically, the various individuals and bodies whose attention is devoted to the ascertainment, illustration, and general exposition of moral and political doctrines. It is very possible that, in the end, this object may be attained, and that the Association which we are met here to open may bear as wide a relation to moral and political science as the British Association, which has now been in successful action for considerably more than a quarter of a century, does to mathematical and physical science."

His Lordship then proceeded to show that five of the most important branches of moral and political inquiry had been singled out—those which especially formed the practical portions of Social Science, comprehending the principles and details of plans, that should further the improvement and secure the stability of the social system, correcting its faults upon rational and temperate views, steering a middle course between those who oppose all change as pernicious, and those whom no change can satisfy. After giving a luminous sketch of the subjects embraced in the five Departments, and dwelling on the great advantage that must arise from mutual intercourse, and the high importance of united action, Lord Brougham illustrated its results by the labours of two Societies,—that for the Diffusion of Useful Knowledge, and the Society for Promoting the Amendment of the Law. He also mentioned occasional concert and union, as having been often productive of beneficial results, seen in Reformatory and Educational Conferences. After dwelling on the importance of diffusing among the various classes of the community the knowledge of such subjects as would come within the scope of the National Association, the eloquent speaker thus concluded :—

"But its diffusion has another and most happy tendency: it leads to the improvement of the system, because it inspires all classes with the desire of promoting measures shown to be safe as well as effectual,—in a word, wholesome reforms. Nor can anything be more groundless than the fears of progress entertained by some, affected by more. It is, in truth, ignorance continued, not knowledge advanced, which they have to fear; nay, which, when we come to an explanation with them, they really do fear. Knowledge is power; but its natural ally is the friendly power of virtue with which its dominion is willingly shared. This is, above all, true of the knowledge which we shall seek to improve and to impart. The Supreme Disposer and Preserver, who 'decketh Himself with light as it were a garment, but defendeth all the earth as it were with a shield,' has provided that the false steps into which we are led by the twilight will be prevented or retraced when the day dawns. If any one is still alarmed at the force which the people seem to receive when their faculties are expanded by cultivation, let him recollect that this happy process cannot be continued, and further knowledge acquired, without a new security being given by that very increase of knowledge, against the delusions and the excesses, from which the peace of the community has most to fear. We are reminded by the subject as well as by the place where we are assembled of the exquisite invention, the happiest, perhaps, in the history of science, which makes the power of steam provide by its expansion for its own control, the one being nicely proportioned to the other. Knowledge is, then, both power and safety; it exercises this self-control; it gives to the mighty social engine both the movement and the governor,—

> 'Unmeasured strength with perfect art combined,
> Awes, serves, amazes, and protects mankind.'

But it is not safety alone that we expect; we fondly hope for more—we confidently look higher. Undaunted by the resistance of adversaries, undismayed by the obstruction which the bias of prejudice, or the conflicts of faction, or the strife of controversy, raise to impede social progress, or to retard, its friends lift up their view to the loftier heights where religious and moral truth sheds an eternal light, piercing the darkness of ignorance that shrouds our region, the mists of doubt that obscure the storms of passion that vex, the instinct of selfishness that chills; and then the eye loves to repose on that bright summit where the same beams dispel all doubt from our opinions toward God, and warm our benevolent feelings toward men,—

> 'As some tall cliff that lifts its awful form,
> Swells from the vale, and midway leaves the storm,
> Though round its breast the rolling clouds are spread,
> Eternal sunshine settles on its head.'"

CHAPTER V.

Reformatories—Cost of Crime—Its Prevention—Reformatory Movement—Meeting at Birmingham—Act of Parliament—Second Report of Rev. Sydney Turner—Prison Discipline—Its Rise and Progress—The Irish Convict System—Public Health—Rise and Progress of Sanitary Reform—Miss Nightingale—Treatment of the Insane—Department of Social Economy—Statistics—Population, &c.—Condition of the Working-classes—Public Baths—Public Parks—Intemperance—Drinking Fountains—Workhouses—The Slave Trade—Conclusion.

"I conclude that all our endowments for social good, whatever their especial purpose or denomination—educational, *sanitary*, charitable, penal—will prosper and fulfil their objects in so far as we carry out the principle of combining in due proportion the masculine and the feminine element, and will fail or become perverted into some form of evil in so far as we neglect or ignore it.'—*Mrs. Jameson.*

MANY years before Great Britain possessed Reformatory Institutions, other countries of Europe, and the United States of America, had taken the lead in this important matter, and possessed Institutions which were effecting the ends desired. Private benevolence in England endeavoured to supply the deficiencies of Government, and accomplish much; but the evil was one of too gigantic a nature to be grappled with, but by the united force of a great nation. The cost of juvenile delinquency at last attracted the attention of politicians; and the philanthropist, who devoted himself to the consideration of the question from the purest motives of Christian benevolence, found he had enlisted on his side men of all parties and principles, who at last saw the necessity, in an economical point of view, of arresting the evil at its fountain head. When it was found that the expense of prosecuting and of imprisonment was far beyond that of educational and reformatory institutions, attention was at last paid to the warnings of those Christian men and women, who had formed a correct estimate of the extent of the evil.*

A heavy debt, accumulating for many long years, had been recorded against Great Britain, for her neglect of the crowds of wretched children who thronged the streets of her great cities; but Christian effort at length prevailed, and the experience of the pioneers in this path of social reform was at last listened to with respect; and when Lord Shaftesbury expressed it as his opinion that "Refuges and Ragged-schools had done more to humanize and civilize the people than any institution that ever was invented, and if they persevered in developing their operation, he believed that, in five years, they would be the means of doing away with every vestige of vagrancy that demoralized and disgraced the metropolis at the present time," there were

* "The Superintendent of Dumfries Prison calculates the expense of prosecuting ten youths, under fifteen years of age, at £1,000, or £100 each. At Liverpool the expense, in similar cases, according to the calculations of the late Mr. Rushton, amounts to £145. 8s. each. The Rev. John Clay, late Chaplain of Preston House of Correction, estimated similar costs for prosecution £50; imprisonment, £65; public works, £50; transportation, &c., £35; or a total of £200 for prosecuting and punishing each prisoner. Alexander Thompson, Esq., at the first Birmingham Conference, said that he has often thought when he saw a little ragged urchin in the street, "My poor little fellow, you are a bill of exchange for £200 or £300, drawn on the public of Great Britain." Dr. Guthrie calculates that every reclaimed child is a saving to the country of £300. Mr. Pierce, of Bow Street, confirmed this when he said, "I never see a criminal boy placed at the bar of the police-court, but I think, "Well, you will cost the country £300 before we have done with you."—*Ragged School Magazine*.

many who shared his confidence, who at a previous time would have ridiculed such an opinion.

It has been well observed that, *in the prevention of crime, God appears, almost without exception, to act in accordance with means.* The criminal returns of our land inform us that there are "so many thousands of criminal offences committed in London each year, and that probably ninety-nine out of every hundred of these offences are committed by the working-classes; and will any one from this infer that any other causes than those which are within our own control lead to so very remarkable a distinction between the different orders of society in their respective numbers of criminals? If the moral state of the middle and upper classes is such that only a solitary individual among them, here and there, ever enters a prison, it is not because the grace of God has converted the soul so much more extensively among these classes, but because so many more preventives to crime of a mere natural order exist among them. Even apart from increased religion, if mere natural means almost entirely exclude one moiety of our population from prison walls, there seems to be nothing very Utopian in the expectation that similar means should exclude also the other moiety, and that, even apart from increased religion, an increased attention to the prevention of crime should render prisons and police-courts almost useless institutions." *

It is deeply interesting to find that the above opinions are confirmed by many other persons. Thus, Charles Dickens, who has spoken of the state of the neglected children of London in these words,—"There is not a country throughout the earth on which it will not bring a curse; there is not a religion that it will not deny; there is no people upon

* "City Mission Magazine."

earth it would not put to shame," has given his deliberate judgment with relation to Ragged Industrial Schools in these words:—"I do not hesitate to say—why should I, for I know it to be true?—that an annual sum of money, contemptible in amount as compared with any charges upon any list, freely granted in behalf of these schools, and shackled with no preposterous red tape conditions, would relieve the prisons, diminish county rates, clear loads of shame and guilt out of the streets, recruit the army and navy, waft to new countries fleets full of useful labour, for which their inhabitants would be thankful and beholden to us."

In an important Paper read at the first Meeting of the Social Science Association, at Birmingham, "On the Possible Extirpation of Regular Crime," by T. B. L. Baker, Esq., a gentleman who has devoted a life to practical efforts for the reformation of juvenile criminals, we find opinions coinciding with those of Lord Shaftesbury and others we have named, and founded on facts that had come under Mr. Baker's notice.

The day is past when it would have been necessary to dilate on the importance of education for the poor. Christian voluntary effort has done and is doing much, and the national system is seconding its efforts. England is rapidly educating her population, who are passing from an uneducated to an educated people. If, as we trust, Christian liberality shall enable our City and Town Missions so to extend their operations, that an accurate estimate can be formed of the further extent to which the Industrial, Ragged and Sunday-school, and Reformatory movement needs development, if the free-will offerings of Christian men and women still increase in proportion to the demand, there will indeed be hope for the future condition of the hitherto neglected children of Great Britain,—hope founded on the experience of the past, and on the promises of Scripture.

"My friends," said the excellent Inspector of Reformatories, the Rev. Sydney Turner, in his sermon, at the Birmingham Meeting of Social Science, "I have seen much and read much about the reformation of the fallen and depraved; and I declare to you that I would rather entrust the criminal to the agency of some plain man, in whose heart the love of the Lord Jesus reigned supreme, and to whose mind the teaching of the Lord Jesus was familiar, than to the ablest philosopher, schoolmaster, drill-master, and work-master in the world, to whom the Saviour is a stranger."

In 1834, the prison of Parkhurst was first designed as an Industrial place of Reformation, and Detention for Boys, and between that period, and the time when an Act of Parliament was passed in 1854, various conferences were held on the causes of Juvenile Delinquency, by persons interested in the subject.

In 1851, a most influential Meeting was held at Birmingham, when some important Resolutions were passed.

The points upon which all seemed agreed were chiefly as follows:—

"That though Ragged Schools and similar institutions were doing a vast amount of good to the neglected class, yet in most localities they were cramped and hindered from want of funds, and that all the agencies yet at work were totally inadequate to check the increase of juvenile crime.

"That voluntary agency was not equal to the work, and that support was needed from some regular and adequate source.

"That where parents neglected, or perverted their offspring, the public should interfere and see justice done; but that parents who could pay should not be permitted to escape doing so.

"That there were large numbers of children in our large towns of the vagrant class, who needed food and some industrial training, (to fit them for future life,) as well as moral and religious teaching.

"That for such children book-learning should be secondary to moral training, and instruction in some trade or handicraft.

"That no great good would ever be done to this class, until some power were given to the police or parish officers to enforce attendance at a school of some kind, and until street-begging should be effectually put down.

"That it is a great mistake to treat juvenile delinquents, especially those under twelve years, in the same manner as hardened criminals; and that Correctional or Reformatory Schools were needed for such classes rather than prisons.

"That the present treatment of this class, through means of prisons, &c., was radically bad, and in every case doing more harm than good, tending to increase and perpetuate crime, rather than to repress or diminish it—the re-commitments being in most cases 40 or 50 per cent., and in some 70 per cent., on the commitments.

"That Reformatories (the few that exist) were doing much good; the reformations being 40 and 50, and in one case 80, per cent. of the admissions.

"That prisons were far more expensive than Ragged Schools, Industrial Schools, or Reformatories, merely considering the current annual expense; and

"That in every respect Prevention was better than Cure.

"In accordance with these views, the following Resolutions were unanimously adopted and confirmed by a Public Meeting held the same evening:—

"1. That the present condition and treatment of the 'perishing and dangerous classes' of Children and Juvenile Offenders deserve the consideration of every member of a Christian community.

"2. That the means at present available for the Reformation of those Children have been totally inadequate to check the spread of Juvenile Delinquency, *partly* owing to the want of proper Industrial, Correctional, and Reformatory Schools; and *partly* to the want of authority in Magistrates to compel attendance at such Schools.

"3. That the adoption of a somewhat altered and extended course of proceeding, on the part of the Committee of Privy Council, is earnestly to be desired for those Children who have not yet made themselves amenable to the law, but who, by reason of the vice, neglect, or extreme poverty of their parents, are not admitted into the existing Day Schools.

"4. That for those Children who are not attending any School, and have subjected themselves to police interference by vagrancy, mendicancy, or petty infringements of the law, legislative enactments are urgently required, in order to aid or establish Industrial Feeding Schools, at which the attendance of such Children shall be enforced by Magistrates, and payment made for their maintenance, in the first instance, from some public fund, power being given to the public authorities to recover the outlay from the Parents of the Children.

"5. That legislative enactments are also required in order to establish Correctional and Reformatory Schools for those Children who have been convicted of felony, or such misdemeanours as involve dishonesty; and to confer on Magistrates power to commit Juvenile Offenders to such Schools instead of to Prison.

"6. That the following Gentlemen form a Committee, with power to add to their number, to adopt such measures as they may think desirable, in order to obtain the requisite Parliamentary enactments, as well as to prepare a Memorial to the Committee of Council, and for the attainment of the specific objects laid down in the foregoing Resolutions.

"By these Resolutions, the Committee, of which a list is appended, was appointed to undertake the necessary preliminary steps in London:—

"D. Power, Esq., Mr. W. Locke, J. Macgregor, Esq., Rev. S. Turner, with the following gentlemen from various parts of the country:—M. D. Hill, Esq., Recorder of Birmingham; Rev. J. Field, Reading; A. Thomson, Esq., Aberdeen; H.

I

J. Grant, Esq., Bristol; Dr. Bell, Edinburgh; Rev. W. C. Osborn, Bath; W. Campbell, Esq., Glasgow; E. W. Nutt, Esq., York; A. M. Caird, Esq., Wigton; Rev. E. Clay, Preston; Joseph Hubback, Esq., Liverpool; J. Conder, Esq., Birmingham.

"At a Committee Meeting held after the Conference, other names were added; and it was resolved—

"That the result of this Conference, together with a list of the principal gentlemen present, be laid before Sir George Grey, by a deputation, consisting of Rev. S. Turner, Rev. J. Field, Rev. W. C. Osborn, D. Power, Esq., J. Macgregor, Esq., and Mr. W. Locke; and that D. Power, Esq., be the Treasurer; also, that the Rev. Sydney Turner and John Macgregor, Esq., be the Honorary Secretaries of this Committee."

One of the results of this Conference was the appointment of a Select Committee of the House of Commons fully to consider the subject, and to report; still further to adopt measures to secure the co-operation of the Legislature in establishing Reformatory Institutions through the kingdom.

A second Conference was held in Birmingham in 1853, which was attended by noblemen and numerous influential persons, and presided over by Sir J. Pakington.

Although the two Bills brought before Parliament by the Earl of Shaftesbury and Mr. Adderley in the previous year had not been passed, yet their discussion had much advanced the question both in and out of Parliament. Lord Shaftesbury's Bill dealt with the large class of children who are found in the streets of London engaged in the demoralizing pursuit of mendicancy; and on the principle that prevention is better than cure, sought to arrest the evil in its bud—while the young were on the threshold of crime. Mr. Adderley's Bill was for the better care and reformation of Juvenile Offenders, by sending them to institutions for their care, training, and discipline.

In 1854, the active supporters of the movement obtained the enactment of a measure for the better care and reformation of youthful offenders in Great Britain, which enabled Judges and Magistrates to send criminal children below the age of sixteen to a Reformatory School for a lengthened period of detention; empowered the Home Secretary to certify schools for this purpose, thus saving the necessity for local Acts; and established the great principle of the pecuniary responsibility of parents for their children when sentenced under the Act.

A Reformatory Union, the idea of which originated in 1855, at the house of Barwick Baker, Esq., of Hardwick, near Gloucester, held its first provincial Meeting at Bristol in 1856, and was attended by many influential persons. Lord Stanley delivered the Inaugural Address, and many important Papers were read on the subject.

The interference of the Legislature gave a great impulse to the movement, and seven Reformatories were certified in 1854, ten in 1855, seventeen in 1856, and ten in 1857. Thus the Reformatory School movement has gone on a course of gradual but steady development, and the Rev. Sydney Turner records it as his opinion, that he does not think "any movement has been characterized by a greater mixture of earnestness of purpose on the one hand, and practical good sense and discretion on the other."

Not less than £150,000 has been contributed by voluntary offerings towards buildings, land, and other expenses of Reformatory Schools, while, for the last year, the subscriptions reached £13,102. 19s. 9d. The second Report of the Rev. Sydney Turner, the Inspector of Reformatories, informs us that—

"On December 31, 1857, forty-five Reformatories, viz., thirty-four for boys, ten for girls, and one receiving both

I 2

boys and girls, had been certified in England; and eight, viz., three for boys, one for girls, and four receiving both boys and girls, had been certified in Scotland.

"The whole number, therefore, of certified schools on December 31, 1858, was—

ENGLAND	Boys	34	
	Girls	12	
	*Boys and Girls	1	
WALES	Boys	1	
			—48
SCOTLAND	Boys	3	
	Girls	3	
	*Boys and Girls	4	
			—10

"In all fifty-eight, of which six are for Catholic children, viz., three for boys and two for girls in England, and one for girls in Scotland.

"There were 1,830 boys and 354 girls under detention in England on December 31, 1858, of whom 1,364 boys and 224 girls were in Protestant, and 466 boys and 130 girls in Catholic, schools.

"The number under detention in Scotland on the same date was 472 boys and 141 girls, of whom four girls were in a Catholic Reformatory.

* It is scarcely necessary to remark, that it is very inexpedient (to say the least) to receive into the same school, however ingeniously divided and arranged, inmates of *both sexes* of the reformatory class. The young offenders detained in Reformatories are often of 15 and even 16 or 17 years of age, and have been necessarily familiar, in most cases, with vice and immorality, as well as crime, before they have been committed to the school. To throw boys and girls of such ages, and with such antecedents, into familiar intercourse in the same institution, and to have them in the close neighbourhood of adjacent departments, only separated by partitions in the house and fences in the playground, cannot but be productive of much moral risk to the inmates.

"The admissions and discharges during the year were—
"For England—

	Protestant Schools.	Catholic Schools.
Admissions—		
Boys . . .	569	119
Girls . . .	105	35
Discharges—		
Boys . . .	329	76
Girls . . .	39	3
"And for Scotland—		
Admissions—		
Boys . . .	125	0
Girls . . .	52	4
Discharges—		
Boys . . .	59	0
Girls . . .	29	0

"The admissions and discharges, however, include 51 children transferred from one Reformatory to another during the year, and five re-admitted.

"The total of fresh cases admitted was therefore 953, and the total discharged 484.

"Of those admitted, 104 were under eleven years of age, and 245 between eleven and thirteen; more than half were thirteen years old and upwards. 395 had not been in prison previously to the offence for which they were committed for detention in the Reformatory; 300 had been previously in prison twice and upwards.

"Of those discharged, 23 had died; 66 emigrated; 67 had enlisted or been sent to sea; 66 had absconded or been committed to prison, or expelled as incorrigible; 262 had been apprenticed, placed in service, or assisted to employment in England.

"Most of the Reformatories now in operation have their full complement of inmates, and the difficulty of getting admission for the young offenders who are thought to be proper subjects for reformatory treatment, and sentenced accordingly, is now considerable. As the admissions will for some time yet exceed the discharges, this difficulty can only be met by the enlargement of the existing schools, or the establishment of others. To what extent additional accommodation will be required can a yet hardly be definitely judged of. But the diminution in the numbers of the class for which the schools are needed has been so considerable, that there seems good reason for anticipating that a very moderate increase of our schools will suffice. This will, however, be best discussed in the latter portion of my Report.

"In reviewing the results and the future operation of the certified Reformatories, there are three points for inquiry :—

"1st. What has been the effect of these Reformatories, up to the present time, on the individuals committed to them, and on the criminal class of juveniles in general?

"It may be safely affirmed that the effect has been more considerable and more beneficial than might have been at all anticipated. Any of those who have had to do with young criminals, and especially with those of our large towns, will agree in the assertion, that to undo the effects of their early initiation into crime, to arouse a real sense of shame, a consciousness of degradation, a desire of rising to a higher standard, a courageous resolution to break off old associations and long-practised habits, a strength to bear the hardships, and struggle with the difficulties, attendant on their entrance or return to honest and industrious life, is no light matter.

"It would not be a matter of surprise, therefore, if a large proportion of the young offenders sentenced to detention in our Reformatories were found after all to have relapsed; if the cases in which the boy or girl lived honestly and soberly when sent back into the common world of industrial life around them, were the exception, and the reverse the rule. It is, therefore, very satisfactory to be able to assert that the result is exactly opposite to this, and that the relapses and re-convictions of the young offenders discharged from detention up to the present time form out a moderate per centage of the whole. I have obtained a return of the number discharged up to the 31st December, 1857, and of the proportions of those that have been again convicted, and that are known to be doing thoroughly well, up to the present time. The results are, that of 410 discharged up to that date, only 50 have been convicted, and 215 are known to be going on satisfactorily. This return is less favourable than such returns will be in future, as it embraces the initial stages of most of the Reformatories, during which the discipline and arrangement of the schools were imperfect, and many of the inmates were lost by desertion. It may be safely assumed that, in their more settled and better organized condition, the per centage of success will considerably increase. I have limited myself to the results as regards the young criminals committed under sentence to these schools. Had I enlarged the inquiry so as to comprehend all the voluntary inmates who have been received on their own application, or by arrangement with their friends, I should have added largely to the favourable side of the account; but it seems fairer to deal only with those whom our reformatory legislation aims at, and for whom the certified schools are more specially opened, as more distinctly criminals.

"But I have long felt that the individual reformation and rescue of the offender is a matter rather of individual religious charity, than of social policy or duty. The Christian is called on to seek and to save what is lost; but the statesman and the community are most concerned with the effects of Reformatories on the general amount of crime, and in reference to the protection of the property and the moral health of society at large.

"The number of juveniles under 16 sentenced at Quarter Sessions, Assizes, or by Summary Conviction, in England and Wales, was—

Sentenced to Imprisonment.	For the Year ending Sept. 30th,		
	1856.	1857.	1858.
Under 12 years.................Boys	1,674	1,622	1,018
,, ,, ...Girls	316	255	144
Between 12 and 16 ....Boys	10,134	9,200	5,551
,, ,, ...Girls	1,857	1,424	909
Total..................	13,981	12,501	7,622

"These numbers in each case include the children sentenced to detention in Reformatory Schools, these being in all cases committed in the first instance to prison.

"The number so sentenced was, for the same years, as follows:—

Sentenced to Detention.	For the Year ending Sept. 30th,		
	1856.	1857.	1858.
Under 12 years ....Boys	54	96	117
,, ,, ...Girls	17	30	22
Between 12 and 16 ... . Boys	403	870	583
,, ,, ...Girls	60	123	127
Total......	534	1,119	849

"The diminution in the number of commitments of offenders under 16, for the year 1858, was, therefore, just upon

40 per cent. as compared with the year 1857, and above 45 per cent. as compared with the year 1856. This fact is the more remarkable as the total commitments of all ages for the year 1858 amounted to 118·162, corresponding to a total of 124,823 for the year 1857, and of 113,736 for the year 1856, exhibiting a decrease of only six per cent. in the one case, and an increase of four per cent in the other.

"It is encouraging to observe that the number of commitments to Reformatories also decreased in 1858 as compared with the year preceding. This shows a steady decrease to the amount of 25 per cent. as compared with the numbers for 1857.

"An important question to be answered is, What rules and principles must be observed in the management of Reformatories, to preserve and increase their usefulness?

"First of all, their tone of discipline, dietary, industrial occupation, lodging, habits of recreation, &c., &c., must be made strict, and must have something of the hardness which St. Paul prescribes as an essential element of the Christian's training, and which, if needful for the formation and development of characters of the common class, must *à fortiori* be still more wanting for the correction and reformation of those who have fallen away in weakness, or been corrupted by vice. Self-will, disorder, idleness, and sensual indulgence have more or less been the causes of their perversion. A manly training to obedience, regularity, industry, and self-control is the needful remedy. It is one of the advantages of the certified Reformatory, in which the offender is detained by law, as compared with the refuge from which he is free to depart, and in which he needs to be induced and persuaded to stay, that the regulations may be more strict, the food plainer, and the work more exact and regular, without endangering your hold on the inmate

I 5

whom you wish to retain and benefit. On this point I think the views of most managers have been considerably modified by experience, as I acknowledge my own to have been also. There is little of the petting and bribery which, at the outset, obtained here and there. The principle of *duty* is more clearly insisted on, persuasion less trusted to. You do not find the really capable masters and superintendents how speak of dealing with boys by some mystical process of individual training. The common law of family and social regulation is more appealed to, and the child shown what it ought to do, and required to do *that* as a *duty* to God and to itself, not as a favour to its friends and instructors. The result is seen in the improved look and manner of children. The coarse, impudent, exacting demeanour once far from uncommon, is now rarely met with. Confidence without familiarity on the part of the child; kindness without weakness or flattery on the part of the teacher; earnestness in everything—work, devotion, school instruction, or play; self-respect, in personal cleanliness and neatness; order in daily habits and meals; these are the marks of a successful Reformatory; and, in proportion as these increase, the school will be more and more useful."

It is right we should mention that, at the time the National Reformatory Union was formed, another Society, called the Reformatory and Refuge Union, was founded in London, which has had a very wide range of usefulness. The two Societies had the same objects in view, though formed, from conscientious motives on the part of the founders, on a different basis.

The National Reformatory Union having now achieved its great object, which was to create such a pressure of public opinion on Government, as to obtain the recognition of

Reformatories as an integral and essential part of the system we should pursue with youthful criminals, has ceased to exist.

We shall now briefly glance at the past and present condition of Prison Discipline. At the time when the illustrious Howard first attracted public attention to the state of our prisons, it was, in the highest degree, disgraceful to our country: while so little care was taken of the physical condition of prisoners, that a most fatal disease, generated by want and filth, was known as the gaol fever, still less concern was felt, or expressed for their spiritual interests.

Before Howard died, he had the happiness of seeing some considerable progress made towards a recognition of the principles he advocated, and the whole system of modern prison discipline may be considered but their complete development.

Yet, notwithstanding all a Howard and a Fry had effected, in 1832 prison discipline was in such an unsatisfactory state, that a Committee of the House of Commons met and strongly reported on the necessity of abolishing the hulk system, erecting prisons on the separate system, and appointing Prison Inspectors. It was shown before this Committee that dancing, drinking, cursing, fighting, &c., were carried on with impunity by the convicts on board the hulks, and that so far from this system being deterring as was supposed, it was quite the contrary.

Of the scenes that often occurred and still happen in the common gaols of our country, conducted on the association principle, an excellent Prison Chaplain, the Rev. J. Kingsmill, has given us a vivid idea: "The day room or night room of a prison of this kind, is a school of thieving and every vice, and a place of torture to whatever right feeling may still linger in the breast of unhappy persons who have fallen from better things. At night the scenes which take place in gaols of this class baffle all description."

Mr. Crawford and the Rev. W. Russell were shortly afterwards appointed Inspectors of Prisons. The former in 1834 visited America, and reported on the Philadelphia system. Both these Inspectors strongly advocated the separate system, calling attention to the circumstance that it had been tried in Gloucestershire for many years, and succeeded.

These recommendations were adopted, and Pentonville prison, in London, was built in December, 1842, on the separate system.

This prison was placed under the management of Commissioners, who exercised great watchfulness over its inmates.

Several experiments were tried in this prison, and, at first, convicts were sent direct from isolation to the colonies: this was found not to answer, and the necessity of some intervening step became obvious. A Board of Directors of Convict Prisons in England was formed under Colonel Jebb.

It was found necessary to have a stage of discipline at the Public Works of England, in which the convicts should be associated at labour during the day, and sleep separately at night. An establishment was formed at Portland under Captain Whitty, in 1848, to carry out these views; thenceforward, instead of convicts being sent direct from Pentonville to the colonies, they were sent from Portland and other establishments formed on its model at Portsmouth, &c.

In reviewing our convict systems, it is impossible not to be painfully struck with the imperfect manner in which England has fulfilled her duty to her erring sons. Till within a few years, the chief reliance of those who took charge of convicts in the penal colonies, was in a system of brute force,—severity, and terror, the punishments of the chain gang, the lash, and the gallows, instead of "the power of that merciful Word, at which the stubborn knee has been known ere now to bow, and the hardened heart to melt." *

A person who had seen the management of convicts thus writes: "I saw and conversed with ten criminals in their condemned cells, on the eve of their execution. They had never heard the Word of God preached since their childhood, some not even then. They had never entered a church or chapel in the colony, or attended a Sabbath service, and they had fled to the bush because their backs had been bared to the bone by repeated scourgings."

Even in 1840, Judge Burton wrote of the convicts in Norfolk Island, "Little indeed has been done by the English Government, and until lately nothing, to render their captivity productive of that improvement in them which should restore the Creator's image in their souls." When this excellent Judge visited the island, he found 130 prisoners in confinement on a charge of conspiracy to disarm, and if necessary murder their guard to effect their escape. In the course of these trials, the horrible condition of the prisoners was fully revealed. One man described their state as a hell upon earth, and represented with others that the crimes of which they had been guilty were not such as should subject them to such a place of wretchedness and despair, and companionship with convicts of the worst character.

Till the year 1838, the system of assignment was in force in the penal colonies. In this system, if the convict were so fortunate as to be placed with a merciful master, his fate was a happy one, but it was often otherwise. These and other objections led to a change in 1838, by which all con-

* Rev. J. Kingsmill.

victs were to be coerced in press-gangs, before being assigned as domestic servants.

In 1842, Lord Stanley introduced the probation system, under which convicts were made to pass through five distinct stages, the first, however, which was detention in Norfolk Island, only being employed in very aggravated cases. This system, however, was found not to answer, for the probation gangs became the nonseries of vice instead of the correctors of it, the intentions of the Home Government not being carried out in the penal colony. Lord Grey's probation system was the next step in convict discipline, by which separate confinement was substituted for congregating the convicts in masses in the first stage; the second stage being penal labour upon public works—both stages being passed at home. The penal establishment at Portland, near Weymouth, of which we have already spoken, was formed to carry out this plan. Portsmouth Convict Prison was opened in 1852, and the good effects of removing convicts from the hulks into a prison, which secured separation at night and at meals, was soon obvious. Dartmoor Prison had been partially opened in 1850, and the arrangements were such as to have a very beneficial effect on the prisoners.

The propriety of employing convicts on public works has been proved, by the success attending the working of the system pursued at Portland; in this case the convict has previously passed through a term of separate confinement.

The plan adopted by Lord John Russell's Government, for carrying out the sentence of transportation, was, after a prescribed number of years' probation in one or other of the prisons we have mentioned, to transport the convict with a ticket of leave. The utmost care was taken to insure the success of this system by the selection of suitable officers and zealous Christian men as Chaplains and Schoolmasters.

A very formidable difficulty soon arose, no less than that the colonists refused any longer to take our convicts, and in 1853 transportation ceased, excepting to Western Australia, which could receive but a very small number.

In 1853 an Act of Parliament passed, enabling tickets of leave to be given to convicts at home, who would otherwise have been sent to a colony.

An improved system, called the "Intermediate System," originated in Ireland, to meet the new state of things, which has been established nearly four years, and has been attended with the greatest success. It is highly approved on the Continent, but has not yet been adopted in England; and as it appears to us to solve the question, "What shall we do with our criminals?" we are most anxious it should be widely known. Matthew Devonport Hill, Esq., who has visited the Irish convict prisons, with a view to test the system, closes a most important Paper read on the subject to the Birmingham Meeting of the National Association with these emphatic words:—

"Thus, then, in my humble judgment, the Board of Directors of Irish Convict Prisons have practically solved the problem which has so long perplexed our Government and our Legislature—What shall we do with our convicts? The results of their great experiment answer thus—Keep your prisoners under sound and enlightened discipline until they are reformed—keep them for your own sake and for theirs. The vast majority of all who enter your prisons as criminals can be sent back into the world, after no unreasonable term of probation, honest men and useful citizens. Let the small minority remain, and if death arrive before reformation, let them remain for life."

How, it will be asked, can such great results be attained? We answer, that by this system the mind of the criminal is enlisted in his own reformation, while the discipline through which he passes deters him from crime.

But its astonishing success is, to us, but another proof of a principle we have long recognised as the only correct foundation, for all reformatory and educational enterprises, and as the key to what Dr. Chalmers calls the prosperous management of human nature. Let the erring child, the prodigal son, the convicted felon, alike know that man is not less merciful than God, that while justice demands punishment for crime, yet mercy may, acting in harmony with offended justice, hold out hope of reformation and restoration to society.

We are too apt to look upon convicts as men abandoned to crime, and irreclaimable; but what says the evidence of practical men who have devoted their lives to the subject? One writes:—"As far as I have learned of prisoners convicted of graver offences, infinitely the largest portion of them are *men of one crime*, whose general tone of feeling and morality is very much on a par with men outside the prison walls." The Rev. John Davis, Ordinary of Newgate, speaks of "having all hope of the effectual reformation of offenders, based upon the blessing of the Most High, in the religious instruction given in prisons;" and mentions "many thoroughly reformed inmates, who are now pursuing their avocations, in many instances, unknown and unsuspected."

In 1856, Mr. Hill repaired to Ireland, to investigate the results of Captain Crofton's system, and thus describes the course pursued:—

"A convict, on coming under the control of the Board, goes first to the cellular gaol, called Mountjoy, situate in the city of Dublin. He is placed day and night in strict separation from his fellows, the only exceptions being the time which he passes in the chapel, the exercise ground, and the school-room. But, even when so engaged, all conversation with his companions is strictly prohibited. The chapels in the convict prisons had formerly stalls, in order that each prisoner might be isolated during the celebration of Divine worship; but it has been found in Ireland, as in England, that the prisoners were stimulated by the physical difficulties of intercourse to find the means of evasion. The result of the change has been favourable to the maintenance of decorum, and has been followed by greater attention to the service.

"In this first stage the ministrations of the Chaplain—ever precious—are all-important. Much time must be separately spent with each prisoner, the object being to acquire a knowledge of his individual character, and, by adapting lessons and exhortations to his peculiar wants, to obtain an ascendancy over his mind and his affections. The value of what is called the separate system (but improperly so called, as separation can only be one feature in any system) is, at the commencement of the prison discipline, very great. It draws a broad line of demarcation between the past and the future. If the opportunity is duly improved by the Chaplain and the schoolmaster, the course of reflection upon which the convict is thrown by the absence of associates will for a time be most wholesome, although (or perhaps because) his retrospections must be painful and humiliating. 'The system of separation pursued in Mountjoy,' says Mr. Cooney, one of the Chaplains, 'joined to religious instruction and reading, tells very powerfully on the minds of the convicts for two or three months. It subdues, and almost invariably leads to a change of sentiment.'

"The first stage at Mountjoy ends at the expiration of nine months, unless the prisoner has misconducted himself. He is then removed to Spike Island, a fortified station in the Cove of Cork. In the daytime the shores of this islet are the limits of his prison. At night he is shut up in a strong building formerly used as a barrack. This contains cellular divisions, and thus he is again, during the hours of rest, separated from his comrades; but he is no longer in solitude, the compartments being so constructed as to admit of conversation,—an indulgence which his desire for rest, to say nothing of the presence of a watchman, secures from abuse. During the daytime he toils at the repair and enlargement of the military works. The poor convict, on his removal from Mountjoy, feels that he has to some extent earned his advancement, and, perhaps, for the first time in his life, looks back with satisfaction on the past. At Spike Island he is employed according to his abilities. If skilled in the labours of the stone-cutter, the mason, or the carpenter, his former training determines his avocation. If able-bodied, but brought up to no handicraft, he works as a labouring man. If his trade is that of an indoor artizan, a tailor, or a shoemaker, for instance, instead of going to Spike Island when he leaves Mountjoy, he is sent to Phillipstown, an inland prison about forty miles from Dublin.

"Even at Mountjoy he has had the privilege of books, not wholly religious works, but, with the entire approval of the Clergy of that prison, interspersed with volumes of a secular character, some intended for his recreation—none, however, being devoid of a moral aspect.

"At Spike Island and Phillipstown, which must be considered as standing on an equality as regards their inmates, these privileges are extended. The schoolmaster now becomes a lecturer also, his choice of subjects being adapted to the wants, the capacities, and the tastes of his hearers.

"In order, however, to exemplify his mental improvement, I pass for a moment to a subsequent stage at the Forts, on the shores of the Cove, which may be called his summer. 'Were the prisoners,' says Mr. Harland, the Head Schoolmaster, 'subjected to a rigorous examination in literary subjects, their progress might appear slow (for many of them are aged men of blunted intellect, and speaking only the Irish language); but this would be an unfair test, for most of them have acpuired much useful information, though incapable of answering correctly for want of expression. Men who cannot read or write, so as to gain much information from books, have been taught, orally and by lecture, something of life in general, and are partially educated. It is both amusing and edifying to hear these old men teaching each other geography, by pointing out on the maps the several countries, under the Irish names, for the different colours that mark them. Those who have learnt to read, and who also speak Irish, very generally translate the subjects and substance of their lessons into Irish, for those who have failed to learn to read. There was a remarkable instance of the effects of application and perseverance in the case of S. C., an old stolid man, scarcely able to utter a word of English, and not knowing a letter in the alphabet; yet such a desire had he to learn to read, that he applied himself day and night to the book, and though extremely dull and slow, in the course of a few months he could read a First and Second Book, and was reading the Sequel when discharged.'

"The Board has devised subsidiary stages, or 'classes,' as they are called. Classification begins on removal to Spike Island, when the prisoners enter the third class, save in exceptional cases, which I pass by. They are eligible for promotion to the second class at periods varying as follows:—

"'If their conduct and character in separate confinement have been recorded "exemplary," they will be eligible for promotion in two months.

"'If "very good," in three months.

"'If "good," in four months.

"'If "indifferent," in six months.

"'The promotion of prisoners, whose conduct shall not be *exemplary* while in this class, will be delayed beyond the periods above laid down, according to the nature and number of the offences committed by them against the rules and discipline of the prison, under such regulations as shall from time to time be approved by the Directors.'

"In the first class, blameless conduct for twelve months raises the prisoner into what is called the exemplary class, qualifies him for his last stage, and transfers him to the Forts, to Lusk, or to Smithfield, according to his aptitude for the employments pursued at one or other of these establishments.

"The scale of advancement being by means of stages, with their sub-division into classes thus minutely graduated, the hopes and fears of the convict are always on the alert, without over-stimulating his mind. A step in progress may soon be taken, while, on the other hand, the danger of backsliding is ever present; yet a single movement in either direction is not usually so important in its results, as to produce an unwholesome disturbance of his spirits. The prisoner, having now worked himself out of Spike Island, passes into what is called the intermediate stage, whether he is placed at the Forts, at Smithfield, or at Lusk. In other words, he gains a position midway between confinement and liberty, one in which his prison (if it must be so designated) is rather a place of enforced residence than of custody. He is, however, subjected to very strict supervision, and he holds his post under arduous responsibility.

"When the prison authorities observe that an intermediate man has acquired some capacity for self-control, he is sent out on messages. It is found in practice that he does not abuse this privilege, but having transacted his business with promptitude, he straightway returns. A number of such men will then be entrusted to leave the establishment, for the purpose of performing some work procured for them at a distance from their home, returning every night immediately on the conclusion of the day's labour. Here, again, instances of abuse, such, for instance, as entering a public-house, are rare, if not altogether unknown. The intermediate man, having now established a character, is entrusted with money to make purchases, or to pay bills on behalf of the prison.

"Intermediate men are striving to acquire such a character as will recommend them to employers, and thus accelerate their discharge; consequently desertion rarely, if ever, occurs. I heard of no instance in which it had been attempted. Bodies of men so trained may surely be turned to the best account. Thousands of hands might be usefully employed on public works of pressing necessity."

Such is a short outline of some of the leading features of a system which has received the unqualified approbation of the National Association, and which, we hope, may be fully introduced into our English prison discipline, making our prisons Reformatories, and sending our convicts back to society as honest, industrious, and orderly members of the community.

Numerous as are the subjects embraced in this Department of the Association, we have given this one peculiar prominence, on account of our anxiety to see it receive suitable attention from those who may have it in their power to promote its adoption on English soil.

We shall now pass on to the next Department, that of Public Health, which embraces many subjects of peculiar interest and importance in the present day to the community. We shall first give a Statement of the Rise and Progress of Sanitary Reform.

In the January number of the *Westminster Review*, for the year 1825, there appeared a long and elaborate article on Quarantine and Sanitary Laws. This was followed in the succeeding number (April) by a second article on the same subject. The object of these papers was to prove that fever in general, and all the epidemic forms of it, have their origin in the bad sanitary conditions of the places in which they rise, and to which they spread, and that the true means of prevention are not quarantine regulations, but the removal of causes polluting the air, such as filth, overcrowding, &c.

The prevention and removal of all such causes are the object of that department of the State which is charged with the care of the public health. But until 1848 no such department existed in Great Britain. No attention had been paid to the subject on the part of the public, the Legislature, or even the medical profession. A striking illustration of the state of the medical mind on this subject, at that time, is adduced in Mr. Howell's Paper on "Sanitary Reform."

A board, comprising all the most eminent and skilful physicians of the day, was assembled in the College of Physicians, under the presidency of Sir Henry Halford; and, after declaring, in opposition to the unanimous opinion of the physicians of Bengal, "that measures of external precaution for preventing the introduction of the cholera morbus, by a rigorous quarantine, have hitherto been found effectual," they issued an official notification to that effect.

This announcement by the English physicians of 1831 was published throughout the land in the form of an Order of the King in Council. But the strong good sense of the public averted many of the mischiefs which these scientific advisers would have produced, had their counsels been carried into execution. The preventive measures which were eventually adopted by them consisted in prohibiting intercourse between one town and another by sea, and permitting it by land; thus, communication between London and Edinburgh by stage coach was perfectly free and uninterrupted, while communication between those capitals by sea was prohibited with such rigour, that no interest, however powerful, could procure an exemption. Francis Jeffrey—at this time holding the high office of Lord Advocate of Scotland, and whose influence, from his personal and official connexions, was very great—was unable to obtain permission for his faithful servant, in the last stage of dropsy, to go from London to Leith by water, lest he should carry with him to his native country, by that mode of conveyance, not the dropsy, which he had, but the cholera, which he had not.

The articles in the *Westminster Review* attracted the attention of the leading men of that day, in and out of Parliament; they were often referred to in the debates in both Houses, and they were followed by a Parliamentary Inquiry into the subject of Quarantine. The articles were, of course, anonymous, but it was no secret that they were written by Dr. Southwood Smith.

After five years' more experience in the wards of the London Fever Hospital, and the wide district included in the Eastern Dispensary, containing, perhaps, the worst fever nests in the metropolis, Dr. Southwood Smith published, in 1830, a Treatise "On Fever," in which he traced and exemplified in a more extended manner the true causes of fever, and demonstrated, by striking examples, that they are strictly local, and capable of removal and prevention. Of the causes of fever, Dr. S. writes:—

"The causes of fever are of two kinds; first, those which

immediately produce the disease, and, secondly, those which bring the system into a condition capable of being affected by the first: the former are called the exciting, the latter, the predisposing, causes.

"I. Of the Immediate or Exciting Cause of Fever.

"The immediate or the exciting cause of fever is a poison formed by the corruption or the decomposition of organic matter. Vegetable and animal matter, during the process of putrefaction, give off a principle, or give origin to a new compound, which, when applied to the human body, produces the phenomena constituting fever. What this principle or compound is, whether it be one of the constituent substances which enter into the composition of organized matter, or whether the primary elements of organized matter, as they are disengaged in the process of putrefaction, enter into some new combination, and thus generate a new product, we are wholly ignorant. Of the composition of the poison, of the laws which regulate its formation, and of its properties when generated, we know nothing, beyond its power to strike the human being with sickness or death. We know that, under certain circumstances, vegetable and animal substances will putrify; we know that a poison capable of producing fever will result from this putrefactive process, and we know nothing more.

In the same treatise is given (page 433) the original table which proved that fever is the great pauperizer of the country, by attacking chiefly the heads of families, the range of its greatest prevalence being there shown to be from the ages of 20 to 40.

In the year 1833, on the motion of Lord Ashley, a Royal Commission was issued to Inquire into the Condition of the Children and Young Persons employed in Factories. The members of this Commission were Thomas Tooke, Esq., Edwin Chadwick, Esq., and Dr. Southwood Smith. The object of it was to ascertain the effects of factory labour on the children and young persons employed therein, and the legislative restrictions, which from the nature of the case were required. The recommendations of the Commission were adopted by Parliament, and made the law of the land, which is now in operation. *Incidentally*, a very important exposition was made, in the course of this inquiry, as to the Sanitary Condition of the Working-classes, and brought directly and impressively under the notice of the Legislature.

An unusual prevalence of fever having taken place in the metropolis during the year (1837), and especially in the Eastern District, the Poor-Law Commissioners appointed Dr. Arnott, Dr. Kay, and Dr. Southwood Smith, to make inquiry into the causes, and to report thereon.

The following is the summary of their Report as to the causes of the various kinds of endemic contagious fever. They are reducible to two classes :—

"1. Circumstances injuriously affecting the well-being of the poorer classes, and arising independently of their habits.

"2. Circumstances injuriously affecting their well-being, and originating to a considerable extent in their habits."

Among the former classes of causes are enumerated,—

"I. Imperfection or want of sewers and drains in the parish or district.

"II. The existence of uncovered and stagnant drains or ditches, containing vegetable and animal matter in a state of decomposition.

"III. Open stagnant pools of water rendered putrid by the admixture of animal or vegetable substances.

"IV. Undrained marsh land.

"V. Accumulations of refuse, either thrown from the houses, or otherwise collected in the streets, courts, and lanes.

K

"VI. Lodgment of filth in large cesspools, in situations where the exhalations are destructive of health.

"VII. The situation of slaughter-houses in densely-peopled districts, among narrow streets, and the bad regulation of these establishments.

"VIII. The state of some of the public burial-grounds in thickly-peopled districts.

"IX. The want of ventilation in narrow alleys and close courts inhabited by the working-class."

Among the second class of causes are enumerated,—

"I. The state of the lodging-houses of mendicants and vagrants, and of a certain class of Irish poor.

"II. The crowded state of the dwelling-houses, which, in certain districts, contain several families under one roof.

"III. The gross want of cleanliness of the person and dress and habitations among certain classes of the poor.

"IV. The prevalence of intemperance.

"V. The habit of dwelling in previously-deserted houses, cellars, &c., &c. Keeping pigs, &c., in dwelling-houses.

"VI. Indisposition to be removed to the hospitals when infected with contagious disease.

"VII. The neglect of vaccination.

In the year following, Dr. Southwood Smith pursued the inquiry commenced in 1837, and gave the results of his examination of twenty Metropolitan Unions or Parishes.

The public generally, and many Members of both Houses of Parliament, having been deeply impressed by this Report, the late Bishop of London moved for an Inquiry into the Sanitary Condition of the People generally, with a view to ascertain whether it was the same in other large towns as in the Metropolis.

In the year following (1840), Mr. Slaney obtained a Committee of the House of Commons to Inquire into the Sanitary Condition of Large Towns and Populous Districts, with a view not only to extend the investigation, but also to bring the facts and results which had been already obtained directly before the attention of the Legislature.*

In this same year a Royal Commission (the Children's Employment Commission) was issued to Inquire into the Condition of Children, and Young Persons employed in Mines and Manufactories *not* under the Factory Act. The Members of this Commission were Thomas Tooke, Esq., Dr. S. Smith, and the Factory Inspectors.

Under this Commission the sanitary condition of the working-classes was made a main subject of inquiry. The effects of particular kinds of labour, on the physical and moral condition of the workers, were traced; and drawings, illustrative of their employments, and more especially of those in mines, were introduced. This pictorial exhibition was quite a novelty in a Parliamentary Report, and caused a great excitement at the time. Some members of the Government were angry; but the public were gratified, and the Lords and Commons impressed, though shocked. It should be borne in mind, that a large proportion of the mines and great manufactories were the property of members either of the Lords, or Commons. It is certain that, without the pictures, the real condition of women and children in mines, and in several other employments, could not have been made intelligible. With them the effect was such, that Lord Ashley was easily able to carry through both Houses a Bill to prohibit altogether the employment of women and female children in mines; and regulating the ages of children, and the hours of their work, in all cases in which large numbers work together.

---

* The Report of this Committee is full of interest. The Bishop of London's was an *official* inquiry under the Poor-Law Commission; Mr. Slaney's a *Parliamentary* one.

K 2

A full and faithful account of the physical and moral condition of the working-classes in the mining and manufacturing districts of Great Britain is found in the Reports and Appendices of this Commission.

The year 1841 was remarkable for the first attempt at direct sanitary legislation. To the Marquis of Normanby, then Secretary of State for the Home Department, the honour of this first effort is due. In his speech on the Second Reading of the Drainage and Buildings Bill, he thus speaks:—

"I believe it may be considered an established fact, that almost all diseases, and more especially those of an epidemic character, to which human nature is subject, where there is no predisposition to disease existing in the constitution itself, arise from malaria—the consequence of the decomposition of animal and vegetable matter. It is ascertained that this is a subtle poison, thus produced, which affects the vital energies, as certainly though not as speedily as arsenic. An experiment has been tried by which it is ascertained that by condensing a certain quantity of air, saturated with malaria, a fluid is produced so charged with animal or vegetable matter in a state of putrescence, that when the smallest particle is inserted in the vein of any small animal, as that of a dog, for example, the animal dies with absolute certainty and with great rapidity, and with the common symptoms of fever. This poison must produce the same effect, though more slowly, when respired by the lungs, and thus introduced into the system.

"If I can once establish the character of this poison, and the extent of its operation, the next point will be to direct your attention to the means by which the evil may be removed. There is an instance in illustration of the facility with which the evil is abated, when the cause is removed, given before the Committee of the Commons last year, and no doubt familiar to your Lordships, from the fact that it has been repeatedly quoted in the newspapers. There was a ditch near the terminus of the Birmingham Railway, at Euston Square, into which a number of drains discharged themselves: the consequence was, typhus fever continually prevailed there. The simple remedy was applied of covering in the ditch, and ever since the place has been quite free from such infectious diseases."

Subsequent inquiry and discussion greatly improved both the principles and the details of sanitary legislation, as compared with the proposals in this Bill. Still, honour to the Lords, who carried the Bill with a cordial and noble spirit through their own House, and sent it down to the Commons! The Session, however, broke up before any discussion could *there* be held upon it.

In the year 1842 was presented Mr. Chadwick's Report on the Sanitary Condition of the Labouring Population of Great Britain, which was, in fact, a return to the Bishop of London's motion in the House of Lords, in 1839. This Report confirmed and extended the results of preceding inquiries, and prepared the way in a signal manner for practical legislation on the subject.

In 1844, Lord Normanby made a second attempt at legislation, moving an Address to the Crown on the Sanitary Condition of the People; but the Administration of which he was a member broke up before much progress had been made with the new and improved Bill which he introduced.

The new Ministry, soon after their assumption of office, appointed a Royal Commission (the Health of Towns Commission), for Inquiring into the State of Large Towns and Populous Districts.

Seeing the difficulty of obtaining any practical result from all the labour that had been devoted to this subject, Dr. Southwood Smith now determined to endeavour to bring together the distinguished men who had taken an interest in Sanitary Reform, and who had exerted themselves to promote it, with a view to induce them to co-operate in promulgating the information, which had been acquired and accumulated, and to force it on the attention of the public and the Legislature. He succeeded in forming the "Health of Towns Association," the action of which proved to be highly influential.

This Association was formed at a Public Meeting held at Exeter Hall, on Wednesday, the 11th of December, 1844, "for the purpose of diffusing among the people the information obtained by recent inquiries, as to the physical and moral evils that result from the present defective sewerage, drainage, supply of water, air, and light, and construction of dwelling-houses; and also for the purpose of assisting the Legislature to carry into practical operation any effectual and general measures of relief, by preparing the public mind for the change."

The Chair was taken by the Most Noble the Marquis of Normanby. The Meeting was addressed by the Chairman; by Sir Robert Inglis, Bart., M.P.; Benjamin Hawes, jun., Esq., M.P.; R. L. Shiel, Esq., M.P.; Rev. W. W. Champneys; Sir W. Clay, Bart., M.P.; R. D. Grainger, Esq.; the Hon. W. Cowper, M.P.; R. A. Slaney, Esq.; Dr. Southwood Smith, and other gentlemen; and a most influential Committee was appointed.

The labours of the "Health of Towns Commission" were embodied in a Report, which was of great service in clearing away the difficulties to legislation, and in determining the principles on which sound and workable legislative provisions should be founded. Just at the close of the Session of

1845, Lord Lincoln introduced a Bill, avowedly for consideration during the recess, founded, for the greater part, on the recommendations of the Health of Towns Commissioners.

In the meantime, the "Health of Towns Association" constantly kept attention alive to the subject; but, among all the services it rendered, perhaps the most important was the publication of its "Report to the Committee of the Members of the Association on Lord Lincoln's Bill." We quote a passage from this Report, equally painful and convincing:—"There is a consideration connected with this subject which deserves the still more serious attention of the statesman and the Legislature, namely, the connexion which is proved to exist between an habitual state of personal and domestic filth and the moral corruption of the people. 'When the drainage and receptacles for filth are neglected,' says another witness, 'it is quite impossible for decent persons to continue the better habits they may have formed in other places: it tends to destroy everything approaching, I will not say to refinement, but to the common decencies of human creatures. A decent and respectable woman coming into such a district cannot maintain the cleanliness and propriety of demeanour to which she has been accustomed. A short time ago I was standing in one of the streets branching off from Rosemary Lane, called Blue Anchor Yard, looking at a stream of abomination that was flowing down from a court into the open gutter in the centre of this Blue Anchor Yard, the open gutter being the common receptacle for the filth from the houses. This noisome stream was flowing close to a house at the door of which there stood a woman with ruddy cheeks, neatly clothed. "Five times this very day, Sir," said she to me, "have I swept this place as clean as I possibly could; but you see the state in which it is again. It is no use to try to keep it clean." Her whole appearance indicated that she was a new comer; in a few days she would naturally give up her hopeless attempt to keep the place clean; and if she remain there, she must necessarily sink into the state of squalor and filth so general among her neighbours.'

"On the husband, the want of a decent and cleanly house is described as still worse. The wretched state of his home,' continues this witness, 'is one of the most powerful causes which induces a man to spend his money on strictly selfish gratifications: he comes home tired and exhausted; he wants quiet; he needs refreshment; filth, squalor, discomfort in every shape are around him; he naturally gets away from it if he can.' 'I have already more than once expressed my conviction,' he adds, 'that the humanizing influence of habits of cleanliness and of those decent observances which imply self-respect, the best, indeed the only real foundation of respect for others, has never been sufficiently acted upon.'"

This Report was drawn up by Dr. Southwood Smith, assisted by notes and suggestions from various members of the Association and others, particularly from Mr. Chadwick, who, though not a member of the Association, and never taking any direct part in its action, materially assisted it on several occasions, and on this especially.

A fourth Royal Commission ("The Metropolitan Sanitary Commission") was now issued (1847) to inquire, Whether any, and what, Special Remedial Measures were required for the Metropolis?

In this year Lord Morpeth introduced a Government Bill for "Improving the Health of Towns in England." Though the provisions of this Bill indicated a great advancement in the knowledge of the principles of sanitary legislation, it encountered great opposition, and was defeated for that Session.

In the year following, Lord Morpeth brought in a second Bill, entitled "A Bill for Promoting the Public Health." Cholera now came to his Lordship's help. After a severe struggle, the Bill finally passed both Houses of Parliament: its provisions now constitute the "Public Health Act," and an administrative body was constituted to carry those provisions into operation, namely, the General Board of Health.

In the next year, an Act supplementary to the "Public Health Act," called the "Diseases' Prevention Act," was passed, giving the General Board of Health additional powers on the outbreak of any serious epidemic.

The course of sanitary legislation, which distinguished this period, was completed in 1850, by the passing of the "Metropolitan Interment Act," the object of which was to prohibit any further interments within towns, and to introduce a more decorous and economical mode of conducting funerals. The latter design failed, but a complete stop has been put throughout England and Wales to any further intramural interment.

Upwards of ten years have now elapsed since the passing of the "Public Health Act." Many towns have adopted its provisions, and in several they have been carried into complete practical operation. It is known that, in several instances, a great improvement in the public health has taken place, but the exact results have not yet been collected.

About seventeen years ago, however, an Association was formed ("The Metropolitan Association for Improving the Dwellings of the Industrious Classes"), the object of which was to ascertain, by actual experiment, whether some of the worst sanitary evils which at present prevail in all our large towns and villages could not be checked by the erection of dwellings for the poor. The plan of the Association was to

K 5

erect a large building divided into suites of apartments, capable of accommodating a number of families, and provided with the following sanitary conditions:—

1. The thorough subsoil drainage of the site.

2. The free admission of air and light to every inhabited room.

3. The abolition of the cesspool, and the substitution of complete house-drainage.

4. An abundant supply of pure water.

5. Means for the immediate removal of all solid house refuse, not capable of suspension in water, and of being carried off by water.

It was no part of the plan of the Association to attempt to assist the labouring classes by offering them any gift or charity. The proposal of the Association was, that the industrious man should pay the full value for his house; but that for the sum he pays, he should possess a salubrious and commodious dwelling, instead of one in which cleanliness, comfort, and health could have no place.

More recently, another Society, instituted for the same objects, and of which Lord Shaftesbury is President, has been instituted,—"The Society for Improving the Condition of the Labouring Classes, incorporated by Royal Charter." And from the operations of these Societies' results, social, physical, and moral, of the most satisfactory character, have been extensively obtained.

The "Common Lodging Houses' Act," also, by enforcing certain conditions of cleanliness, and by preventing overcrowding, has extended to vagrants and others, forming the very lowest portion of the population, the like immunity from disease, which the improved dwellings have secured to the industrious labourer and artizan.

The same may be said of another Act,—an Act conceived

in the same beneficent spirit, entitled, "The Labouring Classes' Lodging Houses' Act," which received the Royal Assent on the same day as the "Common Lodging Houses' Act." The "Labouring Classes' Lodging Houses' Act" is merely an enabling Act, but it gives most important powers to local authorities for raising loans, procuring lands, and uilding or buying houses, in order to provide more convenient and commodious dwellings for the labouring classes.

Space will not permit our illustrating by facts the beneficial results that, from all quarters of the country, have flowed from these Associations and Legislative Acts. We can only, in conclusion, avow our entire and hearty acquiescence in the opinions expressed some years since by Dr. Southwood Smith,* that "it is impossible to over-estimate the practical importance of them." "They show," he says, "the extent to which, under circumstances of the utmost difficulty and danger, it may still be possible to save life; they open a prospect of the physical and social improvement of the people, such as, before these results were obtained, there was no warrant from experience to anticipate; they indicate that the first step in this progress must be the removal of the degrading influence of the present dwellings of the labouring classes; and they prove the practicability, without loss to the capitalist, or additional rent to the tenant, of the universal substitution of houses for hovels."

* We cannot but express our surprise that this eminent sanitary reformer should have been removed from public service in 1854, when the late Board of Health was dissolved, without pension, reward, or provision of any kind; while the only other paid member of the Board retired with a pension of £1,000 a-year for life, and a Companionship of the Bath. Dr. Smith's claims were, indeed, honourably and amply recognised by Lord Palmerston, but he remains to this day unrequited, an oversight which we hope soon to see rectified.

"The physical improvement of these masses, it is now admitted, must precede their intellectual and moral elevation. When the house ceases to be a sty, and possesses the conditions which render it capable of being made a home, then, but not till then, may it receive, with some hope of benefit, the schoolmaster and the Minister of religion

It will be seen, by what has been already said, how important a branch of the National Association is the Department of Public Health.* The sanitary condition of our Army and Navy is a subject that must occupy a large share in public attention, and Miss Nightingale's Papers will have the desired effect of drawing notice to evils which ought never to have existed. Proper drainage and ventilation contain weighty problems still unsolved by social reformers, but in a fair way of being so, by the combined effort of scientific and practical men. Dr. Kebbel, in his recently-published work on "The Climate of Brighton," does not hesitate to say, that "there is no reason why town populations should not be as healthy and live as long as the inhabitants of the country, if all undue crowding were avoided, an abundance of pure water supplied to all classes, and all organic impurities removed from the vicinity of the dwellings as fast as they are formed."†

---

* A labourer in the same cause is the Ladies' National Association for the Diffusion of Sanitary Knowledge, which is becoming daily more known and appreciated. We ask our readers' co-operation in its important objects, by subscriptions and circulation of its Tracts. Office, 14 A, Princes Street, Cavendish Square.

† Dr. Farr, who is one of the greatest authorities in Statistical and Sanitary Science, gives the following valuable suggestions as to the means of promoting public health in Great Britain :—

"I. If Government can do little by *direct* enactments for the diminution of sickness, it is, nevertheless, their duty to determine, by statistical enumerations, the actual state of health, and the extent

Among other subjects brought under the notice of this Department, is that of "Residences for the Insane." Thirty years ago the asylums for lunatics were in a wretched condition, and we are indebted to Members of the Society of Friends at York for awakening general attention to proper

---

to which it is deteriorated in different circumstances. Returns of the diseases and deaths in the Army and Navy, and all bodies of men employed in the public service, should be made annually; and this, with the results of the General Registration, would improve public health, by showing so distinctly the connexion between diseases and their natural causes, that men would either avoid or obviate the evils destructive of health by some invention.

"II. Almost all classes of people of this country are profoundly ignorant of the physiological laws which regulate their own existence; health may, consequently, be improved, by making a knowledge of the nature of the human organization, and of the external agents by which it is influenced, an elementary part of National Education.

"III. It has been shown that external agents have as great an influence on the frequency of sickness as on its fatality; the obvious corollary is, that man has as much power to prevent as to cure disease. Yet medical men, the guardians of public health, never have their attention called to the prevention of sickness; it forms no part of their education. To promote health is apparently contrary to their interests. This may be corrected by modifying the present system of medical education, and the manner of remunerating medical men.

"IV. Public health may be promoted by placing the medical institutions of the country on a liberal scientific basis, by the medical societies co-operating to collect statistical observations, and by medical writers renouncing the notion that a science can be founded upon the limited experience of an individual. The healing art may likewise be promoted by encouraging *post-mortem* examinations of diseased parts. The practitioner who never opens a dead body must commit innumerable, and sometimes fatal, errors.

"V. It has been proved that, in the present state of things, the mortality is greatly augmented wherever large masses of the people are brought together. It will be the duty of the Government, the Municipal Corporations, and all classes of citizens, to render the towns of this country, and every establishment where large numbers are collected together, perfectly adapted to the wants of the human organization, and compatible with the full enjoyment of health."

treatment. A great and marked improvement has taken place both in public and private asylums of late years; but we cannot but feel that there is still wanted a severer scrutiny than is exercised, as to the position and character of persons undertaking the charge of the insane.

The Department of Social Economy defines those branches of Social Science for which no appropriate place could be found in the four preceding Departments. The subjects have been thus classified :—First, Economic Science, under which head Statistics, which, as we have before said, lie at the foundation of Social Science, find their fitting place.

Almost all civilized countries now have statistical departments under Government direction; and Lord Stanley has well said, that statistics are the function of the State in a sense in which no other science is so. "In fact, an executive regularly supplied with such knowledge may be said to have its finger on the pulse of every province, ready, at the first symptom of disease, to intervene with the requisite remedy."

It is a singular fact, that no attempt has ever yet been made to bring the entire British Empire, including our Colonies, under a uniform statistical organization, important and desirable as it must be, that a thoroughly organized system should be in existence.

Statistics have been called the account-books of a nation, revealing the state of its affairs, and the excellencies or defects of its institutions, and, as the President of the Statistical Section remarked at the Meeting of the British Association at Leeds, "that which gave to the recent Report of Mr. Sidney Herbert's Commission on the Health of our Troops in Barracks its resistless force was, the certainty and precision with which statistical researches enabled it to measure the amount of loss sustained, by comparison with the mortality in other classes of the population at the same ages. The Report might have dwelt on sickness, on injudicious diet, on defective ventilation, on want of drainage, and so forth, and all such statements would have been pronounced to be exaggerations or errors; but when it applied the ascertained scale of mortality, so as to prove that there were so many deaths in the thousand, when there ought only to have been half that number, the definiteness of the figures and facts defied evasion, fastened on the public mind and conscience, and compelled immediate measures of reform. Those persons who have ignorantly charged upon political economy and statistics a disregard of moral considerations and of humanity, may now see how egregiously they were mistaken, and how the arithmetic which they thought so heartless is rising up as the most powerful advocate of the value of human life, of health, of domestic comfort, of temperance, of virtue, of proper leisure, of education, and of all that can purify and elevate society."

So fully is the importance, now recognised, of Statistics as a science, that a European triennial Statistical Congress is held by the greatest staticians of the age. This congress has held its Meetings with great success at Brussels, Paris, and Vienna, and is to be held this year in London, when we may hope to realize a suggestion of Lord Stanley's with relation to the objects at which we should aim,—viz., first, the adoption of a method or organized system for the annual collection and publication of national facts, as perfect in itself as possible; and, secondly, the assimilation of that method to those which prevail elsewhere, so that nations may mutually profit by each other's experience.

The second division of the Social Economy Department was Population, Labour, and Capital, which embraced the consideration of Emigration, Associated Labour, and Cooperative Provident Associations. The Condition of the Working-classes forms a third branch, which is of the highest importance, and, we are thankful to know, occupies the earnest attention of the leading philanthropists of the day. Most valuable Papers on this subject have been brought before the Association, and will be productive, doubtless, of practical results. Both in England and Scotland the immorality induced by the condition of the dwellings of the poor is frightful, and, as a Scottish Provincial Paper describes it, "speaks trumpet-tongued to the necessity for improved cottages."

In Mr. Robert's Paper, read at Liverpool, on this subject, he mentions that the difficulty with regard to workmen's dwellings had been met in Paris by a Parisian tailor, who had built an entire village outside the Barrier for his workpeople. "Why," asks Mr. Roberts, "should such an idea not be adopted by hundreds of those who, in our own metropolis, are similarly situated, the daily journey of a foreman being sufficient, for arrangements with the work-people who have their workshops erected in juxta-position to their dwellings?"

We do not believe that in any instance the decent poor will be found backward to avail themselves of opportunities for improving their physical or moral condition. Thus we find that the baths and wash-houses for the working-classes are largely used by them. At Goulston Street Baths and Wash-houses, 150,000 bathers in one year are reported, and "as an illustration of what these establishments are doing in the way of educating the young in cleanly habits, it being the rule to admit two boys under eight years of age into one bath at a single charge, boys are found to unite readily for the purpose, and often on half-holidays or evenings, the boys of the charity-schools come together, when, as the superintendent tells us, 'we don't make a very strict inquiry into their ages—so that they are not evidently too old.'"

Public parks, walks, and recreation grounds are instrumentalities for raising the character of our poor, that we earnestly desire to see carried out in the neighbourhood of every large city, town, and country village. In Nottingham, where large pieces of land are thus appropriated, it is found that the working-classes highly value such means of enjoyment, and appreciate and respect the flowers and shrubs in the arboretum.

But there is another matter connected with the condition of the working-classes, to which we must refer. We have in an earlier part of this little work said a few words on Intemperance, as one of the great causes of crime, and we now recur to this subject as one we consider imperatively demanding the attention of the National Association.

The following important extract is from the Report of the Select Committee on Inquiry into Drunkenness, and deserves the serious attention of our readers :—

"The Select Committee appointed to inquire into the *extent, causes, and consequences* of the prevailing vice of intoxication among the labouring classes of the United Kingdom, in order to ascertain whether any legislative measures can be devised to prevent the farther spread of so great a national evil.

"1st. *Extent of the Evil.*—It appears to your Committee from the evidence taken before them, that the vice of intoxication has been, for some years past, on the decline in the higher and middle ranks of society, but has increased within the same period among the labouring classes, and exists at present to a very great extent in the population of England, Scotland, and Ireland, and in the sea-port and manufacturing towns, as well as in the agricultural districts, including in its victims men, women, and even children.

"2nd. *Remote Causes of its Production.*—The influence of example set by the upper classes of society, when habits of intoxication were more frequent in such ranks than among inferiors in station, and the many customs and courtesies still retained from a remote ancestry, of mingling the gift or use of intoxicating drinks with almost every important event in life, such as baptisms, marriages, funerals, &c., as well as in the daily interchange of convivial entertainments, and in the commercial transactions of purchase and sale.

"3rd. *Immediate Causes of its Extension.*—The increased number and force of the temptations placed in the daily paths of the humbler classes, by the additional establishment of places at which intoxicating drinks are sold, the number now being considered, from the average of several districts in England, Scotland, and Ireland, to be not less than one such place to about every twenty families throughout the United Kingdom."

The testimony of our Judges, Magistrates, Gaol Chaplains, and all connected with the care of criminals, is most striking with relation to the bearing that such habits have upon crime.

Judge Coleridge says, "There is scarcely a crime comes before me that is not, directly or indirectly, caused by strong drink." Judge Pattison observed to a grand jury, "If it were not for this drinking, you and I would have nothing to do." Judge Alderson remarks :—"Drunkenness is the most fertile cause of crime, and, if it were removed, this large calendar would become a very small one." Judge Wightman observes :—"I find in this, as in every calendar, one unfailing cause of four-fifths of the crime is the sin of drunkenness." Judge Erskine even went so far as to say, at Salisbury, in

1844, "that ninety-nine out of every hundred are from the same cause." Nor let us omit to mention the almost dying words of Justice Talfourd, in the charge with which he opened the Stafford Assizes. After lamenting the heavy calendar of crime, and the atrocity of some of the offences, he said that these crimes might, in most cases, be traced to the vice of intemperance. He was dwelling, with deep feeling and energy, on the duty incumbent on the higher ranks to endeavour, by enlightened sympathy, to wean their poorer neighbours from such sensuality, when he was suddenly silenced by the stroke of death. To these facts, with regard to England, might be added others, with relation to other parts of the British empire, showing a similar or worse state.

Can it be that Great Britain is content that her people shall remain the most drunken among the civilized nations of the earth?

The upper and middle classes formerly possessed this character; and now, when improved habits have displaced the former ones of intemperance, the working-classes are the victims of this dreadful vice; and shall Christian legislators and philanthropists be satisfied without doing something to meet this crying evil?

To Temperance Societies is delegated a great work. Both in England and America, some of the leaders of this movement are men of great ability; and, were the Christian public fully alive to the fact that to the habitual drunkard there is no medium between the indulgence of his habit and total abstinence, we should see many, now standing aloof, uniting in earnest effort to promote the foundation of Temperance Societies throughout Great Britain.*

* "Public-houses and beer-shops are seldom or never resorted to by any but the working-classes and mechanics. It is, therefore, a

It has been with no slight pleasure that we have watched the movement going on through the country with regard to Drinking Fountains. When in Rome, some years back, we have often wished we could transfer to our own country the splendid public fountains of that city, and wondered that such did not exist in Great Britain. All honour to the man who took the initiative in this most important step towards the removal of the curse of drunkenness that hangs on our land. We believe the beneficial consequences stretch far into the future, and distant generations will learn to bless the name of George Melly, who first placed within the

---

fair presumption that the magistracy by whom they are licensed are, in many cases, really ignorant of the true nature of the feelings of the poor and labouring classes ; probably the great mass of these very people would heap their blessings upon the magistracy who could and did stop this great curse, as they well know it to be, to themselves and their families. We know a prison where the Chaplain visited, consecutively, 137 prisoners. No less than 86 expressed a distinct wish that there was no such stuff as beer or spirits to be got at ; 32 said they should be glad if there were no public-houses, but they should like a glass at their own homes; and 19 did not care about it. A recent canvass with regard to the Permissive Bill has shown still more decidedly the feeling of working men with regard to their desire to be freed from the presence of temptation. We should mention that the Permissive Bill is designed to enable the ratepayers, in a given district, to veto the licensing of public-houses and beer-shops within that district, on the principle that the people should be empowered to decide for themselves as to the necessity for the traffic in strong drink. We give a few results of the canvass in different places :—

	Adults Canvassed.		
	For.	Neut.	Against.
ENGLAND.			
Manchester, St. George's Ward	2,589	451	246
„ St. John's Ward	1,144	273	94
„ Medlock Street Ward ...	2,978	848	308
SCOTLAND.			
Greenock	14,352	1,858	349
Glasgow (partial)................................	10,964	1,300	798"

"MELIORA."

reach of the weary traveller, and the thirsty sons of toil, the blessing of a timely supply of that pure element provided by bountiful nature for her children.

In glancing at the condition of the working-classes, and the means most calculated to benefit them, we earnestly advocate (united to improved dwelling-houses, recreation grounds, &c.) an abundant supply of a pure literature, and this must be provided in towns by Free Libraries and other appliances, while country districts can be reached by itinerating libraries and colportage. Of the latter as a means most suited to the present condition of our country, we have fully spoken in a recent publication. We are well assured that the wide dissemination of such works as the "British Workman," "The Band of Hope Review," the "Leisure Hour," "Sunday at Home," with popular scientific publications, would go far to supersede infidel and indecent periodicals.

We can only hope that the paper-duty will be removed, in which case education and a taste for literature would be widely promoted, by the increased sale and circulation of the works we have mentioned. Everything that stands in the way of social improvement ought to be made to give way, where practicable, and we hope we may live to see the day, when a penny postage to our Colonies will assist in this great object.*

The fourth branch of the Department of Social Economy is one in which we have long taken a deep interest,—viz., Workhouse Management; and here, as in Reformatories and Prison Discipline, the communion of labour, or the introduction of the feminine element, is largely needed.

* See a Paper, by the Author, on this subject, read at the Leeds' Meeting of the British Association.—"Handbook of the British Association," page 139.

Twenty-four years ago, when the new Poor Law superseded the old, it was hailed by many persons as a remedy for great abuses. If over indulgence was the fault of the old Poor Law, who will venture to deny that some degree of harshness has not attended the administration of the new? Perhaps few of our readers realize the fact that there are in our workhouses upwards of 100,000 inmates.

A large proportion of these are "homeless, houseless, helpless poor—orphan children, lame and blind, aged, who lie down on this, their last bed, to die." * While some have been blameably improvident, how many of these poor creatures have lost health and strength through causes connected with sanitary neglect, over which they had no control!

Should poverty, in such a case, be a crime, and should the national charity of England be dispensed grudgingly and ungraciously, and the workhouse be rendered to the non-criminal poor a less comfortable home than the prison? We agree with sentiments uttered in Parliament, in 1678, by King William the Third, that "Workhouses, under a prudent and good management, will answer all the ends of charity to the poor in regard to their souls and bodies. They may be made, properly speaking, nurseries for religion, virtue, and industry, by having daily prayers and the Scriptures constantly read, and poor children Christianly instructed."

But does the workhouse, when left entirely to the superintendence of paid officials, answer to this character of Christian charity to the poor? Charity, to be such, must possess the element of Christian love; and the experience of City Missionaries, and all who work in the same spirit, attests that the poor appreciate that high principle.

There has lately been formed an Association for im-

* Mrs. Jameson.

proving the religious and moral condition of the inmates of workhouses, through a Visiting Society, the Secretary of which is a lady who has for some years devoted herself to visiting the workhouses of the metropolis. The three special branches of work which it embraces are the exercise of moral and religious influence over the children, the sick and the aged, and the able-bodied. "It will hardly be doubted," to use the language of the Secretary, Miss L. Twining, "that in these several departments or classes of inmates, there is an ample field for the exercise of Christian love and exertion."

This Association is connected with the National Association, and is likely to prove a great blessing to the poor neglected inmates of our workhouses.*

What at the present moment is the condition of the majority of these poor people? In illness, how are they tended? In the London workhouses are 50,000 such persons alone. Is it not a well-known fact, that the nurse is often a person incapacitated, in some way or other, from other employment? Thus, "One nurse in a London workhouse boldly stated that she had been *sixteen* times in the House of Correction, and she was not ashamed of it. She was a woman given to drink, and of a violent, ungovernable temper, causing great misery to the aged people under her control." In the London workhouses the propensity to drink among many of the nurses is so strong, that it is with the utmost difficulty they are kept from indulging it; nor can we wonder at this, when we know that in many workhouses it is the custom to allow the nurses a glass of gin daily, be-

---

* Inquiries respecting this Association and subscriptions may be sent to the Secretary, Miss L. Twining, at the Office of the National Association, 3, Waterloo Place, Pall Mall, London. S. W.

side their beer. "These," it has been well observed, "are the sisters of mercy to whom our sick poor are confided."*

We believe it is now generally admitted, since the practical results of Miss Nightingale's labours, that the union of education and refinement is not incompatible with the exercise of those qualities, that are desirable in the sick nurse; and we cannot but hope that there are, among the single women and widows of our country, many who would, from Christian principle, and from physical and moral fitness for the office, take, as their mission and vocation, the office of nurse. Our hospitals are now earnestly claiming the help of a superior class of women; but be it remembered that it is essential to success, in any work, that there should be special training. It has been observed that "Dilettanti visiting, desultory fits of charity, must give way to serious application, laborious preparation, and long study."

"It would be endless," writes an experienced workhouse visitor, "to give all the cases of harshness and injustice that have come before me. One of the most pernicious *consequences* of all this is the preference of the *prison* to the workhouse. There is generally better diet and better treatment there, the officials being of a superior class. The present Lady Mayoress says, that she was first led to think about the management of workhouses from being one of a Committee for visiting prisons, when she frequently was told by the women and young girls that they would rather be there than in the workhouse.

"Several poor creatures have said to me that they were weary of their lives; they had nothing to live for; they even hoped they might not live to see me again on my return from a short absence, and they longed for death to release them from their trials. Who can wonder at this, after

* Mrs. Jameson.

spending years helpless on a sick bed, with literally no variety from day to day? Others, not bed-ridden, sit round the walls, vacant and dreary, with no occupation beyond the two or three well-worn books, which some tell me they know by heart; and probably in every ward there are several who cannot read at all, from blindness or infirmity. If these things could only be *seen*, it would need little persuasion to touch the hearts of those who hear this, and persuade them to introduce some element of mercy into these abodes."

The Chaplain of a prison says, "It is almost heart-breaking to witness the number of poor creatures who come to prison in preference to the union, or because the obtaining what they require from the latter place is beset with so much difficulty and delay. It is not only an anomalous state of things, but highly discreditable to a Christian country, that a poor man should be led, as is frequently the case, to seek necessary food and shelter in a prison."

Our limited space does not allow us to go further into a subject which possesses our deepest sympathies. Let the women of England who seek a mission, turn to the workhouse, and there, either in visiting the able-bodied or sick, directing the schools, &c., &c., they will find one of the most noble they can fill. Mrs. Jameson has well said,—

"Surely it is worth considering whether the administration of these institutions might not be improved by the aid of kindly and intelligent women sharing with the Overseers the task of supervision. * * * * Can any one doubt that the element of power, disunited from the element of Christian love, must, in the long run, become a hard, cold, cruel machine? and that this must *of necessity* be the result, where the masculine energy acts independent of the feminine sympathies? Since it is allowed on all hands that we want

L

institutions for the training of efficient 'Sisters of Charity' for all offices connected with the sick, the indigent, the fallen, and the ignorant among us, why should not our parish workhouses be made available for the purpose? In such an application of means and funds already at hand, it appears to me that there would be both good sense and economy, therefore it ought to recommend itself to our so-called practical men."

Under the fifth head of Miscellaneous Papers, we find among other subjects that of the "Slave Trade." The question is one of peculiar interest.

Was it wise in England to free her slaves as a question of economy? Were the West Indies ruined by the freedom of the Negro? or is it a fact, that, through this very deed, their ruin has been arrested, and they are growing daily more rich and prosperous?

It is difficult for us to imagine the change which took place on the day when 800,000 British slaves were made free. In many places of worship a watch-night was held, and it was in the house of God, on bended knees, that Afric's sable sons and daughters received the rich boon of freedom. A friend of the oppressed negro thus writes:—

"England's giving freedom to her slaves was an act unique in the history of man. We know not where an example can be found of so noble a sacrifice made by a whole people. As to its prudence, some may think this and some that; but no man can lay it at the door of any selfish feeling. The people of the United Kingdom believed slavery to be cruel. It seemed to them a breach of the law of love which the Gospel had laid down. For these reasons, and for these alone, they made up their minds to be rid of it. But they were not hurried away by their zeal; they chose to pay the cost themselves, and £20,000,000 was paid down by them to get the slaves set free. To us, who saw this done, it may seem an every-day affair; but seen from afar, in the coming ages, it may strike men as sublime.

"Was it after all an act of shining folly? Has it really wrought woe and not weal to the world? It is worth while to find out the true reply to these questions. For if all this were so, then that noble old maxim, that 'Right never comes Wrong,' would be overthrown. Here we have a nation plainly setting itself to do right, 'because right was right,' because it thought more of what was due to God and man than of itself. Has this been a failure? has this done harm and not good? Then it may be unwise to do right. Wrong, perhaps, might as well be kept going. The laws of God, and the rights of man, may be well enough in their way, but should we obey the one, or observe the other, we may find ourselves made fools of."

From reliable authority we learn that our West Indian Colonies are steadily progressing in industry, morality, and education, and that a number export considerably more sugar than they did previous to the emancipation. The Colonies may be considered as recovering rather from the effects of slavery, than suffering from the results of emancipation. We are glad to know that the friends of the Negro are watching his interests, and that attempts to introduce slavery under a new name have been opposed,— free emigration, as it was called, consisting of a revival of the slave-trade (the pretended emigrants being men who were kidnapped on the coast of Africa, and taken on board ship in manacles, or put in them before being many hours on board the vessel).

In looking out upon the broad field of labour occupied by the National Association, we feel how cursory is the glance we have been able to bestow upon its objects, and how difficult it is to assign to each its proper relative position in

L 2

Social Science.* Our work must necessarily be imperfect, but if to some mind, open to the consideration of these topics, it proves suggestive of a suitable and fitting sphere of labour, we shall be amply rewarded. With regard to the National Association, its influence on all subjects connected with Social Science has already been felt throughout our country, and we feel assured it is destined to become one of the most important institutions for social progress in Great Britain.

The Science of Philanthropy now takes its place beside Physical Science, claiming all the help that the latter is so ready to impart; and in this union and communion of labour will be found a pledge of future progress, and a realization of all that can contribute to improve, elevate, and bless mankind. Well has it been said, that "The accomplishment of such great objects, by the voluntary devotion of money, time, thought, and anxious care, are proofs, not simply of the energy of the race, but of the power of a genuine Christian civilization among the middle and upper classes of society. To be insensible to the high significance of such statements, would be as great a stigma on political science as on religious sentiment. Such gigantic facts are, to the statesman, marks of the vast strides of social progress, and, to the eye of faith, signs of the coming of that kingdom for which the believer prays, according to the precept of Him whose promise it will fulfil." †

---

* A Paper, on "Jurisprudence and Amendment of the Law," by G. W. Hastings, Esq., not having arrived in time for this edition of the "Handbook," will account for our seeming omission, which we hope to rectify in a future edition.

† Sir J. K. Shuttleworth.

THE INDUSTRIAL RESIDUUM.[1]

I AM particularly anxious to make it clear from the first that nothing of what I have to say applies to the class of genuinely self-supporting wage earners, to those workers whom we may call the true industrials. At every turn of their daily life the two classes meet and influence each other, they are connected by every tie of mutual service and dis-service; to the casual observer their dress, their food, their living accommodation, even their work, is the same in kind if not in quantity. Yet striking right through this superficial resemblance, and reducing to comparative insignificance (for our present purpose) all social and family alliance, we may find a fundamental distinction which can only be intensified by any attempts to obliterate it by artificial means. It is a difference of character and disposition, and it is to this difference of character and its economic results, rather than to any numerical investigation, that I specially wish to draw attention. The qualities which are characteristic of members of the Residuum are not distributed with any reference to money income, and for this reason it is impossible to base a calculation of their numbers upon any estimate of earnings. Moreover, I know of no important general proposition that can be laid down about all the individuals who are in receipt of small incomes, nor about those in receipt of large incomes; except, perhaps, that the latter will be more comfortable than the former, and this is only an approximate generalization. But by taking as a ground of classification some fundamental characteristic of the individual, some disposition or habit which will determine his actions, it may be possible to mark out a development in human nature which will repay study. The most trivial accident of birth or fortune may enable a true member of the Residuum to conceal himself in that section of society which Mr. Giffen characterizes as the upper barbarians, or may force a respectable man to take temporary refuge in an East End slum; but while such freaks of

[1] Read at the Economic Club, 10th January 1893.

fortune would hopelessly disarrange figures, they will in no way affect our knowledge of how the one or the other will be likely to act under given circumstances.

What then are the characteristics of the class? Measured by the economic standard they are rather negative than positive. The ideal economic man, as we know, is remarkable for his foresight and self-control; in the Residuum these qualities are entirely absent. In place of foresight we find the happy faith which never fails, that 'something will turn up,' and instead of self-control the impulsive recklessness which may lead indifferently to a prodigal generosity, or an almost inconceivable selfishness. The true type of this class lives in the present moment only; not only is he without foresight, he is almost without memory, in the sense that his past is so completely past that he has no more organized experience to refer to than a child. Hence his life is one incoherent jumble from beginning to end; it would be impossible to make even a connected story out of it, for every day merely repeats the mistakes, the follies and mishaps of yesterday; there is no development in it; all is aimless and drifting.

This description may seem overdrawn, but it is based upon an accumulation of experience to which it is difficult to give tangible form. To fully realize the facts it is necessary to live amongst these people, to see them day after day, watch their extraordinary freaks, and feel the burden of their total irresponsibility. But I should like to suggest to those who are more familiar with the wealthy section of the Residuum, whether they do not find exactly the same characteristics amongst people whom mere accident of birth has separated from their natural surroundings. There is the same insuperable aversion to steady work, the same self-indulgence, the same eager devotion to trifles and absorption in the interests of the moment. All that they need to complete their likeness to their poorer brethren are the dirty homes and squalid surroundings, and if they were left for only a week to their own exertions there can be little doubt that these also would appear.

This absence of the economic virtues is of course only one aspect of a very strongly marked type of character; it accompanies a low order of intellect, and a degradation of the natural affections to something little better than animal instincts. It would take me too long to go far into this matter, but in corroboration of the view I may indicate briefly one or two of the more striking facts which we constantly come across in dealing with these people. Take for instance their frequent inability to give the number of the house in which they live, or even the name of the street; when this is combined with their complete ignorance of the points of the compass, and failure to distinguish between the right and left hand, the rational man has dropped very low on the scale towards the sagacious animal, which finds its home easily enough, but has no power of communicating its whereabouts in language.

Or take again the difficulty they have in giving any coherent account of even quite recent events. A little skill in leading questions will elicit almost any statement you please, and this from no wilful unveracity, but from mere confusion of mind. Interesting evidence might also be gained by a student who had time and patience to investigate their vocabulary; it is limited in the extreme, and their power of expression except by means of gesticulation is proportionately small.

It would be hard to attribute this intellectual failing entirely to absence of anything to express; sometimes, I am convinced, there must be actual suffering from the inability to give articulate utterance to the mental chaos within. Nevertheless we are forced to recognize that on the whole these people are as undeveloped—or as degraded—on the side of their affections as of their intellect. The most striking proof of this is the looseness of the family tie, and the absence of all feeling of mutual responsibility between parents and children and brothers and sisters. The children drift away as soon as they become self-supporting, sometimes before, and are often completely lost sight of at an age when they most need the affection and care of their parents. It is very unusual to come across a family of this type where several members are not missing, and very common to find an old couple with a large family scattered about in London, but without communication with any of them.

The economic side of this character may perhaps be best illustrated by reference to the curve in which Professor Jevons expresses the basis of his theory of labour; and which I have copied in the diagram. The horizontal line is the neutral line between pleasure and pain; the upper line represents the increase of pleasure derived from an increase of the reward to labour, the lower curve the increase of pain derived from an increase of labour. Now for the theory which it supports, this curve is of interest mainly when approaching the point where pleasure gained is equal to pain endured; the point, therefore, when the labourer begins to question the advisability of continuing. As descriptive of a process it seems to illustrate the mind not so much of the ordinary industrial of to-day, working under ordinary conditions, as of the member of the Residuum. You notice that it assumes a very rapid fall in the final utility of the reward to labour; but in our modern organization the use of money, and the habit of looking to the needs of the future combine to make the final utility of the reward as nearly constant as possible among the working classes. If a man is accumulating money the final utility will of course diminish, and he may in time reach the critical point where his pleasures and pains balance, and retire from business. But the ordinary wage earner when he receives his pay on Saturday nights is as far off the critical point as when he goes to work on Monday morning. If the idea of leaving his work does cross his mind, it is banished by the thought of not getting it again when he wants it; the mere possibility of a rise in the final utility of the reward is enough to prevent its falling, and for him the upper curve would be far more true if it were almost a straight line.

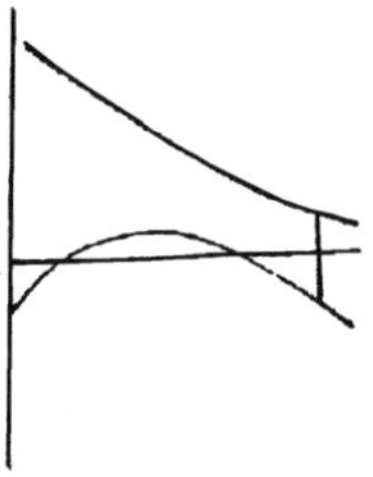

But for the member of the Residuum who has no fears for the future the curve represents a constantly recurring process. With his debts cleared off, and a week's wages in hand, the final utility of the reward is so small that he has absolutely no inducement to work; the smallest temptation will keep him away, the smallest inconvenience cause him to throw up the job; and it is not until he is destitute and his credit exhausted, that he finds himself beginning his curve again, to repeat the process as often as he gets the chance. It is of course only a question of degree. I suppose it occurs once or twice in every man's life to question whether it is all worth while; but an event which is to the normal man a crisis has become with the Residuum a habit, making little or no impression, and leaving no lesson.

In itself, and apart from any special incapacity, this disposition is not altogether an unfavourable one, even from an economic point of view; and the man to whom the future is merely the infinite possibility of something turning up, is so far in a better

R R 2

position for making his bargain in the labour market than the man who is burdened with all the cares of a lifetime. It is the indifferent seller who gets the best price for his wares; and this may partly account for the high wages which clever good-for-nothings sometimes command when they choose to work. But this point is insignificant in view of the facts: *first*, that the disposition is in the highest degree unfavourable to the acquirement of skill, and that though sometimes combined with natural genius it is more often allied to incapacity; and *second*, that though the good-for-nothing is indifferent when the question is of continuing a bargain, he rarely approaches one until he finds the necessaries of life running alarmingly short, until, that is, the claims of the future have become the needs of the present.

From the point of view of happiness there is perhaps more to be said for the disposition. It is difficult to avoid the thought that the facts represented by the curve have possibilities of pleasure about them which are wanting, where the reward of labour has always a fixed and moderate utility; it means an alternation between leisure, excitement, and intense gratification, which might conceivably yield a larger total of happiness than the somewhat low and monotonous level of satisfaction which the regular wage earner gets out of his reward. It must be borne in mind also, that in estimating the happiness of the Residuum, we must leave almost entirely out of account all pleasures or pains of anticipation; and if, as I believe, the worst of pain lies in its anticipation, while the best of pleasure is in its realization, the balance in favour of the disposition in question may tend to become a large one.

Taking this type of character as one of our data, we may now ask about its effect upon the economic position of its possessor. It will be found to result invariably in his permanent failure to maintain himself (and those legally dependent upon him) in that standard of comfort which is considered necessary, and insisted upon by the community. It is, indeed, inevitable that this should be so; want of the economic virtues involves economic failure, and no artificial social arrangements can alter the fact that the man in any rank of life who is not self-supporting is an economic failure. We cannot, however, without some limitations convert the proposition, and say that all who fail to be self-supporting are members of the Residuum.

I have already alluded to those who fall from the ranks of independence through merely temporary misfortune; they owe their failure to the accident of circumstance alone, and not to any inherent defects. It is unfortunately true that long-continued misfortune is only too likely to develop these defects, but until this has taken place there is always hope. There is another class which I should like to exclude, even though their failure to be self-supporting may be more or less permanent. I refer to the large class of women workers, whose earnings have to be supplemented to enable them to live in the standard to which they naturally belong. Their position presents a genuine economic problem, though not quite the one before us now. Looked at from the point of view of exchange, women's labour seems at present to be in the position of what is known as a "bye-product"; it shares a joint cost of production with men's labour, but is so much less in requisition that the latter stands in the position of the main product, and receives by far the greater proportion of remuneration. Some of the histories of joint production have been very interesting; for instance, that of soda and hydrochloric acid, where the latter, originally a waste product, has, through the new uses discovered for it, taken the lead, and reduced soda to the position of a bye-product. It is conceivable that as the most fitting uses are found for women's labour it may advance more nearly to the dignity of being a main product, and thus be able to claim a more equal share of remuneration.

But no such hope can be entertained with regard to the true Residuum; their labour is distinguished by its inferiority alone, and mere inferiority will never find a market; it differs not in kind, but in degree of utility only, and it is inconceivable that a use should be found for it which would not be better supplied by the class from which it has fallen.

Bearing in mind these exceptions—women workers, and the subjects of temporary misfortune—the Residuum seems to fall more or less obviously into two divisions according to the nature of the services rendered. The one consists of those who follow what, for want of a better name, I will call factitious or superfluous employments; the other of men who possess a limited amount of skill, and supplement regular wage earners in the main industries.

The first class is probably the larger in number (at any rate in poorer London), and the most hopelessly excluded from the true industrial ranks. It is a curious product of modern times, and I doubt whether it has its counterpart in history. It is usually assumed that in proportion as labour lacks skill, it falls back upon brute force, mere strength of muscle; but here we have a race living, and to some extent thriving, who have no specialized skill, no 'trade in their hands,' as they will say, and who yet have only the minimum of physical strength. A sort of superficial sharpness you may find in some of them, especially those who get their living in the streets; but it is very shallow, and rarely amounts to more than a ready adaptability of manner and a shrewd facility in saying what is expected. If placed in circumstances which are new to them, or which call for any promptness of action or readiness of resource, their incapacity is immediately apparent. Their mission in life is to pick up the odds and ends of work which are let fall through carelessness or indolence by other people, and their one economic virtue is that of being 'on the spot.' A typical instance of this virtue is found in the protégé of dustmen, who is technically known as the 'follower-up.' Say that the dustman has to empty six dust-bins on a round, and that his cart will only hold the contents of five and a half; here is the opportunity for the follower-up, who saves him the trouble of returning, and gets so much a barrow-load for his pains.

To the immense multiplication of subsidiary employments which is due to the existence of a Residuum, I need only allude; anyone familiar with working London knows them only too well. I do not, of course, refer to the genuine industrial development of subsidiary employment arising from the organization of labour, but to a multiplication of minor services of very doubtful benefit to the community. Compare the legitimate and natural function of the milkman, arising from the perishable nature of his commodity, with that of the oil-man, the coal-man, the wood-man, the coke-man, the coster of every description who haunts the streets of working London, and saves his customers the trouble of going to the shop at the corner. It may of course be argued that this is a legitimate and even desirable service, but those who recognize as the type of this class, not the milk-man, but the tally-man—that evil genius of the poor—will share my doubts. The whole method of retail industry differs from that pursued in higher classes of the community; there the purchaser sends her orders to the tradesman, here the tradesman takes his goods to the purchaser. It is a difference of the imaginative faculty which well illustrates the disposition of the Residuum; for the educated person anticipating her needs, the sight of the stores list is sufficient to provoke a purchase, but for the uneducated person the sight and touch of the commodities themselves is found to be necessary, and these prove so stimulating that debts are frequently incurred for comparatively useless articles.

Other representatives of the class are the girl who cleans steps, the old woman who minds babies, the knocker-up who will waken you at any hour for 2*d*. a week, the self-appointed commissionaire who stations himself outside public-houses ready for odd jobs, and so on with a variety which is to be equalled only by the various forms of indolence which creates the demand for those minor service people. For the most part they are entirely dependent for occupation upon the wage earners themselves, and it is evident that this limitation of their usefulness renders their mode of life in the highest degree precarious. They are exposed to every breath of 'bad times' which excites the smallest desire for economizing in their patrons. Many of them are of no real use, they are even of negative value, for the costermonger who knows his business is as well able to enforce a purchase as the organ-grinder who gets paid to go away. Hence the demand for their services is an unnatural one, and would not make itself felt for a day if it were not artificially fostered. I think it is Roscher who has urged the necessity of taking into consideration the *intensity* of a demand as well as its extent; if I may apply the term in a somewhat different sense, the demand for these services may be described as having the least degree of intensity which is compatible with its being effective. That it is effective at all is due to the peculiar conditions of supply, which we shall have to note presently. It might be thought that in some degree this lack of intensity might be compensated by extent. I believe that in many of the industries which supply the working classes the comparative stability of demand in face of bad times is to be accounted for by their large numbers, cheap goods for the many forming a safer basis for trading than expensive goods for the few. But members of this class never serve a large connection. Some dozen streets will comprise the ordinary coster's round, the charwoman has her half-dozen patrons, the coal and oil-men have their regular customers, and when these fail them they rarely succeed in establishing a new groove.

As a natural link between this class and the next, I should like to refer, in passing, to the charwoman. In the East End she is called in, like other casual labour, when the mistress of a household is unable or disinclined for any reason to do her own work, and the irregularity of an occupation dependent upon such causes quite defies calculation. She probably had her origin in the West, and is a typical instance of the development and results of partial employment. Under any satisfactory arrangement a household will find within its own internal economy sufficient labour power to carry on its necessary and normal work; but the modern system of intermittent cleaning, by which the dirt is allowed to accumulate until the family goes out of town, makes it possible to work with a smaller regular staff, supplementing it from the Residuum upon occasion. Here the charwoman belongs to the class which supplements the labour of regular wage earners; she is intermittently absorbed into their ranks, and rejected again.

The main body of this class consists almost entirely of inferior workmen, or of men who suffer from an exaggerated abhorrence of that regular work which is to all of us more or less of a burden. Here we find a certain amount of skill, but it is either insufficiently developed, or else combined with mental or physical defects which neutralize its utility. It is worth while to distinguish between the men who do inferior work and the men who do their work in an inferior manner. Inferior work is generally, if not always, badly paid for, but it is not necessarily subject to any great irregularity; as we have already noticed, the market for cheap goods has elements of stability about it which are wanting in the smaller markets for highly finished commodities.

It is the work of the inferior man which is subject to the worst forms of irregularity, and which serves as a barometer to indicate depressions and elevations in the industrial atmosphere. These are the men who are always falling into work and out again; they are the first to be turned off as work slackens, and the last to be taken on as it improves. It is only when employers are straining to make the most of a favourable turn that they will employ labour which is dear at any price; and frequently the men will anticipate their sentence, and drop away before the actual dismissal is pronounced; they have worked their spell, and are now ready for their interval of leisure.

According to the ordinary laws of competition we should expect to find this class of labour employed at a lower price than the more highly skilled and reliable; and that this is far from being generally the case, is a fact which requires explanation. There are several causes at work, the action of Trade Unions to a slight extent, but mainly a combination of public sentiment and private sophistry which tend to make employers reluctant to offer a lower than a prevailing wage. It is very noticeable in talking to employers, especially among the minor industries, how they nearly always assume a defensive attitude against any suspicion that they are paying less than an equitable wage. They will urge upon your attention that any apparent deficiency is accounted for by short hours or light work, or compensated by some privilege; facts generally true enough, and deriving their main interest from the proof they afford that an equitable standard is recognized, and that departure from it is not thought to be sufficiently justified by ability to obtain substitutes at the same or a lower rate.

Thus it comes about that inferior workmen will frequently be employed at a rate equal to that of the more skilled in the trade, but the apparent anomaly rights itself in other ways. In piece work, of course, the want of skill tells directly upon the amount of earnings by the smaller quantity of work turned out; but even where the work is by time, the economic laws avenge themselves by the system of partial employment which is creeping into industry. At first sight this would seem to be only an extension of season employment compensated by high wages, such as inevitably occurs in the building trades. But the partial employment of the Residuum exists side by side with regular employment in the same trade, and is a question not of necessity but of convenience. The system is widespread, and may now be found in almost any department of industry. To take an example, it is now quite a common thing for even respectable firms of solicitors, stockbrokers, auctioneers, and so on, to employ a permanent staff insufficient to their needs, relying with perfect confidence on supplementing it from the Residuum when there is a press of work. That is to say, sooner than pay a clerk a comparatively low retaining fee, for the sake of having his services when needed, they will take him on for two or three weeks at a time, giving 30*s*. to £2 a week, and turn him off again as work slackens, with absolute indifference as to what becomes of him in the interval. All the main manufacturing industries, such as boot-making and tailoring, and smaller ones, such as fur workers, feather workers, and trimming makers, have the same fringe of inferior men, only partially employed. (The dockers of course are a case in point, but with them the issue has been so confused that I do not venture to bring them into the question.) That this development is greatly to be deprecated in the true interests of labour, will, I suppose, hardly be questioned. If we must choose between two evils, a low regular wage is infinitely more salutary for the average Englishman than high earnings alternating with periods of idleness; and while the original degradation of a labourer to the ranks of the partially employed is generally due to his dislike of persistent work or want of skill, yet it is evident that the system itself tends to intensify the causes which give rise to it.

In my general dislike of partial employment and its results, I feel myself on fairly safe ground. In certain conclusions to which it has led me I am more doubtful, and will therefore put them in the form of questions, to which I hope to get some answers. The first is: So long as employers have open to them the alternative of partial employment, is it not contrary to the true interests of labour (at least of the Residuum) to bring any pressure to bear, whether by means of trade unions, or through public opinion, towards enforcing a minimum rate of wages?

The second question is: Whether any merely local employment agencies, which without doing anything to break up the immobility of labour enable employers to find an immediate supply at any moment, do not greatly encourage the system of partial employment, and so directly militate against stability in the relation between employers and employed?

Of course it may be urged that whenever an employer finds a workman through such agencies, a workman finds employment, and so the benefit is mutual. In individual cases that may be so to some extent, but we have to look at the tendency of the system as a whole, and that tendency does seem to be towards irregularity and uncertainty. Moreover, for the Residuum the effect is not altogether good, even in the particular case. A member of this class who feels that there is an institution prepared to reinstate him as soon as he drops out will take little trouble to keep to his work, and will become fixed in the unsettled habits which are his ruin.

When we turn to the question how the supply of this reserve labour power is produced and maintained, we are met by the difficulty that we have to deal with causes that are not strictly speaking economic. There could be no greater mistake than to suppose that the Residuum is self-supporting in the sense that the standard of comfort in which it lives is determined by its actual earnings, and for this reason it cannot be handled in the same way as other classes of the industrial community.

Prof. Sidgwick, in discussing the question whether there can be said to be a normal rate of wages corresponding to the cost of production of labour in any class, considers the doctrine to be most applicable in the case of the worst paid labour of which the supply has to be mainly self-maintained. Here a diminution in wage is thought to act as a check to numbers, and a rise to be followed by an increase. But he points out, also, that the worst paid labour of all is that of classes in towns kept up to a considerable extent by the degradation of members of other classes, and therefore unhappily exempted from the economic necessity of keeping up their own numbers. That is to say, in this class the ordinary economic forces which tend to bring about an equilibrium between the wages of labour and the cost of its maintenance, which in the long run is its cost of production, are counteracted by the invasion from other sources. Thus it comes about that to look for any relation between the cost of production of the Residuum and its economic value is—to borrow an illustration—like looking for the relation between the cost of production of cracked bells and *their* value. Members of the Residuum are all cracked bells; in nine cases out of ten they have cost as much to produce as the self-supporting wage earner, frequently more. A child of this class will not cost a farthing less to bring up—at any rate, to the age of twelve or fourteen—than the child of the skilled artizan, for what it lacks in proper food and clothing it makes up for in medical attendance and physic, which the Residuum consumes in really startling quantities. Moreover the degradation into this class is frequently from a standard so high above it as to be practically incommensurable. What comparison can be made between the education of the professional man and the miserable services he can render when he has fallen into the Residuum? Whether we regard the class as reproducing itself, or as largely supplemented from without, it is in either case guilty of an economic blunder; it fails utterly and entirely to regulate its numbers with any reference to its wage-earning powers.

I am inclined to doubt whether this failure is mainly caused by the invasion of its ranks by degraded members of other classes. This upward and downward movement is always going on, and throughout all grades of the industrial organization; no class is so self-contained as to form, strictly speaking, a non-competing group. Perhaps the only real difference is that while other classes give and take, the Residuum only takes. Its members cannot fall lower, and it is seldom indeed that they rise higher. Although the industrial organism is very merciful in the way in which it allows a man chance after chance of proving whether there is any stuff in him, it very rarely succeeds in reabsorbing one of this class; the defects of character are practically ineradicable, at any rate under the present system.

But for the great reason why this class fails in economic elasticity, why it fails to give way before the pressure of circumstances, and why therefore the supply continues to be maintained, we must, I believe, look elsewhere. It is to be found in the fact that it is not self-supporting, that its standard of living is in no way determined by itself, but by the sense of the community to which it belongs, and which for many reasons cannot suffer it to fall below a certain level. And if we need evidence that in England this level is far above that to which the Residuum is capable of falling without danger of actual extermination, we need only question immigrants as to their willingness to return to their native lands.

One reason why they are not self-supporting is to be found in the nature of their employment. It is not so much that their earnings are insufficient to live upon, but that it takes a very high order of intellect to be self-supporting on an intermittent income, and the Residuum is of all classes the least qualified to achieve independence under such conditions.

On the other hand, it is the fact that they are not self-supporting, but are largely subsidized from without, which alone makes it possible for the present system of employment to continue. This, of course, involves us in a vicious circle; but it is characteristic of social problems to be vicious circles, and all that can be done—at any rate on paper—is to point out the links in the chain, and hope that the practical man will some day come along and break through at the weakest place.

One important link we may find in the various sources from which the earnings of this class are subsidized. Of the Poor Law relief and public and private charities (a little within £5,000,000 in London) I need say little here; every one knows more or less accurately that very large sums are distributed among the poorer classes by an expensive machinery, and by far the greater part of this goes to members of the Residuum. Some of them, indeed, the criminal classes, dispense with the machinery, and effect the redistribution for themselves; so far they are less of a burden on the community, but for the most part they are exactly on a level with their weaker brethren; they work occasionally, when they can, and when it is convenient; at other times they help themselves, and live without work.

But large as this recognized subsidy of public and private charity is, I believe it to be unimportant in comparison with the tax levied by the Residuum upon its neighbours. It is very difficult to give any adequate idea of this; it is paid mostly in kind, and comes practically to free board and lodging through a considerable part of the year. So far as concerns free lodging, I can give some actual evidence. I have here a list of twenty-nine families, with a record of their movements during the past three years. During that time two of the families had lived in three houses, eighteen in four houses, seven in five houses, and two in six houses. Of course there is only one reason for these constant removals; that is, arrears of rent. The amount owing in each instance at the last address varies from 15*s*. to £4; further back than that it was impossible to get exact information, but it would very rarely be under 20*s*., and very often over. As an extreme, but by no means an isolated instance, I may cite one which has come under my notice within the last few weeks. Early in 1892 the family took rooms in Holborn, stayed there ten weeks, then left owing £6 for rent; they then took a place in Clerkenwell, stayed there four months, and then moved on into Hoxton, leaving a debt of £7 4*s*. to the landlord; they have been seven weeks at their present address, and already owe over £2, which will certainly never be paid.

It is clear, therefore, that such a list of flittings as I have made represents in itself a large amount of free lodging, quite distinct from the recognized charitable shelters; and I could have increased the list almost indefinitely had it been worth while. The sufferers from this tax are people little better off than the Residuum itself, and the root of the evil lies in the London system of subletting large houses. When a working man with a young family takes a house at £30 a year, and heavy rates and taxes, it is ruin to him to let his extra rooms lie empty, and almost equal ruin to get, as he so often does, a member of the Residuum as tenant. Of course the risk is by this time a fully recognized one, and is covered by the heavy rents which paying tenants have to meet, and which sometimes enable a fortunate householder to live rent free.

The same system of sharing a house among several families is largely responsible for the tax that is levied in food. You will find if you try that it is a practical impossibility to drink your tea if there is someone in the next room who has none to drink. Next door makes all the difference. I have known women live for weeks on the friendly scraps let fall from the landlady's table, and a family of children can always make good an extensive claim. Some striking evidence might also be got from the small general shops which abound in poorer London, if they would only keep books, but they never do. I have no doubt that most of my twenty-nine families have accounts at several shops, of which nothing will be paid until they have exhausted their credit in the neighbourhood, but the shopkeepers themselves are frequently uncertain how much is due to them.

Here is a short statement of the financial position of two representative families.

The first is a man and wife with eight children, the only wage earner at the time in question being the eldest boy. The liabilities of the family amounted to about £6 borrowed in various amounts from a friend, a mother-in-law, a brother and a brother-in-law, £3 15*s*. owing for rent, 30*s*. to the milkman, and an indefinite amount to the provision shop. The family needs for the current week were met by the boy's earnings 7*s*., by pawning the father's boots and the son's best clothes, by a continued free supply of milk from the friendly milkman, and of provisions on credit at the rate of half a pound of butter and four half-quartern loaves per day from the general shop.

The second family consisted of man and wife and four children, whose position was simplified by the fact that the landlady kept a general shop. They owed £4 for rent and 23*s*. for food, and had pawned for £6. Their plan of living was to continue to board and lodge free, the landlady saying that she could not stop them now for fear of losing all they owed; and their only other source of income was an occasional ticket from a neighbouring church.

Now if we take into consideration all these sources, the Poor Law relief, voluntary charity, and the tax levied upon all with whom the Residuum come into immediate contact, we may hesitate to attempt any exact estimate, but we can hardly doubt that the subsidy made towards the support of this class is very large indeed. What are the consequences? The chief, and one which might be most naturally expected, is its almost complete immobility within very narrow limits. No prospect of improvement, not even a promise of regular work, will induce these people to leave a neighbourhood which they have tried, and not found wanting. They have no confidence in themselves, but they have a confidence fully justified in the social arrangements in which they have been developed, and which for them constitutes what they call Providence. Within certain limits, indeed, their life is a constant flitting, but they merely circulate from street to street within a very narrow area, and the causes of their moving have nothing whatever to do with the labour market. No Act of Settlement ever succeeded in establishing an immobility so rigid as this, for Acts of Settlement have at least the merit of kindling a rebellious desire to move.

It may perhaps be said that three years is not very long to test a man's immobility by, but I think it is sufficient for my purpose if we bear in mind that during those three years he has been subjected to a constantly recurring pressure as great as any that ever is likely to be brought to bear upon him; that each removal represents a small crisis, and is the alternative to—if not the actual result of—a forcible ejectment.

Given this immobility of a class of labour of which the earnings are largely subsidized, we have all the conditions which favour the capricious demand for its services which I have noticed. Employers have no need to make sure that their resources are equal to the demand that may be made upon them, for here is an inexhaustible reservoir maintained outside their doors, upon which they can draw at any moment. And that large section of the Residuum, which without skill or strength serve the caprice rather than the needs of their customers, have no need to strengthen their hold upon industry and make themselves indispensable, for they also know that they have inexhaustible resources upon which to fall back.

I cannot leave the subject without alluding to the question of the influence of this class upon those just above it. So far as concerns the labour market I believe that its power for harm has been exaggerated, and that in the long run it competes with the genuine self-supporting wage earners very little, that all questions of wages and hours and employment settle themselves without reference to it. A steady-going efficient workman is never displaced to make room for one of this class, employers know their own interest too well; and it is not until the better supply is exhausted that they have recourse to the Residuum. Like poor land it is only called into use when an increased demand makes it profitable to employ expensive labour, and like poor land it drops out of use again as soon as the demand subsides. And if I may follow the analogy a little further, its utilization is more likely to accompany an increase than a diminution of the return to the more efficient workers.

With regard, again, to the minor service section of the Residuum, we must note that so far as its services are of any value at all, it is the wage earners themselves who reap the advantage. It is their substitute for the domestic servant, and without venturing any opinion as to the desirability of such a substitute we may at least point out that it is harmless in that it does not compete with regulars; it serves a class which must choose between it or nothing, which must either do its own work or employ help fitfully as means will allow.

But though, as I think, the Residuum does not injure the

position of the wage earner by competition in the labour market, it is a drag upon it in a more direct way. I have pointed out to how large an extent it is dependent upon credit and charity, and by far the greater part of this credit and charity is derived from the wage class, either directly or by way of high rents and heavy rates. The cost of living is increased to an extent which is probably far from being compensated for by the services actually endered.

Suppose it possible that by removing the obstacles to the mobility of this class the reservoir of labour could be spread abroad over the country, and gradually reabsorbed into the industrial organism. How would the change make itself felt? In the regular industries there would be less elasticity, less encouragement to season work, more need of organization. The fur trade, for instance, instead of dismissing all but the best hands in the winter, and taking on large numbers of inferior workers for a short summer season, would have to increase its regular staff, and in order to keep them employed to equalise its rate of production throughout the year. A similar policy would have to be observed throughout all industry, and it is not impossible that with greater regularity in production there would be fewer commercial crises.

The alternative to dispersion now urged upon us is organization, but I believe it will prove that the most which organization can do for the Residuum is to define the hopelessness of its position more sharply. We have had a striking illustration of this in the recent movements of dock labour.

After all is said and done, organization is only one amongst many means of self-help; it is impossible to organize dead matter from the outside, and the true Residuum is economically dead. It may be possible to galvanize it into a temporary appearance of life, to raise up a social monster that will be the terror of the community; but the best that can really be hoped for it is that it should gradually wear itself away, or in the coming generation be reabsorbed into the industrial life on which it is at present a mere parasite. And the tendency to this issue must probably be delayed by any scheme, however well devised, which seeks to deal with it in the mass and as a permanent institution, or in any way fosters its reluctance to yield to the pressure of circumstance.

HELEN DENDY

V. Alleged tendency of superior stocks in all classes towards a diminished rate of reproduction.

209. The alleged tendency of the superior stocks in all classes towards a diminished rate of reproduction deserves some notice as one of the possible causes of physical deterioration.

210. So far as this allegation is based upon a diminished birth-rate, a fact general in varying degrees over the whole of Western Europe, there appear in this country at all events to be certain compensating considerations, inasmuch as the Registrar-General's Returns show that among the factors in the reduced birth-rate are: (1) the raising of the age at which marriages are contracted, and (2) a diminution in the number of illegitimate births, both, it will be readily agreed, circumstances tending to the improvement rather than the deterioration of the race.

211. But a more serious view has been taken and put forward by high scientific authority, which, if correct, points to a condition of things which cannot in the long run fail to react adversely on the physical characters of the people.

212. In the Huxley lecture for 1903, Professor Karl Pearson made this announcement:—

"Looking round impassionately from the calm atmosphere of anthropology, I fear there really does exist a lack of leaders of the highest intelligence, in science, in the arts, in trade, even in politics. I do seem to see a want of intelligence in the British merchant, in the British professional man and in the British workman. I believe we have a paucity, just now, of the better intelligences to guide us, and of the moderate intelligences to be successfully guided. The only account we can give of this is that we are ceasing as a nation to breed intelligence as we did fifty to a hundred years ago. The mentally better stock in the nation is not reproducing itself at the same rate as it did of old; the less able, and the less energetic, are more fertile than the better stocks. The only remedy, if one be possible at all, is to alter the relative fertility of the good and the bad stocks in the community. Let us have a census of the effective size of families among the intellectual classes now and a comparison with the effective size of families in the like classes in the first half of last century. You will, I feel certain, find, as in the case of recent like censuses in America, that the intellectual classes are now scarcely reproducing their own numbers, and are very far from keeping pace with the total growth of the nation. Compare in another such census the fertility of the more intelligent working man with that of the uneducated hand labourer. You will, I again feel certain, find that grave changes have taken place in relative fertility during the last forty years. We stand, I venture to think, at the commencement of an epoch which will be marked by a great dearth of ability.

* * * * * * * * * * * * *

The remedy lies first in getting the intellectual section of our nation to realize that intelligence can be aided and be trained, but no training or education can *create* it. You must breed it: that is the broad result for statecraft which flows from the equality in inheritance of the psychical and the physical characters in man."

213. The Committee have not been able to obtain decided confirmation of this view. Professor Cunningham, to whom the point was referred, said:—

"I think that the statement is a pure assumption. I do not know how we can possibly measure this supposed loss of inherited intelligence—we are dealing with inherited intelligence because all his remarks refer to inherited intelligence—and I do not think there is a single solid fact in support of such a view. I am astonished that one for whom I entertain so high an admiration as Professor Pearson should have put forward such a statement, and more especially claim for it, as he does, that it emerges from the 'calm atmosphere' which is supposed to surround the anthropologist." Cunningham, 227(

Adding, in reply to a further question:—

"It should be borne in mind that it is stocks and not classes which breed men of intellect. These intellectual stocks are found in all classes, high and low. No class can claim intellect as its special perquisite. This is a fortunate circumstance, seeing that the conditions which affect the degree of fertility in the higher classes are not as a rule present in the lower classes. The conditions under which genius or outstanding ability appears are peculiar and very little understood. It likewise has a residence, I believe, in no special class and very probably in no special stock. It is not improbable that the physical conditions upon which genius depends may not, in certain cases, be far removed from the domain of pathology." Cunningham, 227

214. Dr. Arthur Shadwell did not think that, beyond the general fact mentioned at the outset, there was ground for the belief that the more capable among the working classes are not reproducing themselves at least in the same proportion as those less capable of putting into existence persons who are physically fit; indeed, vigour seemed to him to imply reproduction—meaning a proper fulfilment of the natural functions of which reproduction is one. On the other hand Mr. Gray, a competent witness, said:— Shadwell, 12230-

"Anything which decreases the difference between the birth-rate and the death-rate among the superior classes and increases this difference among the lower classes tends to produce a progressive deterioration of the average national physique. The tendency of the population in modern industrial communities to concentrate itself in large towns and the increase of wealth appear to have the effect of reducing the birth-rate of the superior classes and of decreasing the death-rate of the inferior classes. It has been established by taking a census of the size of the families of the professional classes in the United States of America, that there has been a great decrease in the size of their families within recent times, and that these intellectual classes are now barely reproducing their numbers." Gray, 3267.

And Sir John Gorst was impressed by the idea that the race is propagated in the greatest proportion by the least fit part of it, the restraints on marriage disappearing as you reached the most unfit.

215. The question is one which can be settled only when brought to the test of definite figures, and in view of the statements made by Professor Pearson, it might be as well if here, as in America, steps were taken to obtain, by means of a proper census, accurate information on the point.

VI. Food.

216. A striking consensus of opinion was elicited as to the effects of improper or insufficient food in determining physique, and this factor was acknowledged by every witness to be prominent among the causes to which degenerative tendencies might be assigned, though in one or two cases its relative importance was thought liable to exaggeration. It is in connection with the young and the conditions affecting their growth and development that the matter must be studied under its most serious aspect, and this will be dealt with in the succeeding section of the report, but in the meantime the Committee desire Rees, 4283-8.

to consider the question in its general relation to the health of the people at large.

Close, 2556-2560, 2636-9, 2774-2780, 2796-2829.
Collie, 4041-4055.
Booth, 1087-8.
Eichholz, 436.
Fosbroke, 6691-5.

Booth, 975-980.
Smyth, 1221-2.
Wilson, 2005-6.
Malins, 3149.
Dowding, 4884.
Ashby, 8686-9.
Garnett, 9265-9274.

217. Into the circumstances affecting this aspect of the subject, the habits of the community enter largely, and these it is obvious are open to a great variety of influences, moral, industrial, and economic. If, as one witness emphatically stated, with the support more or less marked of others, a large proportion of British housewives are tainted with incurable laziness and distaste for the obligations of domestic life, they will naturally have recourse to such expedients in providing food for their families as involve them in least trouble; if, as many contended, the effect of female labour in factory and workshop is to form bad wives and mothers, any changes in the industrial conditions of the people that increase the demand for this class of labour react upon the comfort of the home, and the economic causes that have contributed of recent years to the production of tinned foods in enormous quantities have had the effect of reducing the amount of home-cooking. Changes, moreover, in the relative cost of certain articles of food, whether brought about by economic or fiscal causes, give a direction to popular taste, which it may take generations of educative influence to correct.

218. If competent witnesses are to be believed, the progressive decrease in the price of certain articles of common consumption, such as tea, has not been without consequences of a prejudicial character. Dr. Purdon, of Belfast, is quoted by Dr. Young with entire approval in saying—

Young, 2093.

> "People are living so much on tea and white bread and jam, instead of oatmeal and milk. They are using these cheap jams, and there is the white bread which is supplied so cheaply, which does not contain all the elements of nutrition. In former times the children used to live on oatmeal and butter-milk and potatoes, and the country children still live on that, but the town children live more on tea, and this white bread and jam. Dr. Purdon considers—and I quite agree with him—that that has a very decidedly deteriorating influence."

Hutchison, 10082-6.
Cameron, 10095-7.

Evidence was, however, given that white bread properly made was as rich in nutritive properties as any form of brown bread, and the Committee must not be understood to express any opinion on the relative merits of either.

Hutchison, 9991.

219. Dr. Robert Hutchison, whose authority on this point is unquestioned, said—

> "If I were asked to state the chief fault in the diet of the working classes of this country, I should say it is the excessive use of tea and bread."

Young, 2093-4.
Cunningham, 2298.
Kelly, 11284, 11374-9.
Lyttelton, 5531.
Bostock, 7484-8.

Hutchison, 9995.

It is not so much the actual deleterious effect of tea, though on that point much evidence was given to show that in the form in which it is generally consumed it produces anaemia and neurosis, as that the money spent upon it might with much greater profit be spent on other things. Dr. Hutchison went on to say he would like to see porridge and milk substituted for bread and jam. The Bishop of Ross had the same story to tell of the Irish peasantry who have come to regard bread and tea as a higher and better class of food than potatoes and milk, or porridge and milk; a sentiment which he describes as growing more obdurate every day.

Mackenzie, 6814-6827.
Hutchison, 9957-9978.

220. Both Drs. Mackenzie and Hutchison laid great stress on balance in the formation of a scientific dietary, as well as on quantity. Dr. Hutchison explained that whereas the physiological standard for a man of average weight demands 3,500 units of energy, and for the upkeep of his body 125 grammes of the chemical substance called proteid, that is to say, nitrogenous material found in certain foods, observation showed the diet of the Edinburgh labourer deficient by 280 units of energy and by 13 grammes of proteid; while in York the average diet, according to Mr. Rowntree, was 17 per cent. below standard in energy and 29 per cent. too low in proteids. The consequence of the deficient supply of energy is to lower physical efficiency and reduce the amount of work of which the man is capable. Where the amount available for expenditure upon food is limited, it is of special importance to get the proper materials, and it is in selection that the defects of English habits come out most prominently. As one witness said—

Eccles, 10726.

> "They do not understand how to provide themselves with proper food. They may have plenty of food, but it is not proper food; it is tinned food, easily got, and badly prepared."

221. The question of tea as an article of general consumption requires a little further notice. In a statement furnished to the Committee, relating to the physical condition of the working class children in Ancoats, one of the poorest districts of Manchester, these words are used—

"*Another fruitful and one of the most unsuspected causes of deterioration lies in the long ingrained habit of tea drinking at breakfast and other times in the factories and foundries of the City.* Tea drinking, if it really were so, might not be harmful, but unfortunately the mixture drunk can hardly be called tea at all. More frequently than not boiling water is poured on to too large an amount of poor tea leaves and is left to stand until the tea has become almost a stew, and this dark and nasty mixture is drunk, sometimes three and four times a day, by hundreds of young lads, setting up frequently various forms of varicocele, and is responsible for several kindred evils (excessive costiveness, etc.) We were informed by the late Chief Recruiting Officer in Manchester some time ago that a very large proportion of young men rejected for the Army had been refused on account of ailments brought about by this practice." Appendix XXII.

222. Mr. Atkins also referred to the observations of Colonel Leetham on this point. He found that a large proportion of those rejected for enlistment were heavy tea drinkers who suffered from one form or another of varicocele, and were, in consequence, unfitted for long marches. Atkins, 2954.

223. Dr. Hawkes' experience of female workers employed in factories and workshops in Finsbury pointed to the same abuse of tea. In the case of many of these, tea is the only thing consumed before starting to their work from places in remote parts of the suburbs. During some years' work at a large Metropolitan Dispensary, he found that 80 per cent. of women and girls who came under his notice never touched solid food till the middle of the day: pickles and vinegar were then often the staple of the "solid" meal with tea, and tea again in the afternoon; three or four pints of "tea poison" being thus absorbed in the course of the day. An enormous amount of dyspepsia is thus set up, which rapidly assumes acute forms, with the result that alcohol, at first taken to allay pain, is frequently the final refuge. Hawkes, 13110–13125.

224. With indifference as to the proper distribution of meals and gross ignorance of the right selection of foods required for them, a very general disinclination prevails to spend sufficient money upon food, even among those with ample resources. It is no doubt the case that with greater knowledge the poor might live much more cheaply than they do, but with all classes the tendency appears to be to spend as little as possible on food. Young men are said to take better care of themselves in this respect than the other sex, as they are not under the temptation to spend money in personal adornment, an object for which women will, it is said, stint themselves in food to a terrible degree, and sacrifice many other necessaries of life. Hawkes, 13109, 13126–9.

225. The desire for pleasure is stated to be another fruitful cause of the withdrawal from working class budgets of money that should be devoted to the purchase of food. Mr. Eccles said—

"There is a vast deal more pleasure amongst the working classes now than there was—I mean pleasure which requires money wasted on it." Eccles, 10725.

Other witnesses spoke of the taint with which the love of amusement was infecting large sections of the population, especially amusement in the form of cheap excitement, the desire for some sort of sensation, comparable to the aforesaid dietary of pickles and vinegar. Close, 2569.

226. Betting figures prominently among pleasures of this kind, and much privation in the homes of workmen is due to losses so incurred, while on the other hand, if they win, the family dietary is not improved, but the proceeds of the successful gamble go in drink. The taste for betting is unhappily not confined to men; women, and even children, according to Dr. Hawkes, being very ready to partake of this form of excitement. Loch, 10187–9. Hawkes, 13091–13100.

227. In estimating the causes which contribute to render the poor careless of, or indifferent to, the conditions of proper feeding, it is only fair to remember the extreme narrowness and squalor of their surroundings. Under any aspect of this multiform problem, it is difficult to keep away for long from the housing difficulty, and it enters very largely into the matter immediately under discussion; houses that were originally built for one family have, by the operation of pressure upon limited space and high rents, become occupied by several, but often only one room in the whole house contains a grate of proper service for cooking, with the result that a large number of tenements do not contain the requisite apparatus for the preparation of food, and the culinary art, if practised at all, is reduced to its crudest form of expression. Looking to the large rents that accrue to the proprietors of these premises by their sub-division into so many separate tenements, it does not seem too much to require that every tenement let for the occupation of a family should include a grate suitable for cooking.

Close, 2569.

228. The extent to which tinned food has superseded other kinds was the subject of some emphatic comment by a lady who claimed to know much of the habits of the people at home and abroad. In one farmhouse, notwithstanding the existence of a well-stocked garden, there was nothing but tinned vegetables and tinned fruits besides tinned food for breakfast and dinner. This she attributed partly to laziness and partly to ignorance. The nearer you come to London the more infamous, in her opinion, the food and cookery, young women caring for nothing but pleasure and amusement; their failure to make the home attractive was, she believed, one of the causes of the rural exodus, and while on this subject she described the conditions of squalor and discomfort with which she was familiar.

Lyttelton, 5375–5382.

229. It has already been indicated in an earlier section of this report that for the purpose of bringing home to the people the importance of properly selected and carefully prepared food there is much room for training of a socially educative character among girls and young women. Both Mrs. Close, the witness just quoted, and Mrs. Arthur Lyttelton agreed that the teaching at present imparted in schools was of little value, partly because of the unpractical character given to it by some managers, but mainly because of the early age at which children leave school. The evidence, however, on this point is somewhat conflicting. Mrs. Lyttelton was of opinion that something of a different nature was perhaps more wanted, and instanced the tenement classes that have been established in New York with some success. She would like to see the people instructed in cookery in their own houses or in kitchens fitted up like those within their experience, and employing the simplest utensils; by these means she believed they could be familiarised with processes for using up vegetable and garden stuff, which would tend to economical and wholesome living. Both she and Dr. Hutchison thought much might be done by mothers' meetings and lectures conducted on simple and practical lines, which might be supplemented, after interest had beeen awakened, by the distribution of leaflets, undertaken by health visitors in the employment of the local authority or acting in co-operation with it.

Lyttelton, 5391–5402, Hutchison, 9937–9941, 9979–9981.

230. The Committee believe that all these methods are valuable: nor would the question of cost form an obstacle. In the first place it would be small, and secondly after the lapse of a few years unnecessary, as once a generation of competent mothers and housewives had been brought into being, a family tradition would be created which would contribute to preserve higher ideals of domestic comfort and better standards of life. To these ends the Committee think that a further step might be taken within the near future, and continuation classes for domestic instruction organized, at which the attendance of girls who have left school should be made obligatory twice a week during certain months of the year. The courses of instruction at such

classes should cover every branch of domestic hygiene, including the preparation of food, the practice of household cleanliness, the tendance and feeding of young children, the proper requirements of a family as to clothing, everything in short that would equip a young girl for the duties of a housewife.

231. There is no reason why a judicious discretion should not be permitted in the enforcement of compulsory attendance at such classes: girls in domestic service, for example, might properly be exempt, as the Committee were informed that comfortable homes are as a rule to be found among the working classes where the wives have had the advantage of this training; factory operatives are, on the other hand, said to make the worst wives, and to facilitate their attendance at such classes some modification of their hours of work might be introduced.

232. Further reference to the teaching of Cookery in Schools will be made in the next section, but it may be as well to lay down here the principles by which it should be guided so as to form a preparation for Continuation Classes. The teaching of cookery should be directed to the selection, economy, and preparation of the material best suited to the needs of the poorer classes, including the requirements of young children. It should have for its object the constant repetition of those processes most in request in cottage households, with a view to impressing them as firmly as possible on the minds of the scholars, and care should be taken to use such apparatus and utensils as under favourable conditions are likely to be found in the houses of the poor. The syllabus of instruction should be drawn up by some one with immediate knowledge of the wants of the class from which the children under instruction are drawn; it should vary for urban and rural schools; it should be modest in scope but thorough in application, attentive to detail, but yet based on some broad principle of domestic effectiveness; and, above all, it should provide as much practical work as possible, to the exclusion of mere excursions into theory and demonstrations which tire without exciting interest. Wherever practicable the material cooked should be served and eaten in the presence of those that have prepared it, and in all circumstances the greatest prominence should be given to the utility of the task upon which the scholars have been engaged. For this reason no scientific terminology should be introduced into lessons on the chemistry of food, but the practical value of the different articles of diet should be stated in the simplest and homeliest language.

233. No evidence was taken on Adulteration, but there is no doubt that the opportunities offered to the adulterator by the change from the home production of many articles of food to supply through the channels of trade has had a deleterious effect on public health. The Committee's attention has been called to the fact that the Board of Agriculture, under the powers conferred on it by the Act of 1899, has fixed standards of purity for milk and butter, and they cannot but think it highly expedient that the Local Government Board should, with proper expert assistance, be authorised to fix some standard, which should of course be a reasonably high one, for all foods and drinks. This has now to be done by the Public Analyst, subject very often to prolonged and costly police proceedings, and the standard arrived at necessarily differs according to the views of individuals and the strength of evidence in a particular prosecution.

234. It appears to the Committee that in regard to food, as in other matters, there is something wanting to the ideal of the Local Government Board as a department of public health, and that it is desirable that this aspect of the Board's administrative functions should receive greater attention.

235. The Committee are aware that there is nothing heroic about these remedies, but they believe that by combined pressure in the directions in-

dicated, and the enlistment of an active public sentiment in their favour, much might be done in a few years to reduce evils which are not only a standing reproach to certain classes of the nation, but constitute a serious menace to its general well-being.

VII. Conditions Attending Life of Juvenile Population.

i. *Introductory.—Infant Mortality.*

236. Enough has already been said in the course of this Report to demonstrate the nature and extent of the waste that goes on under the name of "infant mortality," but before discussing the subject matter of the present section under certain particular aspects, the broad facts of this annual sacrifice, and some conclusions to which they point, require a little separate notice.

237. Among the more highly organised nations, where the tendency to a decrease in the birth-rate becomes more or less noticeable, the means by which infant mortality can be averted present a social problem of the first importance. Unfortunately in the volume of vital statistics, from which so many consolatory reflections are drawn, infant mortality remains a dark page. The Registrar-General in his last report lays down that "the mortality among infants and very young children has always been regarded as a valuable test of salubrity." Since the date of that report, Dr. Tatham has caused to be prepared, at the Committee's request, certain interesting statistics respecting infantile mortality in England and Wales, at two quinquennial periods a quarter of a century apart. The tables, six in number, show the mortality among male and female infants separately—(a) in town as distinguished from country; (b) among legitimate as distinguished from illegitimate infants. He has also furnished the Committee with a careful analysis of the tabular returns, and as this analysis is very full, it is only necessary to refer to the appendix, where will be found the six tables referred to, and also Dr. Tatham's remarks upon them.

Appendix VA.

238. Three facts stand out prominently as the result of this investigation: first, that infantile mortality in this country has not decreased materially during the last twenty-five years, notwithstanding that the general death-rate has fallen considerably; secondly, that the mortality among illegitimate children is enormously greater than among children born in wedlock; thirdly, that about one-half the mortality occurs in the first three months of life.

239. The evidence furnished by a variety of witnesses is confirmatory of the conclusions to which these figures point. According to Dr. Chalmers of Glasgow, where the infant death-rate varies in different districts from 63 to 217 per 1,000 births (that for the whole city being 141), "Quite one-third of the infant deaths occur in the first four weeks of life." In certain parts of London, where the proportion of persons living in one or two-roomed tenements exceeds 35 per cent., it has already been seen that this death-rate reaches 223. Preston, with nearly half its female population occupied (30 per cent. being married), has an average rate for the ten years to 1900 of 236, Burnley of 210, and Blackburn, with a still higher percentage of married women employed, of 200. In Sheffield, where the general rate in 1901 was 201, it went up in one district to 234, and there is no doubt that, in parts of the country, during certain seasons of the year, such rates are vastly exceeded.

Chalmers, 5958.

Appendix XIII., 4.

Appendix V., 26.

Greenwood, 8233.

240. Coming to particulars illustrative of these conditions, Mr. Wilson speaking of Dundee, said: "It was quite a common thing to find a woman had had as many as thirteen children, and had lost eleven or twelve out of that number, in some cases the whole of them." Mrs. Greenwood, in a paper submitted relating to Sheffield, says: "One woman I know has buried seventeen out of eighteen children, another has had sixteen sons, of whom only six are living," though neither of them had worked out of their homes since

Wilson, 1952.

Appendix XX.

marriage. Nor, in her opinion, is such waste necessarily connected with poverty; deaths from diseases of the respiratory organs constantly occur owing to the practice of even well-to-do mothers exposing imperfectly-clad children to cold while gossiping with neighbours. Among a certain number of mothers questioned in Hanley and Longton, it appeared that 38 per cent. of the children born to them had died in infancy, and the state of the case in many Lancashire towns is certainly no better; thus in Burnley one woman is said to have had twenty children and buried sixteen, all having died between one and eleven months of age; in this case the father was a collier in good wages and the mother stayed at home. At Accrington things have become so bad as to lead to vigorous action on the part of a conference convened by the Mayor, as the result of which exhaustive enquiry into the conditions under which children are brought up has been demanded.

Appendix V., 35.
Appendix V., 49.

241. In a detailed discussion on the subject of extreme value to those who may undertake further investigation, Miss Anderson was good enough to supplement her original evidence, and the Committee fully endorse the conclusions at which she arrives. The connexion between infant mortality and (i) bad or insufficient feeding, and (ii) the overcrowding of one or two-roomed tenements, is no doubt established, but though the facts seem to point to a strong presumption that it is also connected with the employment of mothers, the information is not so complete as might be desired. Miss Anderson indicates two directions in which this information should be sought: (i) localisation of the infant mortality rates in a systematic way for particular areas in industrial towns where the workers of selected industries live; and (ii) general infant mortality rates for selected industries throughout the country.

Appendix V.
Appendix V, 25-7.

242. The Registrar-General's records, which she points out might be available in throwing light on this important question, are confined to showing the occupations of mothers of illegitimate infants, and the Committee think that it would be well if they could be expanded to show the occupations (if any) of married mothers. In this connexion the Committee desire to call attention to a point of considerable importance. At the Brussels Congress on Hygiene in 1903, the absence of any registration of still-births in Great Britain was noticed, and the consequent difficulty of obtaining the complete figures as to infant mortality. Every witness who was questioned on the subject agreed in deploring the present neglect, and the Committee are emphatically of opinion that still-births should be registered, as apart from the advantages a system of registration would have in making it easier to bring home instances of mal-practice, a knowledge of the facts as to the frequency of still-births would be of great value towards elucidating the causes of infant mortality by throwing light on the ante-natal conditions prejudicial to the survival of the fœtus.

Appendix V, 11.
Scott 1753-4.
Cunningham, 2293, 4
Malins, 3142-4.
Chalmers, 5957-8.
Niven, 6422-3.
Greenwood, 8173.

243. With a view to the protection of infant life, another point arises on which it would be expedient to amend the Registration Laws. By the system under which deaths are registered, a medical certificate as to the cause of death is not absolutely essential to the registration of a death, or the burial of a dead body, although the certificate is always asked for by the Registrar. Having regard to the heavy mortality affecting children of tender years, the Committee are of opinion that the death of no child should be registered in the absence of a medical certificate as to the cause of death. Every witness who touched the subject was strongly in favour of this amendment of the law as the best means of bringing home to the careless parent the consequences of culpable neglect.

Malins, 3142-6.
Collie, 3918-3923.
Dowding, 4915-9.
Greenwood, 8171, 2.

244. In this connexion the subject of infant insurance was also considered. As to the propriety of interfering with this practice different opinions were expressed, though it was the general view that it contributed to parental negligence. On the whole it was thought that if restricted so as to cover the actual expenses of burial its principal abuses would disappear. The evidence of Sir Lambert Ormsby, President of the Royal College of

Booth, 1616-9.
Scott, 1749.
Atkins, 2944-6.
Malins, 3136-3141.
Collie, 3916, 7.
Smith, 8497-9.

Ormsby, 12573-12591. Surgeons in Ireland, upon Irish practice in this regard pointed to the prevalence of a very low view on the part of many medical men in respect of their obligations toward the security of infant life under the conditions touching insurance in that country.

245. The Committee do not think that upon the evidence they are in a position to make any definite recommendation on this point, but they consider that the operation of the practice should be carefully watched.

ii.—*Hereditary Taint.*

246. So far as the Committee are in a position to judge, the influence of heredity in the form of the transmission of any direct taint is not a considerable factor in the production of degenerates.

247. Prof. Cunningham's views, that inferior bodily characters, the result
Mackenzie, 6742-6781, 6900-2. of poverty and not of vice are not transmissible, were confirmed by Dr. Mackenzie, who at the outset of his interesting evidence elaborated a distinction between inheritable characters and their environmental modifications, the result of these last being imposed on the individual by his life history and
Eichholz, 556-560. not therefore transmissible to the offspring. Dr. Eichholz was disposed to go further and sought to explain how some mysterious law of transmitted impulse made for the recuperation of each generation, the unborn child fighting strenuously for its own health at the expense of the mother and arriving in the world with a full chance of living a normal physical existence. This view he supported by the assertion that the number of children born healthy in the worst districts was very great, he himself putting it at not less
Ashby, 8671-7 than 90 per cent. Dr. Ashby thought this was only partially true, as nature too often failed in its effort, and Dr. Mackenzie would not even concede so much, as investigations into the effect of food on guinea-pigs during pregnancy had, he said, shown that the embryo suffers in greater proportion than the mother. He quoted the opinion of Dr. Noel Paton that "the nourishment of the maternal tissues seems to take precedence over the nutrition of the fœtus."

248. The Committee deemed it advisable to hear on this point Dr. Edward Malins, President of the Obstetrical Society of London, and Professor of Midwifery in the University of Birmingham, who thought the testimony of experienced persons was on the whole in accordance with the views expressed by Dr. Eichholz. He would say that from 80 to 85 per cent. of children were born physically healthy, whatever the condition of the mother might be antecedently, so far confirming the opinion that nature intends all
Malins, 3123-9, 3203-5. to have a fair start. Dr. Malins kindly undertook on behalf of the Obstetrical Society to institute an enquiry among the Lying-in-Charities and Hospitals in London which should furnish information on these facts; this enquiry is unfortunately not complete. The Committee were, however, supplied by the courtesy of Dr. Eichholz with evidence which did tend to establish this conclusion from the Medical Officers working for the Royal Maternity Charity, and from the Paddington and Kensington Workhouse Infirmaries.

249. The Committee cannot ignore these opinions though it may well be that the depressing effects of the life-struggle on parents are, nevertheless, in some measure transmitted to the offspring. At any rate some vulnerability towards disease may co-exist with a superficially healthy appearance, and granted unfavourable environment the seeds of degeneration are not long in producing a rank harvest. The consolation of the doctrine lies in the encouragement it gives to working for the removal of the causes which are prejudicial to the health of each successive generation, an encouragement which is immensely strengthened by the concurrent testimony of all concerned as to the immediate effect upon growth and development brought about by the withdrawal of even the most unpromising material from noxious surroundings.

250. It must be remembered that even Professor Cunningham, while denying the influence of heredity in most cases, expressed the firm belief that diseases such as syphilis and alcoholism transmit their effects to the third and fourth generation—and in this opinion the Committee fully concur. Cunningham, 2210.

iii.—*Employment of Mothers late in Pregnancy and too soon after Childbirth.*

251. A very general agreement was expressed that the factory employment of mothers had a bad effect on the offspring, both direct and indirect, but opinions differed as to the extent of the evil and the practical steps that could be taken to remedy it. It is to be found in the most acute form in the Pottery districts and in textile mills. Speaking from an extensive experience of the Potteries, Miss Garnett declared that married women's labour was really the root of all the mischief; the children are born very weakly they are improperly fed, and placed in the charge of incapable people. She admitted the impossibility of interference by any general prohibition, but thought the period during which women are not permitted to return to work after their confinement should be extended. Drs. Scott and Young, certifying factory surgeons of long standing, took the same view. Miss Anderson acknowledged the evil in evidence, but her appreciation of all the conditions of the problem led her to pause before subscribing to the prudence of any legislative change. As the result of further consideration the Committee were favoured with the Memorandum already described, in which the whole subject is discussed with a fulness of detail and wealth of information that have rendered it unnecessary to go further for the basis of the following paragraphs. Garnett, 9015-9034, 9116-9122, 9345-9367. Scott 1834-8, 1888 1901. Young 2169-2172. Anderson 1590-1613 Appendix V.

252. The existing law requires that no occupier of a factory shall knowingly allow a woman to be employed within four weeks after she has given birth to a child. Thus no legal offence arises unless the occupier, with a full knowledge of the facts, is yet responsible for the employment, a situation which, in the ordinary conditions attending factory labour, it is almost impossible to prove. It is needless to say that in these circumstances prosecutions are infrequent or abortive, and though there may be a pretty uniform observation of the law, cases in which it is broken are numerous in some districts, amounting it is thought to general evasion. Appendix V., 3.

253. It has already been indicated that the extent to which the mothers of young children are employed in factories can only be roughly determined, nor are the means forthcoming for an accurate judgment of its effects on infant mortality, but the enquiries of three members of Miss Anderson's staff—in Dundee (jute), Preston, Burnley, and Blackburn (cotton), Hanley and Longton (potteries)—did lead to their noting in those towns "A very striking degree and amount of ignorance of maternal duties, especially of feeding and cleanliness." In Blackburn, Preston, and Burnley it is estimated that of women employed, 37·9 per cent., 30·5 per cent., and 33·5 per cent. respectively are married or widowed. In the several towns of the Pottery district the proportion varies from 10 to 20 per cent. The three ladies quoted by Miss Anderson were unanimous as to the stress and strain involved in the "employment of women from girlhood, all through married life, and through child-bearing"; the decreasing physical capacity of the child-bearing woman brings her at last some relief at the hands of the manager of the mill, and she is sent away, often to take up the equally unsuitable occupation of charwoman or house-scrubber. In this connection Miss Paterson says— Appendix V., 14. Appendix V., 16.

"Great harm is done and suffering occasioned to the women by their remaining at work too long before confinement as well as by their returning too soon after it. Factory managers, doctors, health visitors, and workers themselves are agreed that the four weeks absence is often shortened to three or even less." Appendix V., 31.

254. As an instance of the lack of sufficient care for the children that follows, a subject on which she was able to form very definite ideas, she gave particulars of 144 cases where the health visitors found two, three

Appendix V., 32. or more very young children left alone in the house (in some cases locked in) while the mother was at the mill, with only such food as the mother could prepare over night or in the early morning before leaving. Definite arrangements with another woman to take charge of the children seem less common in Dundee than in Lancashire and the Potteries. It was stated by two doctors in Preston that a large number of premature births were attributable to continued work in the mill during pregnancy, and members of the profession in Blackburn held that the evil of the employment of women during that period was aggravated by their desire to earn as much as possible before they are forced to give up work. In Preston it is found that factory operatives are in the habit of nursing their babies at meal times, and before and after the day's work at the mill; but in Burnley this is exceptional, while greater ignorance and unintentional cruelty in the giving of unsuitable food seem to be common. In the Potteries Miss Martindale notes the effects on infant life of improper feeding and the "appalling ignorance and objections to being taught."

(Appendix V., 34; Appendix V., 35.)

255. Coming to the causes that render the labour of mothers necessary, Miss Anderson enumerated them thus:

Appendix V., 37.

(i.) Death of father, or lack of employment, or inadequacy of father's wage.

(ii.) Desertion of father.

(iii.) Fear on mother's part of loss of future work in factory.

(iv.) Preference for factory over domestic work.

Miss Anderson points out that some of these causes may be traced to the concentration of women's industries in districts where there is absence of men's occupations, and, so far as these are due to economic and social circumstances not immediately alterable, she considers that more may be effected, and the operation attended with fewer counterbalancing disadvantages, by working towards the mitigation of the results of the employment of mothers than by any attempted diversion thereof on a large scale. The great majority of cases may be said to fall under one or other of the three sub-divisions in class (i.).

256. Dundee and Preston resemble each other in scarcity of employment for men.

Appendix V., 40. Miss Squire, speaking of Preston, says,

"The men are said to look out for a wife who is a four loom weaver, and they have the reputation of being lazy."

It was also found that the husbands of cotton operatives were chiefly employed as labourers in intermittent work, while

"in all cases when the husband was in regular employment as weaver, platelayer, painter, bricklayer, etc., the one wage was insufficient to keep the family at the standard of life they expect."

It must be admitted that the standard is somewhat exacting, as the husband insists on having his spending money whatever the household needs may be, and, therefore, the mother's wage, over which she has control herself, proves a valuable accessory.

Appendix V., 45. 257. Miss Squire proceeds,

"The general opinion, among those best qualified to judge, seems to be that the working classes are well off, and that if it was not for the proverbial improvidence of the cotton operatives there would be no poverty. Still, it seems to be the practice for the women to continue their work in the mill as near to the time of confinement as the manager will allow; always the same complaint being made that he had to keep watch and tell the woman that she must cease work."

Appendix V., 47. 258. In Blackburn, Miss Squire finds a high standard of life among textile operatives; comfortable homes, and money to spend on excursions, holidays and amusements, are considered essential. Of Burnley, it is said that

"The infants are of a miserable, debased type in a large number of cases. Whereas, in Preston, the important point seems to be that the infants should be properly fed, in Burnley it seems as if no amount of nourishment could build up a healthy child." Appendix V., 48.

Poverty and desertion are there the causes of the mother's early return.

259. In the Potteries the lack of regular employment for men is a serious element in the situation. According to Miss Martindale, the early return to work is in very many cases prompted by necessity, but she goes on to say, Appendix V., 50.

"It is impossible not to be impressed by the universal preference amongst the women for factory over domestic life, and how depressed and out of health they became if they were obliged to remain at home."

Analysing the causes of this preference for factory life, Miss Martindale adds:

"Surprising as this appears at first, it becomes less so on consideration. At 13 years of age the majority of these women would have begun to work in a factory, to handle their own earnings, to mix with a large number of people with all the excitement and gossip of factory life. They would thus in most cases grow up entirely ignorant of everything pertaining to domesticity. After marriage, therefore, it is hardly probable that they would willingly relinquish this life to undertake work of which they are in so large measure ignorant, and which is robbed of all that is to them pleasant and exciting. Until as girls they have been taught to find a pleasure in domestic work, and until there is a greater supply of healthy and suitable recreations and amusements in the reach of all women, to counteract the prevailing squalor and gloom of these Pottery towns, it is useless to expect them to relinquish factory life."

260. The Committee have no doubt that the employment of mothers in factories is attended by evil consequences both to themselves and their children, and they would gladly see it diminished if not altogether discontinued; but in approaching a discussion of the remedies that have been suggested, they are convinced of the necessity of extreme caution. They have been reminded of (1) the enormous practical difficulties that would accompany any sort of legal prohibition: (2) the existence of a considerable number of unmarried mothers without means of support, whose main chance of rescue from degradation lies in the fact that they desire to labour and know they ought to labour in support of their infants; and (3) the presence in certain populous industrial districts of a large proportion of married mothers who are necessarily the chief breadwinners of their families, and the danger that, if deprived of the opportunity of earning a wage, means will be taken to prevent these families coming into existence; they also feel that the right of married women to the fruits of their own labour, secured to them under the Married Women's Property Act, ought not to be lightly interfered with. Appendix V, 52.

261. Even Miss Garnett, who was most strenuous in condemning the whole practice, admitted that the time was not ripe for prohibition. Among other witnesses, Dr. Scott would prohibit all employment of married women in factories, unless the children are absolutely dependent on their wages. In that case he thought the municipality must make provision for the care of the children while the mother is at work. Mr. Wilson and Dr. Malins agreed that there would then be good cause for such assistance. Dr. Young would extend the period of exemption to two months before confinement (or three in some cases) and three months afterwards. (In Switzerland the period is two months both before and after confinement.) Dr. Ashby would extend it as far as possible, but saw the practical difficulty in regard to pregnancy that the exact date of the child's birth is not easy to predict, and that this would be a loophole of evasion for both employer and employed. He instanced illegitimate children as suffering most, because the mother so often has to go to work. He also discussed the question of licensed women for the care of infants when mothers went to work, but preferred the crêche.

Garnett, 9118. Scott, 1834-8, 1888-1901. Wilson, 2039-2042. Malins, 3210-2. Young 2169-2172. Ashby 8678-8689, 8785-6, 8800-9.

262. The Committee are not able to recommend prohibition, nor do they advocate any uniform extension of the period, but it does seem to them that the application of the law, as it stands, might be made more regular and elastic.

174. G

The extension of the period at present prescribed, which is less than that in many European countries, would probably be made the occasion for more wholesale evasion, and the Committee think the alternative lies between (*a*) maintaining a hard and fast line and placing upon the employer the burden of obtaining proof that the required period has elapsed since the confinement of the woman he employs, and (*b*) exacting in all cases of pregnant or recently delivered woman (1) a medical certificate from an appointed person, say the Certifying Surgeon of the district, that they can be employed without serious prejudice to their physical well-being, and (2) proof that in their absence reasonable provision is made for the care of their infants, which might take the form of a crêche established under municipal or private management, or be secured by the recognition for the purpose of a duly licensed body of women. The latter plan seems open to the least objection, provided a proper licence is assured, but the two might subsist side by side.

Appendix V, 55.

263. Another source of amelioration is suggested by Miss Anderson, who notes the general neglect of voluntary agencies for helping mothers before, during and after confinement, to take care of infant life, even where such agencies exist. In Lancashire, where, it is said, insurances of all kinds abound, no form of provident society exists to which women could contribute while still able to earn wages, nor has any attempt been made to organize a maternity fund, towards which both employer and employed might contribute. The existence of such a fund at Mulhausen is said to have resulted in the reduction of infant mortality by half. The Committee would strongly urge the adoption of such methods of voluntary assistance, and think it not improbable that endowments may be found in many places which could be utilised as the nucleus for a considerable amount of charitable effort in this direction.

iv.—*Decrease in Breast Feeding—Defective Milk Supply.*

Rowntree, 4980-6, Hutchison, 9909, 10018.

Malins, 3131-4. Smyth, 1239. Collie, 3909.

264. A decrease at the present time in breast feeding is generally admitted to be the case in all classes of society, at any rate in the urban districts. With the poor, it seems fair to say that their failure in this respect is due to inability rather than unwillingness, especially in view of the fact that as long as it can be properly continued breast feeding is much the most economical way of nourishing an infant. It is, however, no doubt, the case that women are often unwilling to nurse their own children, because it interferes with their going to work, and witnesses were found to say that in the absence of such excuse mothers were often neglectful of their duty from sheer indifference or weariness.

Cunningham, 2198-9

265. The heavy rate of infant mortality is said to be in large part due to the fact that infants are now-a-days seldom fed from the breast, in which connection Mrs. Watt Smyth gave some figures for the sixteen months ending January 1st, 1902, derived from one of the institutions for the *Surveillance de Nourissons*, maintained by the *Conseil-Général de la Seine* and administered by the *Assistance Publique* of Paris, wherein it appears that among the children breast fed, being 45·4 per cent. of the whole number, the death rate was 2·6 per cent.; among 35 per cent. artificially fed the proportion was 10 per cent.; and among 19·6 per cent. with mixed feeding, 6 per cent.

266. It was also stated by the same witness that in Liverpool, Dr. Hope, the Medical Officer of Health, made an inquiry a few years ago and found that for every death from diarrhœa which occurred among breast-fed infants under 6 months old there were 15 among those fed partly at the breast and partly on artificial food, and 22 among infants fed entirely on artificial food.

Lyttelton, 5368.

267. It must not, however, be understood that the neglect is universal; on the contrary, Mrs. Lyttelton said that country women usually suckled their

own children, with which statement Mr. Fosbroke and Mrs. Watt Smyth appeared to agree; though with country women the process is often continued too long in the mistaken belief that it keeps off pregnancy. Mrs. Greenwood gave some figures showing that out of 725 infants visited in Sheffield nearly 79 per cent. were being fed from the breast alone; but these infants were visited within a week of registration, and it is therefore doubtful for how long they were so fed. Moreover, a medical officer in that town quoted in a lecture by Dr. Jones and referred to in his evidence, has stated that only one in eight of infants born in Sheffield is brought up at the breast. Dr. Eustace Smith, to whom this difference of opinion was submitted, said, "Sometimes they nurse their children for a month or two," and was quite definite that a very small proportion of babies brought to the East London Children's Hospital are suckled. In Dr. Hutchison's opinion it hardly counts unless a child is suckled for at least six months. Dr. Ashby said the women try to suckle their children; but if they are going to work they soon stop. It was on this account that Mrs. Watt Smyth strongly advocated the extension of the period within which a woman was allowed to return to work, as in her view a limit of three months would give a woman time to take an interest in the development of the child, whereas if she feels she must go back in a month she does not deem it worth while.

Fosbroke, 6623. Smyth, 1240. Greenwood 8177. Jones, 10889. Smith, 8435. Hutchison, 9912, 10019. Ashby, 8708. Smyth, 1225.

268. But whatever may be the case in certain districts, or whatever may be the cause, it is quite certain that there is a great decrease in breast feeding, notably in towns, testified to by a number of the most competent witnesses. The evidence of Dr. Hutchison in this regard is of special interest. He says that the fact is common to all highly civilised people, but the medical profession are in ignorance as to reasons for this disability of mothers. He notes a better state of things among the Jews, which may be connected with the abstinence from labour of Jewish mothers. Sir Charles Cameron appeared to think the Irish better off in this respect, but he admitted that the milk was not of good quality and that many of the women were taking a great deal of whiskey. On the other hand, Dr. Kelly, Bishop of Ross, declares, "The practice (of suckling) is fast dying out," and Dr. Browne testifies to its disappearance in manufacturing towns.

Eichholz, 635. Collie, 3908, 9. Garnett, 9207, 8. Eccles, 10724. Hutchison, 10029-10038. Cameron, 10985, 11084. Kelly, 11247. Browne, 9690.

269. It is obvious that in these circumstances the question of the possible alternatives to breast feeding becomes of the greatest importance. The Committee are informed by the highest authority that modified cow's milk is by far the best. The patent foods so largely in use are stated by Dr. Hutchison to present certain marked inferiorities to milk as a means of nourishing children: most of these are deficient in fat, and a considerable number contain unaltered starch, which, being incapable of digestion, produces diarrhœa and irritation of the bowels, a condition which naturally leads to impairment of growth and consequently defective physique.

Hutchison, 9920-

270. It must not, however, be supposed that these foods enter into competition with cow's milk among the poor, as, in fact, they are more expensive and beyond their reach except in a very diluted form. The substitutes they employ are still more deficient in nutritious properties, and much more favourable to the production of every variety of intestinal disorder. In most cases tinned milk is the only expedient. This in a proper state and of a good kind is not deleterious, and there are obvious reasons of convenience why it should be preferred; but at its best, judged by its value as food, it is three times as expensive as cow's milk and is often of inferior quality and always liable to pollution. The vessel once opened is exposed to every kind of bacterial contamination, and by the time it is exhausted all sorts of foul ingredients have been conveyed into the child's system. When tinned milk is not in use, highly diluted and probably sour cow's milk, in which groats or bread have been soaked, is employed instead. It is needless to say that in such circumstances the worst effects of mal-nutrition make themselves manifest very soon, and the condition of the unfortunate child is not improved by the practice begun very early in its career of encouraging it to eat whatever may be included in its parents' dietary.

Lyttelton, 5371. Chalmers, 6027. Fosbroke, 6702. Greenwood, 8238 8245. Eccles, 10755. Hutchison, 1009? Dowding, 4921-4. Eves, 7644, 7793. Smith, 8454.

Defective Milk Supply.

271. The importance of being able to obtain a sufficient supply of good cow's milk being thus emphasised, the Committee are confronted with a great deal of evidence to the effect that it is next to impossible to ensure such a supply, at any rate to the poorer classes. It is not a little curious that, while people in the rural districts have a growing difficulty in obtaining milk because it pays better to send it into the towns, the great mass of the dwellers in towns are in no better case than formerly. There is in fact a great lack of organisation in the distribution of this prime necessity, a great want of knowledge as to its value, and very inadequate means for its preservation from the most obvious sources of pollution. Mrs. Watt Smyth called special attention to this subject, and her evidence was full of the strongest allegations against the cleanliness of the persons from whom the milk comes and of the processes by which it is treated. Thus—

Smyth, 1317. "It is a fact that the milk when it arrives in large towns has been proved to be putrescent. I mean it goes through so many stages, from the country farm to the towns, that it is badly contaminated. But that is not the worst of it. The farms are in such a filthy condition that the milk when it leaves the farm is already poisonous. I have been over many farms, and the conclusion one must come to is that at the bottom of the whole question is the filth of the farms."

And again—

Smyth, 1347. "The cows are in the most filthy condition, standing in manure, and the cow-sheds, the stalls, are covered with manure; and outside the yards are heaped up with it. There is no proper ventilation, the milkers are filthy, their hands and clothes are dirty, and their vessels very often are dirty."

Smyth, 1350. 272. In another case Mrs. Smyth detected hair and fæces in the milk, which the boy in charge took quite as a matter of course because it came off the cow. Smyth, 1348. These allegations Mrs. Smyth declined to admit were only applicable to a few farms. On the contrary she believed that the great majority of farms from which the milk supply was obtained were of this character. It is only fair to Malins, 3220, 3235. say that from his experience at Birmingham Dr. Malins was not prepared to support Mrs. Smyth's indictment, but he could not state whether any precautions were systematically taken to see that the dairies from which the city received Fosbroke, 6731-9. its supply were under proper supervision; Mr. Fosbroke's evidence that the County Council of Worcestershire had been urging this duty on the District Vincent, 12070. Councils implied the presence of much neglect, and Dr. Ralph Vincent's description of the steps taken in the model farm connected with the Infants' Hospital at Hampstead indicated with much clearness the deficiencies of most farms in this respect. These conclusions were borne out by a report on the milk supply of Finsbury in the course of an inquiry conducted by Dr. Newman, the Medical Officer of Health.

The report runs,

"There is evidence to prove that as a general rule the country cowsheds from which the milk is derived are ill-lit, over-crowded, badly ventilated, and badly drained. There is little or no guarantee that the milk is derived from healthy cows."

273. The remedies for this state of things are, first, the general application of the permissive provisions of the existing law under the Contagious Diseases (Animals) Act, 1878. The Local Government Board may make orders for :—

(i) Registration of cowkeepers, dairymen, etc.

(ii) Inspection of cattle in dairies, and for prescribing and regulating the lighting, ventilation, cleansing, drainage and water supply of dairies and cowsheds.

(iii.) Securing the cleanliness of milk stores, milk shops, and vessels.

(iv.) Prescribing precautions to be taken for protecting milk against infection or contamination.

(v.) Authorizing a Local Authority to make such orders.

The Committee think that in the event of the Local Authority not making such orders the Local Government Board, who, it is understood, have already issued model bye-laws for urban and rural districts respectively, should make the orders themselves; or if it was deemed preferable to bring indirect pressure to bear, the power now enjoyed by a local authority under the Infectious Diseases (Prevention) Act, 1890, for prohibiting the supply of milk from an infected dairy might be extended so as to cover exclusion of supply from areas where provisions of the Contagious Diseases (Animals) Act are not in operation. Under either alternative the County Council should be empowered to act in default of the Local Authority, and in either case it should be the duty of the Local Government Board to intervene in the ultimate resort.

274. Nor are the risks of contamination at an end when the milk has left the farm. Owing to the distance from which the supply is drawn and the number of hands through which it passes, these are numerous; to quote Mrs. Smyth again:

"Dr. Priestly, medical officer of health for Lambeth, has stated that much of the milk consumed in the poorer quarters is three or four days old, and probably Lambeth is not the only district in which this state of affairs exists. The milk has passed through the hands of three or four dealers by each of whom a dose of some preservative has been added in order to prevent the actual onset of decomposition. The mischief is caused by the many hands through which it passes. If there were some system of collecting milk in the country and sending it straight to milk depots it would be better." Smyth, 1267.

In Dr. Newman's report it is stated under this head:

"52 per cent. of the milk shops in the Borough were found to have one or more sanitary defects, and 73 per cent. of the milk vendors fail to keep their milk covered or protected from dust. Further, 48 per cent. of the shops where milk is sold are small general dealers, doing a daily milk trade of only a few quarts or pints."

And again,

"There is reason to believe that Finsbury milk and milk generally in London contains great bacterial contamination. Four unpreserved samples of milk, selected from two good class and two poor class milk shops, gave an average of 2,370,000 bacteria per cubic centimetre, which is about 2,000,000 bacteria in excess of what should be present in good fresh milk. Of 25 milks examined in Finsbury in 1903, 32 per cent. contained *pus* and 40 per cent. contained dirt."

275. Things may be better or worse in selected localities, but the Committee have no reason to think this is an overstated account of the general condition. In proof of the extreme ignorance that prevails even among the better class of dealers as to the precautions that should be taken to preserve the purity of milk, Dr. Vincent instanced the practice of a prominent dairy company in keeping a wide bowl of milk standing on the counter of a shop in the West End within a few yards of a continuous and dense traffic. This they label "Pure Milk," whereas from the wide surface exposed to contamination it is imbibing the maximum of bacterial poison, and in Dr. Vincent's emphatic words "should be pitched down the drain." "It is perfectly incredible," he adds, "how milk companies could act in this way." Vincent, 12066

276. The evil is not at an end when the milk reaches the home of the consumer. A pure supply may be rendered injurious by dirt in the house, the proximity of contaminating articles, the general ignorance, in fact, that prevails as to how milk should be stored and the conditions under which it is fit for use. One fertile source of contamination was said to be the use of the feeding-bottle with a long indiarubber tube, which it was impossible to keep clean. Dr. Hutchison would like to see the use of this tube made illegal, as he believes it is in France, and is certainly in some of the States of America. Hutchison, 9984-9

277. The real remedy, however, is to be sought in that social education already described, by means of which, at first in the school and afterwards in continuation classes, and finally by the fostering care of philanthropic and municipal agencies, the foundations of maternal competence may be laid. The Committee had their attention called to a document issued by the Health Department of the Corporation of Sheffield, entitled "Advice on the Feeding and Rearing of Infants," and to another emanating from the Wakefield and Appendix XVIII.

District Sanitary Aid Society, "How to Rear a Healthy Baby," either of which might be made the model of similar admonition elsewhere. Dr. Hutchison thought systematic instruction of mothers in the methods of feed- Hutchison, 9937-9941, 9979-9981. Chalmers, 6076-9. ing and rearing of the first importance; he believed they were in most cases willing to learn, and recommended the issue in all towns of leaflets by the Registrar on the registration of every infant, as was said to be the practice in France, and as had been done for twenty years in Glasgow.

278. The crux of the question lies in the steps that should be taken to bring an adequate supply of pure milk within the reach of the poorer classes. The Committee have had before them schemes for the organisation of this supply by municipal action, the agency of hospitals or charitable societies, or some better adaptation of trade methods. A good account of the system Atkins, 2938. adopted at Battersea was given by Mr. Atkins. It is claimed for it that it has already had an effect in reducing infant mortality, and will be self-supporting when its area of operation is sufficiently extended. The securities Smyth, 1299-1315. taken against contamination seem adequate. Mrs. Smyth described the methods pursued in Paris, both commercially, by an organization under the direction of Dr. H. de Rothschild, and municipally by the *Assistance Publique.* Both these work under medical supervision, and their operations include the periodical weighing of the children fed thereby. In Liverpool the scheme is directed towards "the supply of sterilised milk to poor people at a nominal price," and Young, 2122-7. it was deemed by Dr. Young that the municipality had thus reduced very largely infantile mortality from diarrhœa in the summer. Dr. Collie, 4204-9. Collie was an advocate of such a system and thought it could be made self-supporting, Chalmers 6069-6081. and Dr. Chalmers explained the steps that were being taken to Niven, 6340-5. organize it in Glasgow. On the other hand Dr. Niven was rather inclined to Ashby, 8791, 8852-8874. deprecate municipal activity in such a field, and Dr. Ashby held that the municipality weakened their position as censor if they assumed the functions of trader, and thought much could be done by helping the people to help themselves and by improving the commercial channels of supply. Dr. Eustace-Smith, 8458-8463. Smith informed the Committee that an attempt was about to be made by the East London Children's Hospital to organise a proper milk supply. The milk will be brought from farms kept under strict supervision and it will be kept under supervision the whole time it is in the hands of the Hospital authorities, and will be sold as cheaply as possible. Dr. Smith believes that the scheme will be self-supporting, and anticipates a large sale owing to the willingness of parents to take advantage of the instruction given by the Hutchison, 9925-9. hospital in the shape of leaflets. Dr. Hutchison, however, saw a good deal of difficulty in the general adoption of such a system. Hospitals were for the most part non-local, and he would rather look to improvement in the ordinary commercial supply. On the whole it seemed to him that concerted action might be brought about in which municipalities, hospitals, and charitable agencies might all bear a part and prepare the way for the introduction of better commercial methods.

279. It is of great importance, with a view to enforcing responsibility and guarding against contamination, that the milk supply should pass through as few hands as possible. It has already been shown that milk vendors are often general dealers whose sale of milk is confined to a few quarts, and in the interests of public health this practice should cease. With the formation of a sufficient number of milk depôts, the machinery of registration and of municipal health visitors could be utilised to bring to the knowledge of mothers of young children where supplies of milk could be obtained, and to the spread of this knowledge hospitals and infirmaries in the district could contribute their share. By these means the Committee believe the difficulty of milk supply could be got over without recourse to direct municipal action, but they think that in all improvement Bills promoted by Local Authorities, the insertion of provisions dealing with the milk supply within their area should be insisted on.

280. The problem of providing *ab initio* for the purity of the milk is Vincent 12061-4, 12137-12143. not so easy: most of the witnesses appeared to think that sterilisation did secure this object; but Dr. Vincent, who gave evidence late in the enquiry,

condemned the practice *in toto* on the ground that it reduced the nutritive value of milk to a minimum, and was calculated to cause a scorbutic condition. The Committee do not feel themselves qualified to determine a point on which professional opinion is so acutely divided, but having regard to its importance and the very general belief that by means of sterilisation the best conditions under which a supply can be organised are secured, they think that it should be made the subject of a special investigation by a small body of experts. They have no doubt, however, that Dr. Ashby and Dr. Vincent are right in holding that the first step to be taken with milk on its leaving the cow is prompt refrigeration to a temperature of 40° Fahrenheit.

Ashby, 8874.
Vincent, 12073.

281. It is obvious that the enforcement of this practice would stand in the way of the small cow-keeper sending his milk to the agents of the great collectors who serve the urban market, but looking to the alleged scarcity of milk for the consumption of the poor in rural districts, it might be a good thing if these persons were confined to the sale of what they produce among their own neighbours, where it would be rapidly consumed, and the supply of the urban market thus left in the hands of producers with sufficient capital to take a precaution which the Committee believe to be an essential preliminary to the protection of milk from pollution.

v. *Parental Ignorance and Neglect.*

282. These failings have not escaped illustration in the preceding subdivisions of this section; unfortunately they are not confined to the subject-matter therein treated, but touch every aspect of child-life. The fact of ignorance and neglect on the part of parents is undisputed, and is testified to by a crowd of witnesses. It is not, perhaps, easy to determine how much of this neglect springs from ignorance and how much from callousness; observers are indisposed to take a view that would appear to press hardly on persons the conditions of whose lives are unfavourable to the development of the domestic virtues, and many testify to the willingness of the poor to learn and to a tractable disposition in contact with judiciously tendered advice; but, at the same time, there is no lack of evidence of increasing carelessness and deficient sense of responsibility among the younger women of the present day, which is a discouraging sign, especially as these features are said to exhibit themselves in the country as well as in the towns. Apart from the effects of extreme poverty, alcoholism and love of ease are contributory causes on which many witnesses laid stress. Mr. Booth admitted a growing disinclination to attend to domestic duties, and Mrs. Mackenzie's description of certain classes in the Canongate, Edinburgh, pointed to a degree of indifference which amounted to positive inhumanity.

Booth, 1087-8.
Mrs. Mackenzie, 7034.

(a) Overlaying.

283. In certain overt directions, the disastrous consequences of this neglect are very palpable. Thus overlaying is described as frequent, and is the result of carelessness or drunkenness, the cases generally occurring between Friday night and Monday morning. The practice of placing a small child in bed with older people is, perhaps, sometimes defensible on the score of warmth, but it should be discouraged, and health visitors might properly point out its dangers, while, at the same time, indicating that next to no cost would be incurred by providing a suitable box in which a child could sleep with safety.

Atkins, 3002-5.
Malins, 3147.
Greenwood, 8165
Smith, 8507-8

(b) Fire.

284. The instances in which children are left alone in a room with a fire unprotected by any guard are said to be very numerous. During the years 1899 and 1900 inquests were held on the bodies of 1,684 young children whose deaths had resulted from burning, and in 1,425 of these cases, the fire which caused the accident was unprotected by a guard. Mrs. Greenwood stated that in the leaflets distributed in Sheffield, "On the Care of Infants and Children," warning was given on this point, showing the picture of a fire-guard, but she had never known one bought as the result of the warning, and thought they could only be brought into use by the inclusion of provisions in the building bye-laws.

Booth, 1008.
Greenwood, 816

(e) *Ventilation.* 285. The lack of ventilation and ignorance of the value of keeping windows open—not only in schools but at home—has been mentioned as having a very deleterious effect on child-life. Thus Mrs. Lyttelton said that "not one person in a thousand understands the value of fresh air," and she thinks cottagers get less air than they used because the cottages are less draughty. Dr. Ashby alluded to bad ventilation in the houses as a particular in which mothering was deficient, and classed it with want of cleanliness and badly fitting and insufficient clothing. In this last connection Miss Deverell, Inspector of Schools, confirmed the remark of Mr. Booth that the poor did very little mending, by stating that in many households there is no such thing as a needle and cotton, and mothers made no attempt to mend their children's clothes. Weight, in their opinion, is the only criterion of warmth: children are often without underclothing at all, and their outside garments are both ragged and filthy.

Lyttelton, 5454–6, 5480–5. Ashby, 8700: 8723–5. (d) *Want of Cleanliness.* (e) *Insufficient Clothing.* Deverell, 7998–9.

(f) "*Comforters.*" 286. One of the most noxious practices described is the habit of giving infants india-rubber nipples to suck; this habit appears to be very common, and was most strongly deprecated. It has the effect of causing contraction of the roof of the mouth and the air passages at the back of the nose, which is prejudicial to proper breathing, and is also instrumental in introducing foul germs into the system by virtue of the dirt accumulated. The first of these evils was illustrated by Dr. Young, the second by a number of witnesses.

Young, 2117. Greenwood, 8197–8200. Ashby, 8907–8911; 8918. Garnett, 9211–4.

(g) *Sleeplessness.* 287. Much evil arises from the chronic sleeplessness fostered by the conditions of life so largely prevalent. The lack of sleep from which town children suffer was mentioned by several witnesses as a cause of degeneration. "One of the great contributory causes of physical deterioration," says Mr. Edwards Rees, "is the fact that the children do not get enough sleep." Children in the slums are habitually up till very late at night. This is attributed by Mr. Edwards Rees more to parental carelessness than to the fact that the children would not go to sleep, even if they were put to bed, by reason of the noise and overcrowding, and so on; "for," he says, "children easily get accustomed to such surroundings;" but the conditions under which many of them are compelled to exist must render sleep very troubled and broken. Dr. Mackenzie attributes to sleeplessness the prevalence of irritability and nervous exhaustion.

Rees, 4418. Mackenzie, 6835–8.

(h) *Indifference to slight ailments.* 288. Very little knowledge is moreover found among mothers of the slighter ailments to which children are prone, and precautionary measures are rarely taken until they have reached an aggravated stage. The improvidence of the poor is another circumstance responsible for much juvenile suffering. Money, when plentiful, is spent on luxuries or absorbed in the pursuit of pleasure, and when the pinch comes the children are the first to feel it. Mr. Loch drew attention to the higher rate of wages and also to the decrease of child pauperism; but he said that "when the higher wage has not been combined with a better power of using the wage the children have not come off better," and Dr. Chalmers stated "that there is no relation between the total earnings of the individual householder and the quantity or the quality of the food going into the household."

Deverell, 8006. (i) *Improvidence.* Loch, 10185–9. Chalmers, 6051.

vi.—*Feeding wrong in time, in kind, and proportion.*

289. The general aspect of the feeding problem is discussed under a previous heading of the report, and the questions of milk for infants and feeding of children at schools under sub-sections of this heading, but looking to the extreme importance of nutrition in relation to the young, the subject requires some special notice among the factors that make for juvenile degeneracy. With the single exception of Mr. Edwards Rees, whose panacea is fresh air, all the witnesses concurred in claiming the first place for food. "Food," says Dr. Eichholz, "is the point about which turns the whole problem of degeneracy." There is, first, the want of food, secondly, the irregularity in the way in which children get their meals, and, thirdly, the non-suitability of the food when they

Eichholz, 436, 7.

get it; and these three circumstances, want of food, irregularity and unsuitability of food, taken together, are, in his opinion, the determining cause of degeneracy in children. Describing their dietary, he says:

> "Their breakfasts are nominally bread and tea, and the dinner nothing but what a copper can purchase at the local fried fish shops, where the most inferior kinds of fish are fried in reeking cotton-seed oil, and this often supplemented by rotten fruit collected beneath coster's barrows."

290. Meat may be had in small quantity and of a poor sort on Sunday, but according to this witness the absence of milk and meat is most important in determining degeneracy in the poorer areas. The tea it must be added is probably in the dangerous form already mentioned, having been brewed in the morning and been allowed to stew throughout the day. Dr. Scott instanced a Scott, 1744.
child in Arran who died the victim of his parents' special kindness in this respect. Mrs. Close thought children would be better brought up on beer than the tea Close, 2592.
they get. Dr. Collie considered it most deleterious, and evidence to the same Collie, 3931-3.
effect was given by other witnesses. The growing consumption of tinned foods and its coincidence with the decrease of cookery at home press with exceptional severity on the young. Sir Frederick Maurice laid stress on this as a Maurice, 331, 2.
serious factor in the situation. Few of the mothers in the poorer districts, according to Mrs. Bagot, do much cooking; tinned foods or bad fried fish are Bagot, 4554.
the alternatives, and if the mother does do any cooking it is only for the father, and the children have to eat what he does. Thus Miss Garnett cited a case Garnett, 9110.
where a baby, who was very ill, had been fed on tinned salmon and orange juice, and Mrs. Lyttelton described the regrets of a country holiday child for Lyttelton, 5388.
its "tasty" supper, which on enquiry proved to vary between bloater and a halfpenny saucer of pickles.

291. In the course of Sir Frederick Maurice's investigations, he gathered Maurice 275.
"universal testimony" that it was the habit of parents to feed children off their own plates, and this would ordinarily include raw herrings, pickles, fried fish, and the like. "They live as we do," was quoted by Mr. Fosbroke as a Fosbroke 6624.
common saying.

292. With the prevalence of such ignorance, it is needless to say, no balance or proportion enters into the calculations of those who cater for the wants of the young. The greater cheapness of many articles of consumption —meat, for instance—has brought them much more largely within reach of the poor but there has been no corresponding increase of knowledge as to the economic expenditure of money on wholesome food; indeed, the general consensus of opinion collected from every variety of witness points to the conclusion that in no branch of domestic life is the English housewife so deplorably destitute of the necessary equipment.

Remedial Measures for the above Adverse Conditions.

293. It is clear that the evils, which it has been the object of the preceding sub-sections to summarize, can only be dealt with as part of some great scheme of social education, to which many agencies must contribute, legislative, administrative and philanthropic, and by which the people themselves must be induced to cast off the paralysing traditions of helplessness and despair.

294. Later sub-sections will deal with what, in the Committee's opinion, may be done in school and during the period between the close of school life and adolescence to raise the standard of domestic competence and ideals of home life, and in this they believe lies the principal hope for future generations; but it is not too late to do something to rescue the present generation from the consequences of past neglect, and to that end the subjoined observations and recommendations are directed.

295. An enactment placed on the Statute Book so recently as 1902 offers an opportunity to the Local Authority that should not be missed. Under the Midwives Act of that year, which places in the hands of County Councils, or any District Council to which they may delegate their powers, the general supervision of all midwives practising within their area in accordance with rules to be laid down under the Act, rules have been made and approved

by the Privy Council by which every midwife is made responsible for the cleanliness and is enjoined to give full directions for securing the comfort and *proper dieting* of the mother and child during the lying-in period, which, for the purposes of the regulations, is held to include a period of ten days after delivery. It is obvious that as women trained under the Act take the place of the illiterate, incompetent persons that very generally look after the poor at this critical point of their lives, a great opportunity is offered for giving mothers the information wanted and providing the newly-born infant with a fair start. If the Local Authority make use of this agency, which is entirely under their control, for the dissemination of proper knowledge and practical advice, a considerable step in advance will have been achieved. Models for maintaining and improving upon this step already exist, and only require careful study and a genuine faith in the value of social education, to be brought into general use.

296. Direct evidence on the value of crêches as places of instruction for mothers was given by Miss Eves. She thought education of young women in the care of infants in the strictest sense technical education, and advocated the establishment of centres for the purpose under the Local Education Authorities.

Eves, *passim*.

297. A large body of evidence was tendered as to the organisation and operations of the Manchester and Salford Ladies' Public Health Society, and the Committee had the advantage of examining on the subject Mrs. Worthington, one of its principal members, and Mrs. Bostock, one of the Health Visitors it employs. The Society, which has been in existence for more than twenty-five years, has for its object the discovery of all those conditions that are adverse to public health, and especially the bringing within the knowledge of the mothers among the poor such information as will enable them to do their duty by their children. The poorer parts of both towns are divided into districts, each under the supervision of one or more of the ladies who constitute the Society, and, subject to their directions, a number of Health Visitors, who are in part paid by the Corporation, undertake the duty of visiting every house in which the birth of a child is reported, with the object of educating mothers in the best methods of bringing up young children. By these means, Mrs. Worthington stated that a good deal of influence has been brought to bear upon them to adopt regular hours and not be quite so miscellaneous in their feeding operations, and it is said that they have now acquired some settled notion of what is the best type of food to give children. Incidentally and very largely the labours of the Health Visitors in this connection bring to their knowledge all sorts of insanitary conditions, arising from overcrowding, stopped drains, and structural defects, which they proceed to report to the municipality on a form provided for the purpose. In the result, an inspector is at once sent and the evil is put right before very long. In a recent report of the Society's work, it is said that the Health Visitors have made 30,364 inspections of houses, and have reported 1,500 cases of insanitary conditions, and the Medical Officer of Manchester testifies that the effect is marked in the poorer districts of the city, and that "an improvement on former conditions can now be generally discerned." The Report goes on to quote from one of the superintendents that the poor "look upon the Health Visitor as their best friend, and there are few homes where she is not made welcome."

Worthington, *passim*
Bostock, *passim*

298. The Report proceeds:—

> "In addition to teaching personal and household cleanliness and giving advice and sympathy, much practical help is given in ways such as giving food and clothing to specially needy families, finding work for men and women, getting recommendations for different hospitals and institutions, sending children into the country or to the seaside, making the beds of sick patients and cleaning their houses."

On the whole, however, the work is preventive rather than remedial.

> "The women in the district are shown the evils of dirt and the dangers of living in unhealthy dwellings; they are taught to prevent the spread of disease, and the laws generally which will enable them and their families to lead moral and healthy lives."

Miss Squire, one of the Factory Inspectors employed by Miss Anderson for the purposes of the Memorandum which has been extensively referred to, visited some houses with one of the Health Visitors and was

"favourably impressed with the effect she seemed to produce upon the mother or nurse, as the case might be. The serious proportion of infant deaths is a matter of common knowledge in the town, and the mothers and nurses seemed to take it as quite reasonable that the Medical Officer of Health should prescribe to them what they might and might not do, and to be impressed with the fact that what their mothers did before them would no longer be allowed to be their guide in the treatment of their children." Appendix, V. 60

These visits are supplemented and the lessons they give enforced by the distribution of short leaflets (as concise and pointed as possible) in order to keep the advice given constantly before the mother's mind, and it is found that these leaflets, following upon personal visits, are read and treasured.

299. Fuller testimony is borne to the value of the system by the extent to which it has already been adopted in other places. Mr. Rowntree advocated municipal leaflets associated with visiting; he described the work of the Health Society in York and also the standard of knowledge and training to which visiting ladies ought in his opinion to attain. Mr. Fosbroke testified to the good results of the lady visitors employed by the Worcestershire County Council. Dr. Smith said that the distribution of leaflets by the East London Children's Hospital had had great effect, and he thought that direct work among parents was more valuable than the training of children and young people with a view to their becoming parents, a view also taken by Dr. Hutchison, because in his opinion these things do not come home to girls until they have a baby of their own. Rowntree, 4994-; 5158-; 5252-; Fosbroke, 6613-; Smith, 8464-7. Hutchison, 993 9953, 9980-9990

300. The Committee believe that enough has been said of the value of the system, and the testimony paid to its success by competent judges, to justify them in urging upon every locality the adoption of similar methods. The system has the advantage of linking individual and philanthropic effort with municipal responsibility, in a way that regularises the one while energising the other, and appears to give to each its proper influence in dealing with social wrongs, at the same time tending to check the overlapping and misdirection so often characteristic of purely charitable impulse.

301. Other agencies to the same end, such as lectures and mothers' meetings are not excluded; indeed, they can very well be worked into the system and be utilised to expand and develope its scope. There is no step, in short, towards training mothers in personal, domestic and infant hygiene, with which it cannot be associated, and towards the effect of which it cannot be made to contribute.

302. The Committee desire to press these considerations with all the earnestness at their command upon the most serious attention of the community, and they would further suggest to the Local Government Board the desirability of issuing to local authorities a circular explaining the objects to be sought and the means by which they can best be attained. It is in connection with steps of this sort that the Committee believe an Advisory Council formed on the lines recommended in the earlier part of the Report might prove of great use to the Department usually charged with the interests of Public Health.

vii. *School System—Medical Inspection of School Children—Feeding of School Children.*

303. In a country without compulsory military service the period of school life offers the State its only opportunity for taking stock of the physique of the whole population and securing to its profit the conditions most favourable to healthy development. It has been shown in the first part of this report how the occasion may be utilised in furtherance of the objects of an anthropometric survey, and it remains to be considered what are the conditions of school life adverse to physical well-being, and how the opportunities which it presents can be made to realise the best results for the rising generation, at, perhaps, the most sensitive period of its growth.

304. It may be said at once that the general effects of school life are not prejudicial to health. The civilising influence exercised by the school during

Eichholz, 526-530, 578-582. Niven, 6246-9. the last thirty years was noticed by both Drs. Eichholz and Niven. The untamed savagery characteristic of certain types has disappeared, and Dr. Eichholz further thought that there was nothing in the conditions of school life which it was not in the power of existing authorities to improve; but popular opinion required educating on such elementary though important points as the seating of children, the arrangement of light, and the supply of fresh air. Evidence was given of children being kept too long at desks which do not fit them, in an attitude of strained attention, either writing or with their eyes concentrated on a blackboard, which, if the class happens to be a large one, perforce occupies a position in which some have a difficulty in seeing it. Very young children are sometimes observed sleeping in postures calculated to develop curvature, and infant occupations are criticised as often much too fine for their eyesight. It was also alleged that the eyes of scholars often suffered from the effects of a strong cross light, and that defective ventilation counted for a great deal among the unfavourable circumstances with which children had to cope.

Cameron, 11053. Gorst, 11852, 11872. Kerr, 801, 812, 818. Collie, 4019. Garnett, 9237.

Greenwood, 8213, 8265-8280. Kelly, 11366-8.

305 While with the last-named exception schools on the whole seem to be in a good state, Mrs. Greenwood drew a sad picture of the dirt and darkness in some of the Sheffield schools, and Dr. Kelly, Bishop of Ross, taxed the National Board with indifference to the warming of schools, from which children suffered acutely. It appears that whatever fuel is used in schools in Ireland has to be procured by voluntary contributions or brought there by the children themselves, and it is not an uncommon thing for children to take a sod or two of turf to school on a winter's morning. Dr. Kelly goes on:—

> "I might set it down as one of the causes of the poor physical development in Ireland that the school children are unfairly, in fact I might say cruelly, treated in the schools themselves. I see now many of these little children going to school all the winter barefooted, and in some instances they go to a school where there is no fire. The country children have to travel a couple of miles to school; a great many of them have no cloak or shawl, or anything to cover them. Ireland is rather a rainy country, and they go wet into the school and sit down there shivering all day."

The Committee think that a system under which the infliction of such suffering on poor children is possible requires amendment.

Eichholz, 579-583. Kerr, 808-810, 946-950. Collie, 3982, 4024-6, 4096-4101. Chalmers, 6164-9. Horsfall, 5687-5693. Gorst, 11811-3. Brunton, 2426, 2443-9, 2470-9, 2528-2537.

306. On the point of ventilation, there was some difference of opinion between the advocates of open windows on the one hand and mechanical appliances on the other, but as in the case of the latter the inlets are usually under the control of the teacher, it is of the first importance that he should be alive to the value of fresh air, and in this respect among others there appears to be something wanting to the education of teachers in the principles of practical hygiene. Sir Lauder Brunton was perhaps the most emphatic critic of the school system, and believed that both in regard to breaks in school work and the substitution of physical exercises for prolonged mental application the Swiss practice was preferable, but his actual knowledge of what goes on in English schools did not appear to be very close, and the Committee are convinced that the regulations in force, if properly observed, are sufficient to cover both points of attack.

Lyttelton 5415-5428. Fosbroke 6598-6608.

307. Infant schools as conducted in urban districts can no doubt be worked to the great physical advantage of the children attending them, but there is evidence that a handful of small children in a rural school necessarily suffer a good deal from neglect or are taught under conditions from which no advantage can be derived. So impressed were Mrs. Lyttelton and Mr. Fosbroke with the unsatisfactory results that may be expected from keeping tiny children at school that they would not permit attendance under five at all. Mr. Fosbroke made this one of his chief points and had evidently given it a great deal of attention. He thought the effect of a later attendance at school on the children themselves, both physically and mentally, would be most salutary, and it would have the indirect result of keeping mothers at home and compelling them to attend to domestic duties. Sir J. Gorst went even further, and would be glad to see country children excluded from school up to seven years of age.

Gorst, 11816.

> "That is the practice in Switzerland, which is perhaps almost the best educated country in the world. They do not let their children come to school till they are seven. They run about in the villages and mountain sides, and they are often employed in looking after cattle, goats, and so on. They do not go into school at all till they are seven years old, and therefore when they do go to school they are sturdy and strong, and their observation is awakened."

While the Committee are unable to accept this statement as entirely correct, they think that school attendance in the *rural* districts should not be compulsory till six or possibly till seven, and should be discouraged, if not absolutely prohibited, under five.

See Appendix III to the Report of the Committee on Employment of School Children (1901, Cd. 849).

308. The organisation of games and the provision of accommodation for outdoor exercises were strongly pressed by some witnesses, but no scheme of games alone can ever be made general enough to supply the place of methodical physical training. A model course of physical exercises, more thorough, systematic, and complete than any previously in use, has just been produced by the labours of another inter-Departmental Committee, and in connection therewith this Committee desire to express their opinion that wherever practicable, the prescribed exercises should be performed in the open air, and the interval utilised to obtain a thorough ventilation of the school premises, and particularly the class-rooms, which as a rule suffer most in this respect. Inasmuch, however, as the effect of physical exercises depends in no small degree on their regularity, and the climate of this country is such that exercises in the open air are necessarily interrupted for days at a stretch, the provision of play-sheds or rooms for physical exercises, other than the ordinary class-room, should be insisted on in ordinary circumstances as an almost indispensable part of the school equipment.

309. Mr. Atkins avowed himself a great advocate of games as an educative as well as a health-giving factor, and he thought there might be a much more systematic effort towards their proper organisation, both for children of school age and young people. Mr. Rowntree called attention to the same want, more especially in towns, and cited Boston as an example of the American practice in this respect, where there is no open space without some one to organize the children's play. Mr. Horsfall and others emphasised the need for physical recreation and exercise, through the lack of opportunities for which Mr. Edwards Rees said the children of Salford were losing the instinct of play. To this end playgrounds should be utilised, not only during the mid-day recess but after school hours. The Committee heard a good deal from Sir Lauder Brunton of an organization now in process of formation, under the name of a National League for Physical Education and Improvement, and they cannot do better than commend to its attention the expediency of placing itself in communication with the educational authorities throughout the country, with a view to their intelligent co-operation in a scheme of so much importance and utility. The Committee have read with great interest the observations of the Royal Commission on Physical Training (Scotland) on this subject, and think that paragraphs 85 and 86 of that Report deserve the special study of the organizers of the League. In the Committee's judgment a full consideration of the subject requires that due prominence should be given to the part that ought to be played by skilled physical instructors, and the need of securing a constant supply of such. In this regard, the suggestion in paragraph 314, as to the training and recognition of teachers and instructors, may be useful.

Atkins, 3000, 3041–3060, 3075–3086.
Rowntree, 5129–5140.
Horsfall, 5673–5683.
Rees, 4318–4330.
Fosbroke, 6648, 6681.
Loch, 10131–5.
Brunton, 2430, 2454–7, 2537–9.

310. The teaching of cookery and household management assumed particular importance as bearing directly on the subject of this enquiry. The principles that should govern the teaching of cookery have already been laid down in paragraph 232 of this Report. These principles are to all intents and purposes embodied in the directions of the English and Scotch Central Education Authorities for the conduct of Cookery Classes, but there is a considerable body of evidence to show that, for various reasons, in practice the teaching of cookery has not had all the beneficial results that might have been expected.

311. Dr. Eichholz summed up the fruits of his experience in these words :—

Eichholz, 499.

> "Many reforms are necessary in the teaching of cooking before it will become sufficiently useful in the home. The children begin to learn it too early to retain much by the time they leave school. They do not follow the work up sufficiently at home or at school to become conversant with the process taught them. It is the exception to find a girl repeat a lesson on the cooking of meat at home. The knowledge which the child gains is too insecure to tempt the parent to run any risks with the modest domestic allowance of meat. So it comes about that the girls become great adepts at making cakes. I think the work would be all the better for being concentrated upon the last six or twelve months of school life."

Lewis, 8579-8581, 8624, 8631.

This testimony is confirmed by other witnesses such as Mr. Lewis, teacher in the West Green School, Tottenham, who pointed out that teachers were apt to substitute a standard limit for the age condition very properly inserted in the Code, and that one cause of the disposition to devote too much attention to the making of cakes and other attractive dishes is the difficulty of getting rid by sale—as the teachers are expected to do—of the plainer and more useful dishes. On the other hand, Mrs. Bostock and Miss Maude Stanley bore testimony to the beneficial effect in the homes of the teaching at present given in cookery classes in school.

Bostock, 7516-7522. Stanley, 13448.

312. It seems to the Committee that certain prevalent defects in the teaching of cookery are, in part at all events, due to the absence of an efficient system of inspection, but on a consideration of the evidence they are disposed to doubt whether any system of teaching cookery, however well organised, is likely to be effective or practical, which is confined to the period of school life and is not continued and supplemented by instruction of the girls at a more mature age, when they are more appreciative of the need of such instruction and more likely to have opportunities for the practical application of what they have learned. It is for this among other reasons that the Committee have elsewhere (paragraph 230) recorded their opinion in favour of compulsory Continuation Classes.

Smyth, 1277-1288. Wilson, 1958-1960. Close, 2604-2619, 2662-2681. Rees, 4366-4383. Lyttelton, 5435-7, 5375-5382. Niven, 6295-6, 6379-6386, 6526-8.

313. As regards the instruction of girls of school age, however, the Committee are disposed to agree with those witnesses who urged that instruction in Cookery, Hygiene and Domestic Economy, should, as far as possible, be made compulsory on the older girls, that such instruction should be concentrated in the last year or so of school life, and that room for it should be made by the omission at this stage of certain other subjects from the school curriculum. In this connection, they would direct attention to the scheme of supplementary courses for girls under the Scotch Code, which virtually carries out these suggestions. Girls are admitted to these courses at or after 12 years of age, on giving proof of reasonable proficiency in the three R's, and thereafter devote most of their school time to *practical* instruction in cookery and laundry work, and all that appertains to the management of a home. As girls must remain at school till 14, unless specially exempted, there is thus opportunity for concentrating the attention of the girls upon domestic subjects for a period of a year or eighteen months immediately prior to their leaving school.

314. It is obvious that if instruction of this kind is to form part of the normal school course, questions of school and personal hygiene and of the proper conduct of a home must receive much greater prominence than appears to have been the case hitherto in the normal curriculum of Training College students, and that further, in the large centres of population, there will be opportunity for the employment of a special class of teachers, of more moderate attainments, perhaps, as regards the ordinary subjects of a general education, but who have a thoroughly expert knowledge of the class of subjects in question. It follows from this that, in the opinion of the Committee, there should be greater diversity of type in Training Colleges and greater elasticity in the conditions upon which recognition as a Certificated Teacher is granted. This status, subject, of course, to sufficient safeguards as regards general education, should be conferred not only upon the teachers of book subjects but also upon those who have received a comprehensive and thoroughly satisfactory training as teachers of housewifery, or, to take another example, teachers of physical exercises, and the training of such teachers should be equally an object of State support.

315. Moreover, the Committee think it worth considering whether the present law as to school attendance could not be so modified as to render the partial exemption from attendance at school, which in England and Wales is at present granted only on a certificate of proficiency or of previous due attendance, obtainable without further condition than that the obligation to attend school, which at present ceases altogether at the age of fourteen, shall in that case be extended to a later age. By this means the knowledge acquired during

attendance at elementary schools, which is too often forgotten on leaving school, might, it is hoped, be retained and extended. Legislation would, of course, be required to effect this, but the principle has already been to some extent recognised by the agricultural provisions of Robson's Act, and the Education (Scotland) Act, 1901, and it seems capable of extension in other directions; it might, for example, be made a valuable means of fostering that sense of domesticity in girls, the decay of which several witnesses deplored. In this, as in many other respects, the co-ordination of education which is now being effected by means of the new Local Education Authorities should be of great service in rendering a scheme of the sort feasible.

316. The problem presents features of too great minuteness and intricacy to be adequately discussed within the limits of this Report, but enough has been said to indicate the lines of advance upon which the Committee think that something may be done in order to equip the daughters of the people for domestic life.

317. Evidence was also forthcoming as to the unsuitableness of a stereotyped school system for children who are not up to normal school standard and are yet not so defective as to warrant treatment as "mentally deficient." Dr. Eichholz advocated the establishment of special schools, to which he gave the name of "Schools of Industry," something on the lines of day industrial schools for "retarded" children; that is, all children other than those actually defective who, by reason of insufficient feeding or whatever cause, are backward and not able to profit by the ordinary school curriculum. The condition of many of these children is the result of parental neglect, so that there are two sets of associated circumstances, educational and domestic, which, in his opinion, make it desirable to extend the industrial school system under Section 16 of the Education Act, 1876, with the same liability on the part of the parent to contribute. In his own words— Eichholz, 486-494, 680-708.

"I should like to see schools, either urban or rural, schools of industry, not punitive industrial schools, which shall recognise the uselessness of much of our present curriculum for retarded children, which will, in the first instance, create self-respect through cleanliness and decency, and aim at re-establishing an enfeebled constitution through suitable regular feeding. As these are secured, physical education becomes possible, and the final aim comes into the foreground, which is to provide a curriculum based largely on manual occupations and manual instruction, and which shall endeavour to implant in the minds of the children a respect for the dignity of work —a fact which their homes have never impressed upon them, and which the elementary schools likewise fail too often to accomplish." Eichholz, 486.

Dr. Eichholz would like to see the thing done through the magistrate rather than by a simple drafting of children to and from such schools by means of the School Attendance Officers and the ordinary authorities concerned. Eichholz, 486-494, 680-708.

318. The Committee may take this opportunity of expressing their opinion that, wherever possible, in cases touching the young where the assistance of a magistrate is invoked, he should be a person specially selected, sitting for the purpose.

319. Dr. Kerr was also in favour of a system of special schools for retarded children, though he had not considered the matter in relation to the food problem. He considered that as many— Kerr, 783-794, 846-869-880, 888-902, 918-941.

"as ten per cent. of the ordinary school populations all over the country, at any rate in the first half of school life, require some consideration on account of debility and backwardness from various causes—some simpler education than the ordinary Board School attempts to give them."

In his view the method of selection for such schools would depend on the results of medical and educational inspection, the great point to be kept in view being "to suit the curriculum to the child instead of the child to the curriculum." It would, of course, be an essential feature of the scheme that children who under special treatment became fit for the normal type of school should be transferred there for the remainder of their school life. Besides these "retarded" children, Dr. Kerr thought there was a group of semi-invalids for whom something between a school and a hospital in the country should be started.

320. It is obvious that underlying all these schemes there is need of a much more complete system of Medical Inspection in schools than has yet been attempted, and it is not therefore surprising to find that all the medical witnesses and others laid great stress upon its introduction on some recognised basis.

Eichholz, 456-461, 550, 1, 615-9.

321. Dr. Eichholz thought it was the greatest need in school organisation. On the ground of expense he would confine a general examination to the poorest schools, and considered that in London the work could be done by ten young men at £250 each. The School Board had nothing like a big enough staff and confined their attention to cases of defective eyesight, feeble-minded children, and the medical examination of pupil and assistant teachers. Dr. Eichholz explained to the Committee the conditions observable in the children attending these poorer schools, and described the means by which medical inspection might be utilised to note and check degenerative tendencies. Children who were thought to need medical examination in better class schools would be examined on special notification by the teacher, but in the schools classified as poor the inspection would cover all new admissions and the re-examination of old cases where necessary. What follows on such a system is thus described:—

Eichholz, 463.

"The business of the school doctor is to examine children at admission when necessary, and periodically later; to make recommendations to the school authority which shall reach the parent without delay. It is for the school authority to determine how to make those recommendations effective. It is impossible that the doctor can enforce anything that he says. The way I suggest this should be done is as it is done in Frankfort and in German schools generally. They issue a slip of paper, which I suggest should be in duplicate, which says, 'Your child is suffering from so-and-so, kindly exclude him until he is fit to attend.' In England this would not suffice, as it would give the parent an excuse for not sending the child to school. We want the certificate in duplicate so that the school attendance officer may keep its contents before the parent. The authority would issue one to the officer and the other to the parent, and the parent in this way might be dealt with according to the measure of his culpability or his powerlessness, if he did not or could not act upon it. Many cases would still need charitable aid and whatever help managers could procure. I do not anticipate that the need for punitive expedients would often arise. A few wholesome examples would effect a very rapid cure of culpable neglect."

Among the further duties of the medical inspector would be:

Eichholz, 464.

"recommendations to the local authority on general and special points of school curriculum, length of lessons, apportionment of intervals for recreation, organisation of recreation, ventilation, lighting, artificial and natural, the use and abuse of needlework, desking for children of various ages, use of slates, towels, local variations in curriculum to meet special needs."

Kerr, 853-9.

Collie, 3989-4006, 4120-6.

322 Dr. Kerr thought that with the assistance of intelligent teachers trained in hygiene, there need not be a very great augmentation of the existing medical staff. Dr. Collie went into the matter in some detail and thought that in the first instance it would be sufficient to entrust to the teacher the duty of reporting to the medical inspector the case of any child who on admission appeared to him to be suffering from any physical or mental defect; and the medical inspector would attend occasionally to receive such reports from the teacher and examine any child who required it. He advocated the employment of doctors of experience rather than young ones, and said the late School Board staff was quite inadequate, but well qualified general practitioners might be entrusted with the inspection of eyes and teeth (and presumably ears) without having recourse to specialists. No very large staff would, in his opinion, be required. Dr. Chalmers quoted the Report of the Royal Commission on Physical Training (Scotland), and declared a medical inspection the necessary accompaniment to any system of general physical training in schools. He thought medical inspectors might be useful in matters of ventilation and health conditions generally. Dr. Niven, like Dr. Collie, would utilize the teachers in the first instance and would require them to have special training. This last indeed appears a necessary corollary to the introduction of any system of the sort. Dr. Mackenzie considered "the systematic medical inspection of schools as one of the things that is most called for at the present time," and explained with some fulness the methods he advocated. He thought the

Chalmers, 6183-6.

Niven, 6346-8.

Mackenzie, 6787-6813, 6858-6869

superintendence of the system might be entrusted to the Medical Officer of Health, and the inspection made on admission and periodically afterwards. The assistance of specialists on certain points would probably be required, but he agreed with other witnesses that properly trained teachers could render most useful assistance. Mr. Loch would not neglect the medical inspection of the home, "the fulfilment of an already recognised public sanitary duty," and he would have the two go together. Mr. Murphy was also among those who emphasized the need of a general medical inspection of schools.

Loch, 10328-1033

Murphy, 10403.

323. For the reasons, then, that appear in the discussion that has occupied the preceding paragraphs of this sub-section, the Committee consider that a systematized Medical Inspection of School Children should be imposed as a public duty on every school authority, and they agree with the recommendation of the Royal Commission on Physical Training (Scotland) that a contribution towards the cost should be made out of the Parliamentary Vote. The value of such an inspection is well illustrated by the particulars given in a recent report by Mr. George Andrew, of the Scotch Education Department, on the *Gemeindeschulen* of Berlin and Charlottenburg. From that report it appears that of 2,547 children examined on admission to the schools of Charlottenburg, 321 or 12·3 per cent. were rejected as being unfit for the work of ordinary schools, and of all the children examined, 63 per cent. were described as not completely normal *(nicht völlig normal)*.

324. The Committee believe that, with teachers properly trained in the various branches of hygiene, the system could be so far based on their observation and record, that no large and expensive medical staff would be necessary. The general inspection may safely be limited to that class of school which from its character and surroundings affords clear evidence of the type of which its scholars are composed, and it should be repeated as often as the medical officer thinks necessary. In other cases it will be enough for him to examine such children as may be submitted to him by the teachers, and it would be for him to determine whether circumstances called for expert assistance. In no instance should the inspector do more than state the facts that require the attention of the parent, cases of poverty or neglect being left to the proper authorities to deal with.

325. The Committee further think that the services of the medical inspector should be at the disposal of the school authority in respect of all matters where the advice of a person skilled in the hygiene of child life may be wanted, and they look in the result to a much more intelligent handling of many of the problems with which that authority is charged.

326. Owing to the healthier conditions commonly prevailing in regard to the nurture and growth of the country population, the Committee do not think that these recommendations will have any appreciable effect in augmenting the charge for education that now devolves upon the rural ratepayers: on the contrary they are of opinion that bringing the sanitary administration of the county into touch with the schools may lead to the anticipation and prevention of many of those epidemics which are now such a fertile source of local expense.

Feeding of School Children.

327. The Committee have reserved to the end of this subsection the discussion of the question of feeding elementary school children (1) because it lies outside any direct obligation that has hitherto been recognised and (2) because some of the matter that precedes has a bearing upon its consideration.

328. Besides seeking to elicit the opinion of a large number of general witnesses on the point, the Committee sought to equip themselves for the task of formulating some conclusion by the examination of certain persons who were either called or tendered themselves to give special evidence thereon. These included Mr. W. H. Libby, Secretary to the East Lambeth Teachers' Schools Dinner Association, Sir John Gorst and Dr. Macnamara, Members

of Parliament, and Dr. Osmond Airy, one of H.M. Inspectors of Schools, who for many years was chairman of the organisation for providing dinners to poor children in Birmingham.

329. Before, however, proceeding to examine the evidence on the point, it may be as well to state that the general trend of opinion was to the effect that the ultimate means of dealing with the difficulty lay in the development of those forces of social education which have been described above, and in the operation of the great body of ameliorative tendencies which would raise the general condition of the poor, and foster a sense of parental responsibility, spreading knowledge and enlightenment in their train.

330. It was nevertheless acknowledged that the evils arising from underfeeding were so widespread, and in certain localities so pressing, that some authoritative intervention is called for at the earliest possible moment to secure that the education of the children who are obliged to attend school shall not be hampered and retarded by the physical conditions thereby engendered.

331. The evidence on the specific question of the feeding of children compelled by law to attend a Public Elementary School can conveniently be treated under the following heads:—

(*a*). The extent to which underfeeding prevails at present.

(*b*). Existing voluntary methods of providing food.

(*c*). Proposals in regard to the more systematic feeding of school children.

(*a*). *The Extent of Underfeeding.*

332. On this point Dr. Eichholz, one of H.M. Inspectors of Schools, gave very interesting evidence. In pursuance of his view, already noted, that the whole question of bad physique practically centres round feeding, he made a special investigation into the conditions of the Johanna Street Board School, Lambeth, as a type of school in a very bad district, and he considers that 90 per cent. of the children are unable, by reason of their physical condition, to attend to their work in a proper way, while 33 per cent. during six months of the year, from October to March, require feeding. He gave the Committee an estimate, based on the figures of the voluntary feeding agencies, of the number of underfed children in London. Some time ago he furnished the Board of Education with a similar estimate, in which he set down the number as 60,000; but he has since gone into the question in more detail and now estimates the number as approximately 122,000, or 16 per cent. of the elementary school population of London.

Eichholz 435-7.

Eichholz 471, 476.

333. The London School Board expressed the opinion that there were not more than 10,000 children suffering from malnutrition; but, in Dr. Eichholz's view, this estimate is belied by the actual figures given in the report of the Joint Committee on Underfed Children, which gives the average number of children fed per week during the weeks the feeding centres were open in 1902-3 as 22,206. His argument is that in the first place this figure only refers to children attending Board Schools; and secondly it is very fallacious as an index of the total number of children who are underfed, because the same children are not being fed all through the season; the result of investigation shows that on an average the centres feed three times as many children during the season as they feed on any one day, and that the number furnished as the weekly average of children fed may be taken as an index of the daily average of meals provided; and therefore we get 66,000 as the rough total of underfed children attending the London Board Schools. Besides this there are the children attending voluntary schools. And, as the amount spent per annum in London by voluntary feeding agencies is £6,100, and £100 spent is said to represent 2,000 necessitous children fed per season, the total estimate of underfed children in London is, roughly speaking, 122,000, or 16 per cent of the whole.

Eichholz 472, 475.

334. This calculation obviously proceeds on the assumption that all the children fed at schools and centres would otherwise be underfed, but such observations as those of Miss M. Horn, manager under the late London School Board, in a paper recently read to the International Home Relief Congress, throw considerable doubt on the validity of the assumption.

335. Dr. Eichholz also made enquiries about Manchester, and was assured that 15 per cent. of the Manchester children are underfed. As to this, Dr. James Niven, the Medical Officer of Health for Manchester, did not give any figures, but expressed the opinion that there was a very large number of underfed children, that the child must be fed at any cost, and that no voluntary agencies could possibly cope successfully with the evil. On the other hand, Mr. Edwards-Rees, Vicar of Pendleton, Salford, and a member of the Salford Education Committee, stated distinctly that not more than 2 per cent. of the children in Salford and Manchester come to school underfed; and in his opinion the question of malnutrition is not nearly so urgent as that of lack of proper physical training and pure air. Dr. Henry Ashby, of Manchester, nominated by the Royal College of Physicians to give evidence on the subject of nutrition, was inclined to think that most children go to school having had sufficient breakfast; so that there is some apparent conflict of evidence as to the conditions in Manchester. Eichholz, 476. 47[illegible] Niven, 6290-1, 6372-8. Rees, 4285, 4286, 4384-9. Ashby, 8740-1, 8748-9.

336. To revert to London, Dr. James Kerr, Medical Officer to the London School Board, "does not feel as strongly on the point of nutrition, or rather want of nutrition, as most people"; but he admitted not having considered the matter as fully as Dr. Eichholz. Dr. R. J. Collie also, a Medical Inspector to the late London School Board, stated that there is "not a very large proportion of children who are actually half-starved; it is only in some districts"; but he did not give any statistics. Kerr, 942-5. Collie, 3947.

337. Of other witnesses who spoke on the extent of underfeeding, Mrs. Close, a lady who considers herself familiar with rural England, said that children are constantly half-starved when they get to school owing to the laziness and neglect of the parents. Dr. Chalmers, Medical Officer of Health for Glasgow, is now making enquiry into the proportion of underfed children in Glasgow, but had no figures at the date of giving evidence. Close, 2636-2642. Chalmers, 6171-[illegible]

338. Dr. W. L. Mackenzie, Medical Officer to the Local Government Board for Scotland, said that in the slums of Edinburgh a large proportion of children were half-starved, and he agreed that to subject a half-starved child to the routine of school would be the height of cruelty, and the educational result would be poor. Mrs. Mackenzie, the wife of Dr. Mackenzie, gave the same sort of opinion: "the child must be fed." Mr. W. H. Libby said that a feeding agency in Lambeth coped with from 12 to 15 per cent. of the elementary school population, and in the poorest districts with from 25 to 30 per cent., which, so far as that locality is concerned, bears out Dr. Eichholz's figures. Miss Garnett, the head of the Diocesan Women's Settlement in the extremely bad district of the Potteries, attributed the faulty nourishment of the children to the neglect of the mother who goes to work in the factories; the teachers in the various schools notice that children from such homes are less able to do their morning's work; but they do not admit that the proportion of such cases is very large. Mr. C. S. Loch, Secretary to the Charity Organisation Society, did not think a large number of children were sent to school half-starved, but gave no figures. Mr. Shirley Murphy, Medical Officer of Health of the Administrative County of London, regarded the question as by far the most important before the Committee, and said "the child has got to be fed." Finally, Dr. Kelly, Roman Catholic Bishop of Ross, said that in the South of Ireland it was commonly the case that children came to school underfed. Mackenzie, 6977 6980. Mrs. Mackenzie 7030. Libby, 7835-7. Garnett, 9042-[illegible] Loch, 10191. Murphy, 1040[illegible] Kelly, 11380.

(b). *Existing Voluntary Methods of Providing Food.*

339. There has not been a great amount of definite evidence on the voluntary agencies in existence. As regards London, Dr. Eichholz mentioned the following agencies which spend about £6,100 per annum collectively:— Eichholz, 473, 476.

London Schools Dinner Association.

Mr. G. R. Sims *Referee* Fund.

Destitute Children's Dinner Society.

East Lambeth Teachers' Schools Dinner Association.

Southwark Children's Free Meals Fund.

The work of these agencies has been coordinated, so far as the Board Schools are concerned, by the Joint Committee on Underfed Children, which was worked under the ægis of the School Board and existed to prevent overlapping.

Eichholz, 482.

340. In regard to the sum contributed by the parents, Dr. Eichholz said that it amounted to 5 per cent. in the case of the London Schools Dinner Association, and 25 per cent. in the case of the Destitute Children's Dinner Society.

Libby, *passim*.

341. Mr. Libby, who is Secretary to the East Lambeth Teachers' Schools Dinner Association, gave a description of the working of the Free Meal Fund in connection with the Association. The fund has been running about twelve years, and is conducted on business lines; it is worked by the teachers through the attendance officers, and careful enquiries are made as to the circumstances of each family before a child is given a meal. There are breakfast centres and dinner centres. A child can be given a pint of vegetable soup and a piece of brown bread and a piece of cake at the cost of 1d. About 2 per cent. of the children pay the full 1d., some pay part of the cost, and the rest nothing; not more than 7 or 8 per cent. pay anything at all, but Mr. Libby thought this small percentage was due to the thing not being sufficiently worked. In spite of the fund there are still many underfed children, because there is not sufficient organisation; it is difficult to get sufficient organisation by voluntary methods.

Atkins, 2067–2074. 2081.

342. Mr. J. B. Atkins, the London Editor of the *Manchester Guardian*, gave a description of the Free Feeding System in Manchester. There the funds are derived entirely from voluntary sources, but the School Board recognised the charity and the teachers helped in distributing the food, &c. The movement has grown steadily, and in 1902, 139,000 free dinners were given, at the cost of a little over £400. In this case also the circumstances of the family are carefully investigated by the attendance officers, but no attempt is made to recover payment from the parents.

Chalmers, 6173–7.

343. In Glasgow Dr. Chalmers said there is an institution called the "Poor Children's Dinner Table," by means of which the condition of every child is enquired into, and meals are given to the underfed; the system is very comprehensive apparently, so that "no child in Glasgow ought ever to go to school starving," but no detailed account was given to the Committee.

Maurice, 279–283.

Sir Frederick Maurice, however, referred to the Glasgow system and stated that the difficulty as to parental responsibility had there been solved by means of a very thorough system of investigation, and that the applications for gratuitous food have diminished rather than increased.

Mackenzie, 6977. Mrs. Mackenzie, 7005–7013.

344. Free meals are given in Edinburgh to about 2,000 children, but the evidence given by Dr. and Mrs. Mackenzie makes it doubtful whether this number by any means exhausts the number of children who are underfed.

345. The most complete scheme described was started by the late Mr. George Dixon, and has now been in operation for 20 years in Birmingham with Dr. Airy, H.M.I., as chairman of the organisation. In considering the principles on which they would act it was decided in the first place that only those should be helped who could expect practically nothing if it was not given to them, and secondly that only such a meal should be given as would not compete in any way with the meal which could be provided even in a very poor home. It was next decided that cases for help should be selected

Airy, *passim*.

with the greatest care. This is done by three different people—by the head teacher of the school, by the class teacher in whose class the boy or girl is, and by the visiting officer. The co-operation of these three, Dr. Airy states, has been so successful that he does not believe there has been 5 per cent. or anything like it of abuse. The number of children fed in normal times is 2,500 and the plan pursued is thus described:—

Airy, 13253.

"We began with ten centres. We had large coppers for soup at ten centres, to which the children came from all outlying schools. The School Board allowed us in each of those centres to canvas off some twenty or thirty yards of playground, perhaps five yards wide, and the cooking was done at one end. There the soup, a good lentil soup with some animal stock, and the bread and jam, were prepared. The process was simply this: we had to do everything to simplify matters. It had to be a rough business, but it was an effective one. The children come, and form file, and then they walk up, and as they walk up they take a spoon out of a basket and go up to where the voluntary helpers are distributing the soup. They take their bowl of soup and go on to benches on the other side of the canvassed shed and sit down and eat their soup. The moment they have done they put their basin and spoon into another basket, and as they go out they take a large slab of bread and jam, and eat that in the street. The School Board allowed us to do this without any rent, and they gave us the gas. Then the cooking of those meals is done by paid labour, but the distribution is done by the voluntary help of ladies at each centre. There is a rota at each centre and there are two ladies who attend each day. Our manager I will refer to directly—he is a most capable, suggestive man. We were very much distressed at the fact that the children would come a mile or a mile-and-a-half to eat this poor dinner, and they would come through slush and snow and wet, and we wanted to prevent that. A system of baskets was invented. There is a system of baskets at present in use by which the soup can be kept absolutely hot for more than an hour. I have tried it at both ends, and I find it is almost as hot as when it comes out of the copper. We reduced the number of centres to four or five, and now all the outlying schools send their baskets with a paper saying how many dinners they want. Those dinners are put into the baskets at a quarter to twelve or twelve o'clock, and then the staff of the school help in distributing the meals at that school."

346. The Committee have thought it worth while to print this part of Dr. Airy's evidence in the body of the report because of the remarkable economic fact with which he concludes:—

"We give that dinner, a large bowl of soup—in fact, they have two or three bowls if they like—and a large slab of bread and jam, for less than a half-penny, and in that expense is included £150 or £100 a year to the manager." Airy, 13258.

347. He attributed this result (1) to the concentration of the population that has to be helped, (2) to the good will and assistance of the local authority, and (3) to the organizing skill of "a heaven-born manager," a retired naval officer, and to the co-operation of volunteers. In addition to those dinners, which provide for some 2½ per cent. of the children of school age, breakfasts, consisting of cocoa, milk and bread, are supplied by the bounty of a private individual to the necessitous children in about twenty schools in Birmingham, as Dr. Airy believed, under similar conditions. The testimony of the teachers is unanimous, that the system pursued enables the children to do the ordinary school work, and they report that the difference is perfectly extraordinary. Airy, 13272.

(c) Proposals in regard to the more systematic feeding of school children.

348. With scarcely an exception, there was a general consensus of opinion that the time has come when the State should realize the necessity of ensuring adequate nourishment to children in attendance at school; it was said to be the height of cruelty to subject half-starved children to the processes of education, besides being a short-sighted policy, in that the progress of such children is inadequate and disappointing; and it was, further, the subject of general agreement that, as a rule, no purely voluntary association could successfully cope with the full extent of the evil. Even those witnesses who were inclined to think that its magnitude had been much exaggerated, did not question the advisability of feeding, by some means or other, those children who are underfed, provided it could be done quietly and without impairing parental responsibility. The only witness who appeared absolutely to dissent from that view was the Bishop of Ross, who, while admitting an enormous number of underfed children in Ireland, deprecated any steps being taken to remedy the evil, on the ground that it would weaken the sense of self-respect and self-reliance both of parent and child. Kelly, 1130-88.

349. The purely medical view was well put forward by Dr. Robert Hutchison, a well-known authority on nutrition, nominated to give evidence by the Royal College of Physicians. He said, "looking at it purely scientifically, it would be an extremely important thing to ensure to every child at school a sufficient and proper sort of meal"; and again, "I feel certain that the provision of meals would do a great deal to improve the health and growth and development of the children of the poorer classes." It is worth noting, also, that he considers the ages of ten to fifteen as the most critical period, rather in opposition to the view expressed by some witnesses that the period of infant life is the most important. This witness further gave the Committee to understand, that a child ought to have a certain amount of nourishment during the twenty-four hours, but that it does not very much matter how you divide it up; except that it is more important to have a meal before physical exercise than before mental; and, therefore, mid-day dinner is more important than breakfast. This view is interesting, as differing from a very general opinion that no child should be allowed to come to school without sufficient breakfast. Dr. Hutchison further Hutchison, 9973, 10001-10015, 10067, 8.

thought that most children require a certain amount of animal food, and pure vegetarianism is not sufficient. He spoke entirely from the medical point of view, and refused to discuss the economic question.

350. The general trend of opinion is in favour of some sort of regularised feeding in school, or at centres, exacting payment from the parents where they are in a position to pay, but giving the meal free where they are not. Thus, Mr. Charles Booth is in favour of some sor of school restaurant in every school, or connected with every school, where food could be obtained at a minimum cost, but with no intention of giving it away; though he admitted that "in some cases it might be a mere charitable assistance." He is also in favour of some special industrial schools, where children, whose parents culpably neglect them, could be fed and boarded, the cost to be charged upon the parents.

Booth, 992-6, 1128-1133.

Brunton, 2428, 2450-2.

351. Sir Lauder Brunton thought it might be necessary to provide food at schools, and "in cases where the parents are absolutely unable to pay, food might be provided out of the poor rates;" but "every effort must be made to force the parents to pay properly for the food."

352. Mr. Atkins said,

Atkins, 3000, 3006-3018.

> "We have got to the point where we must face the question whether the logical culmination of free education is not free meals in some form or other, it being cruelty to force a child to go and learn what it has not strength to learn."

But he agreed that the parents should be made to pay, if possible.

Collie, 3955-9.

Rowntree, 5039-5041.

Mrs. Mackenzie, 7030-4.

Garnett, 9215-9.

353. Dr. Collie thought that underfed children should be fed by means of school kitchens, and that the parents should be prosecuted for neglect; a few prosecutions would have a salutary effect. Mr. Seebohm Rowntree advocated the judicious feeding of school children, so as not to pauperise the parents. Mrs. Mackenzie thought the child must be fed, but there ought to be power to "arrest the wages of the parents." Miss Garnett thought the children must be fed, and there were other opinions to the same effect.

Loch, 10192-9. 10242-6, 10271-6.

354. The opinion of Mr. C. S. Loch is worthy of consideration, as being presumably the official view of the Charity Organisation Society. He found fault with the existing systems of voluntary feeding, as "purely a movement against destitution without regard to education;" he stated his belief that no child should ever be fed without thorough investigation into the circumstances of its family, and no free meal given except in special cases, and then only as secretly as possible; but he admitted the necessity in special cases. The feeding should not be at the school, though it does not appear from his evidence where it ought to be. He instanced the difficulty in former days, before the Free Education Act of 1891, of getting educational fees out of parents, and argued there would be similar difficulty in getting feeding fees. Both Mr. Loch and Mr. Shirley Murphy thought that in cases of real destitution the Poor Law Administration should always be brought into play, and not kept out by any system of free feeding.

Murphy, 10402-10411.

Niven, 6349-6378, 6508-6510.

355. Dr. Niven propounded a definite scheme for ascertaining the fact of malnutrition, for feeding the child, and enforcing parental responsibility. The teacher would be taught to note all children obviously suffering or undersized, and to report them, if diseased, to a medical officer; if underfed, to the educational authority. The director would then instruct the attendance officer to ascertain the circumstances of the family; and the subsequent feeding of the child would depend upon the facts thereby elicited. It is worth noting that any such system would in the first instance be rendered easier by an organised medical inspection of schools, a proposal, as it has been seen, very widely and generally advocated.

Gorst, 11834.

356. The most uncompromising advocacy of public responsibility came from Sir John Gorst and Dr. Macnamara, and as the first-named appealed to the authority of the other, it is fair to treat his proposals as put forward

in the name of the two. Dr. Macnamara has based his recommendations on the following proposals of a Committee of the London School Board in 1898, which, however, do not appear to have commended themselves to the School Board of the day:— Macnamara, 12376.

Children attending School unfit for School Work.

(i.) It should be deemed to be part of the duty of any authority by law responsible for the compulsory attendance of children at school to ascertain what children, if any, come to school in a state unfit to get normal profit by the school work—whether by reason of underfeeding, physical disability, or otherwise—and that there should be the necessary inspection for that purpose.

Provision for Children sent to School "Underfed."

(ii.) That where it is ascertained that children are sent to school "underfed" (in the sense defined above), it should be part of the duty of the authority to see that they are provided, under proper conditions, with the necessary food, subject to the provision contained in Clause (vi).

Supervision of Voluntary Organisations by Authority.

(iii.) That existing or future voluntary efforts to that end should be supervised by the authority.

Voluntary Effort to be Supplemented if inadequate.

(iv.) That in so far as such voluntary efforts fail to cover the ground, the authority should have the power and the duty to supplement them.

School Dinners available for all Children, and Method of Payment.

(v.) That where dinners are provided it is desirable that they should be open to all children, and should be paid for by tickets previously obtained, which parents should pay for, unless they are reported by the Board's officers to be unable by misfortune to find the money; but in no case should any visible distinction be made between paying and non-paying children.

Prosecution of Parent for culpable Neglect.

(vi.) That where the Board's officers report that the underfed condition of any child is due to the culpable neglect of a parent (whether by reason of drunkenness or other gross misconduct) the Board should have the power and the duty to prosecute the parent for cruelty, and that, in case the offence is persisted in, there should be power to deal with the child under the Industrial Schools Act.

357. Dr. Macnamara was good enough to explain to the Committee how a scheme on these lines could be brought into practical application, and gave interesting evidence as to the success claimed for similar schemes in Brussels, Vienna and Paris. Upon the plan followed in the last-named city he estimated the outside cost of its application to London to amount to £120,000. He did not believe that any serious difficulty would arise in determining what children should be fed without paying, as "the attendance officer and the teacher together could bring you very near to the actual state of facts," and he was prepared to press the case against the parents who neglected their children with all the force at the disposal of the law. He advocated dealing with them by an extension of the Prevention of Cruelty to Children Act, or under the Education Acts by enabling the school authority to recover the cost of food in the same way as fines for non-attendance are recovered. The permission to parents to have their children fed at school, if they desired it, would, it was believed, greatly facilitate the operation of such a scheme, and Dr. Macnamara agreed that where parents made use of it as a convenience it would be fair to make them pay something in excess of cost price, which would to some extent diminish the charge that would in the last resort fall upon the community. The weak point in the system was admitted to be the difficulty of maintaining voluntary effort and providing public help at the same moment. Sir John Gorst, it is true, said the British public is very fond of doing some part of the work of the Government for them, instancing the lifeboat service, and thought private benevolence would still come in, though the school authority should be responsible for its efficient working; but Dr. Macnamara was obliged to acknowledge that in those towns abroad where the municipality had stepped in, the flow of charity had been arrested, and he was not prepared to think his scheme could be adopted without a substantial public charge. This he would allow, subject to increased powers of dealing with a parent who could and would not pay, which he regarded as an essential part of the scheme. Macnamara, 12429, 12438. Macnamara, 12425. Macnamara 12438. Macnamara, 12468 Gorst, 11915. Macnamara, 12384.

358. On a general survey of the evidence, and bearing in mind the considerations which form the subject of Paragraphs 329 and 330, the Committee think that a large number of children habitually attend school ill-fed, but this

number varies locally with the time of year and with the conditions of employment, and is not likely to increase —indeed they look, as they have said, with confidence to the operation of many causes towards its diminution.

359. It seems, further, that in a large number of cases voluntary organisations with the support and oversight of the Local Authority are sufficient for the purpose, and as long as this is so the Committee would strongly deprecate recourse being had to direct municipal assistance.

360. Circumstances, however, do arise which call for more immediate aid, and in which the School Authority, taking into account the difficulty in the way of home provision of suitable food, and the number of children who attend school habitually underfed, are willing to provide regular and sufficient meals, and in such cases the Committee agree with the opinion of the Royal Commission on Physical Training (Scotland), that "the preparation and cooking of these meals, where it is found necessary to provide them, "ought to be regarded as one of the charges incident to school management."

361. By a differentiation of function on these terms — the School Authority to supply and organise the machinery, the benevolent to furnish the material—a working adjustment between the privileges of charity and the obligations of the community might be reached.

362. In some districts it still may be the case that such an arrangement would prove inadequate, the extent or the concentration of poverty might be too great for the resources of local charity, and in these, subject to the consent of the Board of Education, it might be expedient to permit the application of municipal aid on a larger scale. As a corollary to the exercise of such powers—which should be by scheme sanctioned by the Board—the law would have to be altered so as to furnish means, as was suggested in evidence, to compel the neglectful parent to take his full share of responsibility, and the Committee are sanguine that a few prosecutions to this end would have a most salutary and stimulating effect.

363. It seems probable that the best way of dealing with many of these children, whose antecedents place them definitely in the category of "retarded," is by means of special schools of the Day Industrial Schools type, in which feeding would form an essential feature, and the choice between establishing such schools or merely treating children as underfed must be largely left to the Local Authority, after considering all the circumstances, to determine.

364. The Committee, moreover, do not think that children should be made the subject of either experiment without the concurrence of the Poor Law Authorities, and the funds should be found through the machinery of the Poor Law, with all due precaution against affixing any unnecessary stigma upon the deserving parent.

365. The Committee deem that by these means the community may be protected from the consequences of the somewhat dangerous doctrine that free meals are the necessary concomitant of free education. Education is a great social need, which individual citizens are, as a rule, not able to provide for their children on a sufficient scale, but food, like clothing and lodging, is a personal necessity, which in a well-ordered society it is not inherently impossible for parents to provide; and the effort to supplement their deficiencies, and to correct the effects of their neglect, should aim, in the first instance, at the restoration of self respect and the enforcement of parental duty.

viii. Risks of Contamination during Adolescence.

366. The Committee are impressed with the conviction that the period of adolescence is responsible for much waste of human material and for the entrance upon maturity of permanently damaged and ineffective persons of both sexes. The plasticity of the physical organization, the power it possesses of yielding rapidly towards degenerative or recuperative influences, appears

to terminate at eighteen, and the records of the years preceding that age are in the great majority of cases decisive for self-improvement or the reverse. Unfortunately, it is a period of which too little account is taken. With the classes under consideration education in the ordinary sense of the word is over just when in its full significance it becomes most necessary. Parental direction is almost entirely absent, and in lieu of it very little supervision is exercised in any other quarter over physical or moral development.

367. The Committee are not prepared with any comprehensive scheme of social regeneration, but there are agencies at work whose efforts deserve recognition and support, and there are ways in which the State might still exert some restraining and guiding influence in the interests of the adolescent of either sex.

(a) *Girls.*

368. Taking girls first, it is said, "the conditions under which they work, rest and feed doubtless account for the rapid falling off in physique which so frequently accompanies the transition from school to work." The secretaries of Girls' Clubs were quoted "that as soon as a girl leaves school she does not fall off immediately, but between fifteen and eighteen she begins to suffer, and if her work is too hard she does not recover." Hot rooms, unhealthy surroundings at work, bad food, late hours, excitement and stress of work, superadded to functional anæmia, are described as most detrimental during these years. The extension of the mid-day restaurant to meet the needs of adolescent girls was recommended, but the difficulty, in the opinion of one witness, was to get girls to feed themselves properly, however much restaurants might abound.

Eichholz, 435.
Eichholz, 566.
Hawkes, 13113.

369. Testimony, however, was forthcoming that girls, as a whole, during these years showed improvement. Prof. Cunningham noted a conspicuous diminution in the number of young women whose figures were distorted by tight lacing. Whether due to "improved æsthetic taste" or the greater prevalence of outdoor exercises, the important effect of the change is recognised by all familiar with the anatomy of the thoracic and abdominal cavities. Moreover, it was remarked as a common impression that girls were actually bigger than they used to be, a statement which, in the absence of anthropometric observations, it is of course impossible to prove.

Cunningham, 2220-5.
Eichholz, 551.

370. The opinion of the Hon. Maude Stanley is valuable on this point. From an experience of thirty-four years in connection with Girls' Clubs in Soho, a district which Mr. Charles Booth marks as one of the blackest on the map, this lady was able to assure the Committee that the physical conditions of girls had improved. Though 50 per cent. of those that come under her review are working as tailoresses under circumstances which cannot be particularly favourable, she is satisfied that they are better grown and likely to do better towards their children than were their mothers. Even in a poor district in Walworth, where a club was recently started for the benefit of a most demoralised class, the girls who joined it were not physically deteriorated. Miss Stanley attaches great value to the contact with helpful and sympathetic women of a superior class which these clubs establish, and to the ten days or a fortnight in the country which most of the members enjoy through the medium of their organization. Miss Stanley was able to add, owing to maintenance of relations with members after they married, that they proved competent housewives and were greatly in request upon that ground.

Stanley, 13381-8.
Stanley, 13408.
Stanley, 13465.

371. Unfortunately, the number of these clubs is small and the sphere of their operations far too limited to produce any general effect. Mr. Douglas Eyre, who spoke with authority on this point, confirmed the evidence of Miss Stanley that much less was done for girls than boys in this respect. Mr. Eyre enumerated the principal organisations that have for their object the strengthening and development of the work of institutions for the welfare of girls, but, so far as it was possible to collect reliable returns on the subject, he did not believe there were more than 5 per cent. of the youthful portion of the industrial population who were materially touched or assisted by

Eyre, 3555-3563.

anything in the shape of a well organised recreation agency out of school or working hours.

372. The Committee have already (paragraph 230) touched upon the subject of obligatory evening continuation classes for girls beyond school age, which should have for their aim the instruction of girls in the objects of personal and domestic hygiene, and a large body of opinion was in favour of the experiment being made; and if physical exercises of a recreative character were included in the curriculum, the Committee believe it would add to the value and ultimately, it may be, to the popularity of such classes.

373. Wherever it was thought desirable, owing to the employment of married women in factories, or for other reasons, to establish municipal crêches, girls over 14 might be made to attend occasionally, and the teaching of infant management should rank with other forms of technical instruction in the way advocated by Miss Eves.

Eves, 7708–7719.

(b) *Boys.*

Eyre, passim.

374. Turning to boys of the age under discussion, it is not too much to say that in their case physical training contains the most fruitful germ of moral and material well-being. The evidence of Mr. Douglas Eyre, who has been for years connected with the organisation of boys' clubs and other associations which rest on a basis of physical exercise, affords striking testimony to their value, but indicates at the same time how small a field is covered by existing effort owing to lack of funds, scarcity of personal service, and want of systematised co-ordination.

Bagot, passim.

Eyre, 3574–8.

Bagot, 4693–9.

Bagot, 4564–7.

375. The work of a cognate character described by Mrs. Josceline Bagot, who for many years has maintained a Boys' Club in Lisson Grove, shows what may be effected by the single-handed exertions of an intelligent and sympathetic individuality towards giving lads of a very poor class that motive for self-preservation from contaminating influences which is the very bed-rock of self-respect, and it is not a little singular that so obvious an outlet for philanthropic energy should have received such scanty support, either personal or pecuniary. One reason appears to be the reluctance of some of the clergy to associate themselves with anything that is not primarily religious; as Mr. Booth puts it, "Obligatory attendance at a Bible class being administered medicinally, with cricket and football to take the taste away." Boys entering the club physically feeble are greatly improved by a few months' systematic training, and Mrs. Bagot's experience shows that the interest a lad acquires in physical fitness instinctively weans him from idleness and temptation to drink, while teaching the roughest to box gives them a chivalrous sense of obligation to the weak. Material improvement thus becomes the natural ground-work upon which moral and religious impressions are afterwards built up; but all these results are too apt to be lost by prematurely forcing religious interests upon the attention.

Rees, 4318–4341. 4427–4442; 4506–8. Horsfall, 5667–9.

Rees, 4413–8, 4498–4501. Bagot, 4618. Atkin, 2918–2922.

Eyre, 3642–3652, 3692–5, 3710–9.

Eyre, 3532–5, 3548–3564, 3578–3584.

376. The evidence on this subject was not confined to London; the Rev. W. Edwards Rees and Mr. Horsfall testified most forcibly to the need of similar efforts in Lancashire. Mr. Rees considered the lack of physical training to be one of the two main causes of degeneration, and would make such training universal for boys and girls of all ages; both gentlemen deemed the provision of facilities for physical exercise and enjoyment as the most pressing question of the hour, and Mr. Rees in common with Mrs. Bagot and others was emphatic in attaching great importance to swimming. Mr. Douglas Eyre commented on the lack of encouragement given to physical education in continuation schools, and the difficulty, in the absence of some State system of training, of finding competent instructors. He also described the organisation and objects of the Twentieth Century League, which he hoped in due course would provide the machinery for intelligent co-operation among the various voluntary agencies in the field, and thus give the State a firmer foothold for direct intervention and support.

377. It is not a little curious, however, that a league with these objects should not have come to the knowledge of the persons engaged in the formation

of the National League for Physical Education and Improvement, which is apparently being promoted by Sir Lauder Brunton and others, with similar aims. That two leagues should be in the field for the same ends appears to the Committee a signal instance of that waste of energy and overlapping which the existence of each league is presumably designed to counteract. Brunton, 2430, 2454-7, 2537-9.

378. Mr. Murphy was impressed with the number of boys employed on errands and as messengers (by which means they begin to earn money younger) who have no technical training and therefore subsequently drift into the ranks of unskilled labour. He threw out suggestions as to making them join cadet corps, with a view to their being fitted for military service, and at the same time teaching them a trade upon which they could engage after leaving the Army, if they enter it. In this connection it is interesting to note that, in Mr. Edwards Rees' opinion, the Lancashire working men at all events would not be afraid of some form of general military service which did not imply absence from home. Murphy, 10440. Rees, 4336-4341, 4503-8.

379. With a view to giving the people every possible opportunity for physical recreation, the Committee think it should be the duty of municipalities to provide and maintain open spaces in some proportion to the density of the population, and that such spaces, or some of them, should include shelters fitted with gymnastic apparatus, these last being put in the charge of competent instructors; and having regard to the paramount importance of the national physique to the community at large, they are also of opinion that some grant should be made from the National Exchequer in aid of all clubs and cadet corps in which physical or quasi-military training, on an approved scheme, is conducted, subject to public inspection, which grant might be provided without any additional charge to the Treasury, by insisting that a certain proportion of the sums already paid to the local authority under the Local Taxation (Custom and Excise) Act, 1890, should be devoted to physical education.

380. The Committee also think that the obligation should be laid on boys to attend continuation classes, in which drill and physical exercises should take a prominent place; and with a view to the encouragement of clubs and cadet corps, exemption from the obligation might be granted to all enrolled and efficient members of such organisations as submitted to inspection and conformed to the regulations qualifying them for public aid. By these means, without recourse being had to any suggestion of compulsory military service, the male adolescent population might undergo a species of training that would befit them to bear arms with very little supplementary discipline. The older lads could actually be familiarised with the use of the rifle, an exercise of no inconsiderable value from the point of view of general education, and a great deal might thus be done which would not only provide partly prepared material for absorption into the army or reserved forces, but would give a tone and a carriage to all that came under the influence of the system.

381. Direct information on the working of a cadet corps was furnished by Mr. Bennett, Captain of the 1st Cadet Battalion "The Queen's" Royal West Surrey Regiment. His letter on the subject, and some figures illustrative of his experience are printed in the Appendix. Their general effect, so far as it goes, is to negative the presumption of progressive deterioration in the class concerned. Appendix XI

382. In order to organize existing efforts on a comprehensive and effective basis, the Committee would like to see a central body (whether a branch of the Twentieth Century League or another), in touch with municipal activity, established in every large town, and charged with the duty of supervising and directing voluntary agencies of all kinds and bringing them up to a minimum standard of efficiency. One extremely important field of operation for such an organisation in each town would be the collection of information as to the various kinds of employment open to young people on leaving school—the conditions of employment, rates of remuneration, and relative permanence—with a view to advising them in their choice of an occupation, and thus minimising the evil effects of the kind of temporary and casual employment referred to in paragraph 378.

(c) *Juvenile Smoking.*

383. The question of Juvenile Smoking at the period of life dealt with in this and the preceding sub-section has been given some prominence in evidence, and the Committee have received communications on the subject from the late London School Board and one or two Anti-smoking Leagues. The evidence submitted on the point represents a practically unanimous opinion that the habit of cigarette smoking among boys is a growing one, and that its consequences are extremely deleterious. No actual testimony was forthcoming to prove that early smoking diminishes growth, but Prof. Cunningham mentioned it as one of the causes of physical deterioration, and Dr. Scott was of opinion that scarcely two per cent. of cases of undergrowth had not been habitual cigarette smokers : the experience of a schoolmaster at Longton was quoted to show a distinct inferiority of physique in the case of boy-smokers as against non-smokers, and Mr. Atkins adduced evidence from Colonel Leetham, the late Chief Inspector of Recruiting in Manchester, who has said that "perhaps a third of the rejects from the Army in Lancashire might be attributed to smokers' heart." This is, no doubt, an excessive estimate, but it shows one bad consequence of early smoking, and it is common knowledge that smoking affects the wind and general physical capacity.

Cunningham, 2346–9, 2369–2372. Scott, 1766–9. Garnett, 9261–4. Atkins, 2954–9.

384. There appear to be two ways of dealing with the matter which might be of good effect and could be made the subject of a very simple Act of Parliament.

(1.) To prohibit the sale of tobacco and cigarettes to children below a certain age.

All the witnesses were agreed in advocating this step, but there was some difference of opinion as to the age which should be selected. The Committee think that it should at least be sixteen.

Eyre, 3603. Fosbroke, 6642–6. Chalmers, 6202–4. Garnett, 9134–5.

(2.) To prohibit the sale of tobacco and cigarettes in sweet shops and other shops frequented by children.

This appears to be a growing practice, though it is impossible to discover the number of such shops licensed for the sale of tobacco. The Committee communicated with the Inland Revenue Office on the subject, and their reply was to the effect that "a licence to deal in tobacco is granted to anyone who asks for it, and no record is kept of whether the holder of such a licence is solely a tobacconist or whether he combines with that business any other trade." Mr. Eyre was of opinion that sweet-stuff shops should not be so licensed, and called attention also to the gambling propensities encouraged by such shops. The Committee agree with the weight of the evidence that licenses for the sale of tobacco should not be given to these shops.

Eyre, 3605–8, 3678–3687. Rees, 4410–2. Lamb, 11754–6.

APPENDICES.

APPENDIX I.

(a) ORIGINAL MEMORANDUM PREPARED BY SURGEON-GENERAL SIR WILLIAM TAYLOR, K.C.B., DIRECTOR-GENERAL, ARMY MEDICAL SERVICE.

1. A deep interest has been aroused, both in the lay and medical press, by the writings of Sir Frederick Maurice and others, who have brought into prominence certain observations pointing to the fact that there is an alarming proportion of the young men of this country, more especially among the urban population, who are unfit for military service on account of defective physique.

The questions naturally arise as to whether this impeachment of the national health has a solid foundation in fact, and as to whether the condition is true of the population as a whole, or only of a certain section of it. The teaching of public health statistics would appear to show that progressive improvement of the national health has steadily followed the improved conditions of life which have been brought about by the advance of sanitary knowledge and its practical application. It has also been pointed out that athletic records are constantly being broken for all sorts of feats of strength, agility, and endurance, facts which would seem to indicate that the physique of the well-to-do classes, at least, is improving rather than deteriorating. It is nevertheless true, and the fact is a disturbing and disquieting one, that a very large proportion of the men who offer themselves for enlistment in the Army are found to be physically unfit for military service.

2. In an article on the National Health, which appeared in a recent number of the *Contemporary Review*, Sir Frederick Maurice states that, according to the best estimate he had been able to arrive at, it has been for many years the case that out of every five men who wished to enlist, and primarily offer themselves for enlistment, you will find that by the end of two years' service there are only two men remaining in the Army as effective soldiers. Of the men who offer themselves, some are rejected by the recruiting sergeant or recruiting officer, some by the examining medical officer, and some, though enlisted, are found after three months to be unlikely to develop into effective soldiers and are summarily discharged. According to General Maurice's experience, at the end of two years not more than 40 per cent. of the men who wished to become soldiers will be found serving; or, in other words, 60 per cent. of the men offering themselves are physically unfit to serve as soldiers. He points out that it is no good talking of conscription or of any form of compulsory service if we already have five men offering themselves for every two men who are fit for the work; no one has suggested that we should increase our Army in the proportion of two to five, *i.e.*, make it two and a-half times as large as it is now. He then goes on to say that no nation was ever yet for any long time great and free, when the army it put in the field no longer represented its own virility and manhood.

3. But the want of physique, thus shown to exist with regard to a large section of the community, is not only serious from its military aspect, it is serious also from its civil standpoint, for if these men are unfit for military service, what are they good for? As Sir Lauder Brunton says—"Poor in physique as they all are, and poor in mental capacity and power of application as many of them must be, what becomes of them? Many of them probably marry girls as weak as themselves, and have children, some of whom go to swell the lists of infant mortality, some to join the criminal classes, while others grow up more weak and incompetent than their parents." Inquiry is wanted, and it is vital for us to know the truth. Whether part of the physical deterioration is the result of unskilled labour flocking to the towns and there failing to find means for properly rearing a family, or whether it be on account of causes which are attackable, such as early marriages and the ignorance of mothers; the result is that the rising generation of all below the artizan class includes a vast number of men of a very low standard of health and physique.

4. Regarding the condition of the poorer classes, Mr. B. S. Rowntree read a paper at the British Medical Association Meeting, last August, on Poverty and Disease (*British Medical Journal*, 16th August, 1902). His observations refer to York, which has a population of about 75,000. In discussing the question he speaks of poverty under two heads, primary and secondary poverty. He defines "primary poverty" as the condition when the total earnings are insufficient to obtain the minimum necessaries for the maintenance of physical efficiency in the family. "Secondary poverty" is when the earnings would be sufficient if some part of them were not wasted. He found that in York 10 per cent. of the total population were living in "primary poverty," and that of this 10 per cent. just one-half (52 per cent.) is due to the fact that the chief wage earner, though in regular employment, has wages which are insufficient to maintain a family of moderate size in a state of physical efficiency. 18 per cent. of the population were living in a state of "secondary poverty," so that, adding the two classes together, 28 per cent. of the population in York are living in poverty. Some years ago Mr. Charles Booth estimated that in London 30 per cent. of the people were living in poverty. If then the same conditions prevail in other large towns, it would appear that more than a quarter of our town populations are living at, or below, the poverty line. Now in England and Wales, at last year's census, 77 per cent. of the population was urban; namely, 25 millions out of 32½ millions; the town population having increased 15 per cent. during the last decade.

5. Every year a table is published, in the Army Medical Department Report, which classifies the recruits examined according to their previous occupations. The following is the table for 1900. It has not been selected for any special reason, but is given as an average example of our recruiting experience:—

Occupation of recruits.	Number inspected.	Number rejected.	Ratio rejected per 1,000 inspected.	Proportion of each group in 1,000 recruits inspected.
1. Labourers, servants, husbandmen, etc. -	52,022	15,025	288·82	616
2. Manufacturing artizans (cloth-workers, weavers, lace makers, etc.) - - -	11,971	3,478	290·54	142
3. Mechanics employed in occupations favourable to physical development (smiths, carpenters, masons, etc.) - -	11,201	2,923	260·96	133
4. Shopmen and clerks - - - - -	5,950	1,826	306·89	70
5. Professional occupations, students, etc. -	827	220	266·02	10
6. Boys under 17 years of age - - -	2,431	273	112·30	29
TOTAL - - - - - -	84,402	23,745	281·33	1,000

6. Examination of a series of these annual tables shows that the proportion of the different classes remains remarkably constant from year to year, and the figures indicate that the bulk of our soldiers are drawn from the unskilled labour class, and consequently from the stratum of the population living in actual poverty or close to the poverty line. As might be expected, the highest ratio of rejection is shown for men who have been following indoor occupations.

7. The impairment of vigour and physique among the urban poor is easy to understand when we reflect that, in addition to their only being able to provide themselves with food insufficient in quantity and probably poor in quality, their poverty also usually entails unhealthy environment, *e.g.*, defective housing, overcrowding and insanitary surroundings. Add to this the distress resulting from such causes as want of thrift, illness or death of the breadwinner, and alcoholic excess. Further, the physical deterioration caused by inherited or acquired disease must not be forgotten, and in illustration we need only instance the part played in this way by tubercle and syphilis.

8. In his Annual Report for 1902, just issued, the Inspector-General for Recruiting remarks that the one subject which causes anxiety in the future as regards recruiting is the gradual deterioration of the physique of the working classes from whom the bulk of the recruits must always be drawn, and, when it is remembered that recruiters are instructed not to submit for medical examination candidates for enlistment unless they are reasonably expected to be passed as fit, we cannot but be struck by the percentage considered by the medical officers as unfit for the service. In the reports from all the manufacturing districts, stress is invariably laid upon the number of men medically rejected for bad teeth, flat feet, and inferior physique.

9. Examining next the Army recruiting statistics in relation to the assertion that practically 60 per cent. of the men offering themselves for enlistment are unfit for military service, the following table has been compiled from information given in the Army Medical Department Reports, supplemented in some particulars by data obtained from the reports of the Inspector-General of Recruiting. A period of 10 years (1893–1902) has been selected, as of course, the greater the number of observations dealt with, the nearer will be our approximation to the truth.

Year.	1. Number of recruits inspected (A.M.D. Report.)	2. Number rejected on inspection. (A.M.D. Report.)	3. Number rejected within 3 months after enlistment. (A.M.D. Report.)	4. Invalids discharged during the year under 2 years' service. (I.G.R. Report.)	5. Ratio per cent., column 2.	6. Ratio per cent., column 3.	7. Ratio per cent., column 4.
1893 - - - - - -	64,110	25,999	342	962	40·6	0·5	1·5
1894 - - - - - -	61,985	24,705	369	770	39·9	0·6	1·2
1895 - - - - - -	55,698	22,548	368	952	40·5	0·7	1·7
1896 - - - - - -	54,574	22,698	413	999	41·6	0·8	1·8
1897 - - - - - -	59,986	22,370	575	997	37·3	1·0	1·7
1898 - - - - - -	66,502	22,983*	387	983	34·6	0·6	1·5
1899 - - - - - -	68,087	22,071	433	1,003	32·4	0·6	1·5
1900 - - - - - -	84,402	23,105	640	1,514	27·4	0·8	1·8
1901 - - - - - -	76,750	21,522*	1,014	3,825	28·0	1·3	4·9
1902 - - - - - -	87,609	26,913*	1,308*	2,254	30·7	1·5	2·5
1893–1902 - - - - -	679,703	234,914	5,849	14,259	34·6	0·9	2·1

* Does not include men enlisted in 1902 and discharged under 3 months' service in 1903.

10. It will be observed that during this decennial period the number of men medically examined for enlistment was 679,703, and of those 234,914 were rejected as medically unfit for service, giving a rejection ratio of 34·6 per cent.; of the men passed fit, 5,849 broke down within three months after enlistment, being at the rate of ·9 per cent. for this class; while 14,259, or 2·1 per cent. more, were discharged as invalids under two years' service. The smallness of the rate of the rejections within three months of enlistment, varying as will be observed, between ·5 and 1·5 per cent. speaks well, I think, for the thoroughness of the primary medical examination of recruits. But the rejection of one out of every three men examined by the recruiting medical officer points clearly to the poorness of the human material available for army purposes, as a writer in the *Lancet* puts it. Adding together the rates for the three classes of rejections referred to in the table, we find that 37·6 per cent. of the 679,703 men examined during the decennial period proved to be unfit for military service. This percentage of rejections does not, however, represent the whole extent of the physical unfitness existing among men wishing to become soldiers, and offering themselves for enlistment. The Inspector-General of Recruiting states in his report for 1902, that it must be borne in mind, when examining these totals, that they do not represent anything like the total number of the rejections of candidates for enlistment into the Army. A large number of men are rejected by recruiting sergeants and recruiting officers and such men in consequence are never medically inspected and do not appear in any returns. In the decennial period under consideration we have only been able to account for 37·6 per cent. of rejections from official statistics; but according to Sir Frederick Maurice's estimate, 60 per cent. of the men who offer themselves are unfit for service. This indicates that the number of men turned away by the recruiters themselves as unlikely to have any reasonable chance of passing the medical examination is an appallingly large one. As already stated no official record is kept of the number of men turned away; but there is reason to believe that the number is a large one, though whether it is sufficiently large to bring up the figures of the rejected to 60 per cent. we have no means of saying.

11. Examination of the statistics dealing with the causes of rejection by the examining medical officers of men seeking enlistment, brings into prominence the fact that the majority are rejected on account of causes indicating poor physical development, namely, under chest measurement, under height, and under weight. Defective vision may also be regarded in many cases as resulting from developmental defect. The following table gives the statistics relating to the principal causes of rejection for the twelve years 1891–1902:—

CAUSES of Rejection of Recruits on Inspection, with ratio per 1000.

1891–1902.

Cause of rejection.	1891.	1892.	1893.	1894.	1895.	1896.	1897.	1898.	1899.	1900.	1901.	1902.
Under chest measurement -	93·03	95·90	108·55	110·27	126·38	139·64	89·44	73·88	65·84	59·84	49·88	56·72
Defective vision- - -	40·35	42·35	41·51	42·90	39·88	40·72	41·15	42·64	41·99	36·42	35·84	39·23
Under weight - - -	32·47	27·62	39·99	39·61	36·58	35·95	45·58	34·82	33·84	28·52	25·15	21·72
Under height - - -	26·76	32·71	33·24	28·67	28·72	28·77	24·86	21·79	20·21	15·18	13·56	11·59
Imperfect constitution and debility	18·40	9·87	9·47	5·00	3·57	4·44	4·45	5·49	5·82	4·94	3·36	3·91
Disease of veins- - -	16·39	16·24	17·11	15·84	15·85	15·72	15·42	15·74	14·22	11·69	13·98	12·30
Disease of heart- - -	16·06	13·87	17·74	19·62	20·71	18·76	17·67	17·26	15·69	13·15	16·74	17·33
Defects of lower extremities	15·57	17·09	14·40	17·44	18·16	18·14	18·12	17·72	13·98	10·53	10·35	12·27
Varicocele - - - -	12·93	11·85	12·85	14·25	12·28	13·07	13·07	12·29	12·16	11·21	13·89	12·59
Flat feet - - -	11·04	9·83	12·45	14·71	13·16	17·81	16·79	12·24	12·31	9·02	11·66	12·44
Loss or decay of teeth -	10·88	14·56	15·33	16·26	17·95	19·75	24·16	26·34	25·29	20·02	26·70	49·26

It will be observed that bad teeth and flat feet, causes of rejection to which considerable importance was attached in General Maurice's paper in the *Contemporary Review*, occupy a comparatively low place in the list. But with regard to loss or decay of teeth, it must be pointed out that the numbers rejected on this account during the past four or five years have shown steady increase, until this cause of rejection has come to regularly occupy a high place on the list. Whether the increase in the rejections for bad teeth is an indication of increasing prevalence of physical unfitness is open to question, the increase may partly, at least, be due to the more common use of articles of food which readily undergo acid fermentation, and partly also to examining medical officers having gradually come to place a high value on soundness of teeth as a matter of the greatest importance in its relation to the maintenance of the physical efficiency of the soldier on service. The main inference to be drawn from a consideration of the figures in the table is that the bulk of the rejections arise from causes indicating the operation of agencies antagonistic to healthy physical development.

12. It has already been stated that a large proportion of the population live in towns, and this has been estimated at 77 per cent., or 25,000,000. Of this town population about 25 per cent. (probably at least 6,000,000), appear from trustworthy investigations, to be not only poor, but living in actual poverty, so as to be unable to rear their children under conditions favourable to health and physical fitness. The bulk of the men who seek enlistment belong to this section of the population, and a very large proportion (but probably not quite three out of five as stated by General Maurice) of the men who wish to join the Army prove physically unfit for military service. It is not claimed that the same proportion of poor materia will be found generally distributed amongst all classes of the community. In an earlier paper in the *Contemporary Review*, on "Where to get men," General Maurice remarks regarding this point, "Of course, as yet, that stern and brutal fact of the two in five does not mean that of the whole male population of the kingdom there are only two out of five who are fit to bear arms."

In the concluding paragraphs of his paper on "National Health: A Soldier's Study," Sir Frederick Maurice states: "My object is to call upon the great profession, whose immediate concern is health, to give us the guidance and leading we need, and primarily it seems to me that we ought to call upon the Councils of the Colleges of Physicians and Surgeons, as *ex officio* the great national boards of health to help and guide us." I should suppose that they have not at this moment, despite the census, sufficiently comprehensive data on which to pronounce, but if that be so, no Government could or would wish to resist an appeal from them for assistance in getting at the truth on the tremendous question which has been raised by the investigations of Mr. Rowntree: "Is it or is it not true that the whole labouring population of the land are at present living under conditions which make it impossible that they should rear the next generation to be sufficiently virile to supply more than two out of five men effective for the purposes of either peace or war."

The question at issue constitutes a problem by no means easy to solve. Were all classes of the community able to provide their offspring with ample food and air space, a healthy race would be produced, and the proper material to fill the ranks of the Army would probably soon be obtained.

13. Information is wanted as to the causes of physical deficiency and as to the best available methods of remedying defects and improving the national health. Such an inquiry might fitly be undertaken by a commission as to the composition of which the advice of the Colleges of Physicians and Surgeons might be asked. As the matter is one of the utmost importance from the recruiting point of view, it is suggested that the Secretary of State might take the initiative in the matter of getting the opinion of the Councils of the Colleges with regard to—

(*a*.) The necessity for such an inquiry.

(*b*.) The ground to be covered by a commission if appointed.

(*c*.) Composition of commission.

W. TAYLOR, Director-General,
Army Medical Service.

War Office,
2nd April, 1903.

APPENDIX III.

SOME RECENT INVESTIGATIONS AS TO THE NUMBER OF THE "POOR" IN THE COMMUNITY.

By Mr. C. S. Loch.

1. There are two counts of the "poor" which have attracted much attention and which, in my opinion, require more close study than they have received, if their real nature and limitations are to be understood. The counts to which I refer are those made by Mr. Charles Booth in regard first to the East of London and Hackney, and then to the whole Metropolis, and those made by Mr. Seebohm Rowntree in regard to York.

It is not possible to ascertain the limitations of those counts and the qualifications which must be borne in mind in making use of them without a very attentive regard to the text which accompanies the figures and tables. The figures and the text must, so to speak, be read together as one expression, a not very easy task in any case and a task quite beyond the energy or perhaps indeed the ability of the general reader who is not specially interested or well versed in the details which underlie the calculations. Consequently such readers, if they are desirous of arriving at their conclusions quickly and by way of reference rather than slowly, and by the difficult progress of accurate and proven data, seize the figures and forget the text. The text thus becomes little better than a footnote left unread or hastily skimmed, though in fact, if the conclusions are to be tested, it is even more important than the figures.

Such a result cannot but appear serious to those who are dealing personally with questions of social life and habit, and are anxious that the people at large should take a sane and accurate view of social conditions. Above all is it necessary in estimating the improvement or deterioration of the national life, during the past generation, to analyse these counts and classifications of the population—if their evidence is to be admitted as having any exceptional validity.

I have used the word "counts" in the above paragraph. In fact the calculations in these books are only in part counts. To a very large extent they are estimates.

2. To ascertain the nature of these counts and estimates the following points have to be considered: (1) the general method; (2) the methods of collecting the particulars on which the classification is based; (3) the classes—the utilisation of these particulars in forming the estimates and conclusions which are submitted, and (4) the conclusions and recommendations themselves.

I begin with Mr. Charles Booth's book on the "Life and Labour of the People."

"*Life and Labour of the People.*"

THE GENERAL METHOD.

3. Mr. Charles Booth states his object to be: "To show the numerical relation which poverty, misery and depravity bear to regular earnings and comparative comfort, and to describe the general conditions under which each class lives." "I have throughout leaned to the safe side," he says, "preferring to paint things too dark rather than too bright."*

He aims at giving a picture, partly verbal and descriptive, partly statistical. He counts and estimates the people as they are as nearly as may be, at a precise moment, as if he were reporter and photographer combined. He makes no comparison between past and present. He collects no data for such a comparison, nor are his conclusions based on any observation of the changes in the development of the people, their present with their past, numerically or otherwise. He thus passes by all questions of progress or decadence, for in regard to them the comparative method is necessary. He works on a flat surface, so to speak; he cannot measure past and present or account for the present by the past, he can only measure quantities as they appear side by side on a certain date and from that comparison can only presume the conclusions.

*"Life and Labour of the People" I. pp. 5 and 6 (First Edition).

Now this is obviously in itself a very misleading method. Without the ascertainment of causes, conclusions cannot rightly be drawn; and causes do not lie on the flat surface, the momentary view of things. They can be ascertained only by the observance of their growth. In this method the observation of growth or development is wanting. Hence it does not supply the data which are necessary if any conclusions are to be drawn.

4. Perhaps I may be allowed to give an instance of this in reference to the statistical side of the question, though the principle affects the descriptive as much as the statistical elements of the method.

Mr. Charles Booth, the late Dr. Hunter, and some other writers adopted the method to which I refer in the controversies as to Poor Law relief some ten or twelve years ago. Each took a particular date and argued from the number and proportion of in and out paupers in unions grouped together in various ways. The instances of the fallacy are perhaps more striking in Dr. Hunter's paper "Outdoor relief—is it so very bad?" *Contemporary Review*: March, 1894) than in some of the other documents. I will therefore refer to it more particularly.

Dr. Hunter, like Mr. Booth, took one particular date—that of Mr. Ritchie's return "Poor Law (Indoor and Outdoor Relief)" No. 265, 1 Jan., 1892. He grouped the poor law unions according to their population and in other ways. The unions thus grouped he arranged in two sub-groups or sections—one which he called "outdoor relief unions" and another which he called "indoor relief unions." To form these sections (1) he calculated the percentage of outdoor paupers to total paupers in each union of the group; (2) he calculated the mean of these several percentages; (3) he classed as an "outdoor relief union" and placed in that section each union in which the percentage of outdoor paupers to total paupers was *more* than this mean; (4) he classed as an "indoor relief union" and placed in that section each union in which the percentage of outdoor paupers to total paupers was *less* than this mean.

Thus to take an instance:

Group: Unions in Lancashire.

Mean of percentages of outdoor to total paupers in the general Lancashire unions* (as above) 66·9.

Section "Out-door Unions."

Rochdale (and others—with percentage of out-door paupers to total paupers in excess of the mean) 70·4.

Section "Indoor Unions."

West Derby (and others—with percentage of outdoor paupers to total paupers less than the mean) 56·7.

From figures thus compiled—with unions arranged in groups and the groups in turn compared with other groups, he drew various conclusions. Amongst others for instance that "the policy of indoor relief drives into the house." This was an inference assumed from the fact that according to his arrangement of unions there appeared at a particular date in a group or groups a larger number of paupers in the section of "indoor" unions than in the section of "outdoor" unions.

But from this momentary view of pauperism no such inference could be drawn.

First of all, the policy of Boards of Guardians is not so clear and settled that they can be placed in sections on the basis of outdoor to total paupers. Unions both below and above the mean are working usually by haphazard decisions and with no determinate policy at all. So far as such a policy is concerned, it is a mere chance whether they come above or below the line of the mean.

Next when a Board has a definite and declared policy, the inference that indoor relief drives into the house is

* No detailed calculations appear in Dr. Hunter's paper. The above figures are worked out on his method and to explain it.

falsified, not by the figures of the moment which afford no basis for judgment one way or the other, but by the test of the comparative method.

Take, for instance, the case of a union which is classed "indoor." It is not mentioned by Dr. Hunter by name, but it is included in his groupings. Thirty years before the particular date the paupers numbered 1,019—in two groups—one indoor of 107, one outdoor of 912, and the policy of the guardians had been to reduce outdoor relief. Then in the course of the thirty years many recipients of outdoor relief grew old and died, and in accordance with the policy of the guardians their places were taken by very few new recipients. A few of the old recipients went into the workhouse or infirmary as age and helplessness came on them, and also a few others were admitted to these institutions. Then at the particular date the figures stood thus: 100 paupers in all instead of 1,019; of these sixty-nine in receipt of indoor relief instead of 107 and thirty-one in receipt of outdoor relief instead of 912.

Thus the fact as noted on the particular date in this (and in other instances) was taken as indicating a policy and the conclusion to be drawn from that policy. The fact indeed was taken as equivalent to the cause; because on the particular date there were more indoor than outdoor paupers; therefore indoor relief drives into the house. But the fact is itself a resultant of causes which have previously been at work and which only an analysis of previous facts and conditions can indicate. Isolated from them it has no real meaning. To affirm that to be a cause which is only a statement of fact, as disclosed on a particular date, is really to assume the cause, not to ascertain it—to assume the cause without having collected, analysed, or considered the evidence which would show what the cause or causes were.

In this (and in other instances) on the application of the comparative method it is evident that the inference from the figures of the particular date is quite incorrect. The paupers have not been driven into the house but the result has been quite the reverse.

A generation which began with 1,019 paupers concludes at thirty years later with 100; and the potential pauperism is reduced to that extent. The paupers who, on another policy of administration, would have taken the place of the (1,019—100) 919, are not driven into the house, but into the world of self-support.

Mr. Charles Booth's book on Pauperism is full of instances of conclusions of many kinds drawn from one day counts of paupers "in" and "out" in relation to population, and they are, as can be shown again and again, quite fallacious for the reasons given above.

In the "Life and Labour of the People" the method, so far as the statistics are concerned, is the same—though only to a small extent, comparatively, are problems of pauperism there treated. But so far as figures are given of the "poor" and "very poor" etc., it should be remembered that, putting aside for the moment other points in the method, no conclusions can be drawn from the figures themselves. No one can conclude from them in any way whether the people are more or less independent, or more or less poor. They throw no light on the relative condition of the people, and thus Mr. Booth's object "to show the numerical relation which poverty, misery, and depravity bear to regular earnings and comparative comfort" remains unachieved. In the classes A,B,C,D, there is poverty and clearly to some extent misery, using that word to mean extreme poverty, as presumably in A and B, but depravity as a cause of poverty is like other causes necessarily omitted everywhere from the enumeration. Nor is "the relation between poverty, depravity and regular earnings" worked out. His classes "C intermittent earnings" and "D small regular income" are separated only to be merged altogether and finally; so that the distinction of "regular earnings" itself disappears, just at the point where its influence on social habit should be most carefully observed. On the non-comparative one-day method of count and estimate this could hardly be otherwise.

The Method of Collecting the Particulars.

6. Information has been obtained from School Board visitors. Of the total area dealt with part is called "tested." The tested area covered "fully one-half" or "from one half to two thirds" of the whole population.

The whole area consists of seven districts, Shoreditch, Bethnal Green, Whitechapel, St. George's, Stepney, Mile End, Poplar and Hackney. The population consists of 891,539 persons. The figures in regard to the non-tested part of this area are arrived at by a proportion as against the figures in the tested part. The latter thu becomes the gauge and standard of comparison. It is not stated however, what is the area or areas of this tested part, or what streets or families make it up. From that point of view the validity of the proportion cannot be checked by the text.

7. The School Board visitors gave particulars of the heads of families with school children. Of these particulars samples are furnished in the specimen streets. "Every house and every family with children is noted with such information as the visitors could give about them. Here are specimens of each class of street" [21]. [21] I., p. 7.

8. The heads of families with school children are scheduled by *sections* according to employment; and then it is assumed that all male adults in the area, (*i.e.* apparently those in the same district) who are not in the schedule follow the same employments as those of the men, who have been scheduled, in the same numerical proportion. Thus, if there are, say, thirty in the "dress and food" trades section among the scheduled male adults (*i.e.* those with children at school), it is taken for granted, there is a proportional number of male adults engaged in "dress and food" trades among non-scheduled male adults [22] [22] I., p. 4.

9. The adult males whose employment is unknown are thus allocated to employment sections proportionally according to the employments of those adult males whose employment is known. The children not of school age have then to be dealt with on similar lines. The elder children (thirteen to twenty) are assumed to belong proportionally to the several employment sections, the totals of which have been themselves settled by proportion. Thus supposing that in a district the tested families, the families with children attending school, say about half of the whole number, produce 100 railway servants, the other, the untested adult males without children attending school, are taken as producing another 100 railway servants, and then the children and young persons who are in the district according to the census are placed out, section by action, so many to our 200 railway servants, so many to each of the other sections.

10. The relative "poverty" of the section (*i.e.* the *section* by employment) is thus partly settled by the scheduled cases—the cases of the adult males in the employment-section who have children at school. But it is also settled in great degree by proportional calculations and assumptions. Thus, if in the district there are, say, 100 railway servants, scheduled as fifty in B. and fifty in C., all men with children attending school, it is taken for granted that there are among the unscheduled populations, as the census returns of occupations may suggest, say another fifty B's and fifty C's railway servants, whose circumstances are not known; and though these at the time are at least not maintaining children of school age, or indeed may not have had any children at all, they have to be counted as poor as their brethren, and must for their sake, therefore, be themselves also consigned to B. and C.

11. According to the assumptions that underlie these proportional calculations the "non-tested area" has been scheduled "by *other means or in proportion*," for "as is the condition of the tested part so is the condition of the whole population."[24] [24] I., p. 5.

12. The speculative nature of the method is obvious. It is not stated what is the area "of the tested part"—or what are the areas in the different districts which taken together make up "the tested part:" the quantitative distribution of the classes of poverty and the sections of employment is assumed. This is an extremely hazardous assumption, unless the similarity of the population class by class and section by section in different parts of the area is tested and "agreed."

But of this no evidence is given.

13. To complete the population "it was necessary to add 68,451 adult women." These are "distributed amongst the classes in the proportions shown for the rest of the population to which they stand in the relation of sisters or daughters." "It may be doubted whether class B should contain its full proportion of these women or of girls from thirteen to twenty." This remark shows the very speculative nature of the process. The number of adult women added to Class B is 7,799,[23] in a total of 100,062. p. 62 [23] Table IV.

O

14. Many of the phrases which are used show how difficult it is to check the figures and to give them any real statistical value.* The known and, so to speak, counted families may be placed in the several classes rightly or wrongly. But in addition, to account for the whole population many adjustments have to be made "by other means or in proportion."

But even so, it may be doubted if Class B has its right number of girls and women; and other similar cases of unsettled approximation to the figure ultimately entered will be found later on.

15. Though the classes are selected by the rule of "poverty," etc., judging from the published schedules of cases, wages are not entered in them, nor do wages appear to have been verified in the individual cases; nor are statements of rental or ages supplied. Nor is the number of children above school age or working away from the home noted—though their payments are often a very important item in the family budget. All these are indispensable details.

The Classes.

16. The classes into which Mr. Booth divides the population are—

"A. The lowest class, which consists of some occasional
[1] I., p. 37. labourers, street sellers, criminals, and semi-criminals.[2] Their life is the life of savages with vicissitudes of extreme and occasional excess. Their food is of the coarsest
[3] I., p. 38. description; and their only luxury is drink.[3] This class is now hereditary 'to a very considerable extent.'[3] At the same time it is recruited with adult men from all others, and the number of children left in charge of it are 'proportionately small' and the number of young
[4] I., p. 39. persons is proportionately large."[4]

B.—Class B is "very poor"—"casual earnings, very poor." "The labourers of Class B do not, on the average, get as much as three days' work a week, but it is doubtful if many of them could, or would, work full time and for long together, if they had the
[5] I., p. 43. opportunity."[5] "Class B, and especially the 'labour' part of it is, not one in which men are born and live and die, so much as a deposit of those who from mental, moral, and physical causes are incapable of better
[6] I., p. 44. work."[6] "From whatever section Class B is drawn, except among the sections of poor women, there will be found many of them who from shiftlessness, helplessness,
[7] I., p. 43. idleness, or drink are inevitably poor."[7]

C.—Class C, "intermittent earnings poor." "Those who work by the job, who are in and out of work according
[8] I., p. 44. to the season or nature of their work[8] are more than any others the victims of competition: on them falls with particular severity the weight of recurrent depressions of trade." There is a passion for drink "among some of the highly paid irregular workers."[9]
I., p. 46. The employment section "irregular" in C and D taken together "has a very bad character for imrovidence."[9]

D. "Small regular earnings, poor." "It must not be understood that all of these have quite regular work; but only that the earnings are constant enough to be treated as a regular income, which is not the case with
[10] I., p. 48. the earnings of C."[10] "Of the whole section none can be said to rise above poverty, unless by the earnings of the children, nor are many to be classed as very poor. What they have comes in regularly, and except in times of sickness in the family, actual want rarely presses, unless the wife drinks." As a general rule, these men have a hard struggle to make the ends meet, but they are, as a body, decent steady men, paying their way and bringing up their children respectably. The work they do demands little
[11] I., p. 49. skill or intelligence."[11]

The other classes are E regular standard wages. "As a rule the wives do not work, but the children all do." "This class owns a good deal of property in the aggre-
[12] I., p. 51. gate."[12] F. "Higher class labour." G. "Lower midde class, shopkeepers, and small employers, clerks, etc., and subordinate professional men." And H. "Upper middle class—all above G. lumped together," "the servant-keeping class."

* So in vol. ix., p. 13, dealing with "crowding and earnings," on certain statements Mr. Booth bases the *average* earnings for an *average* number of employees "not necessarily or probably the true average for the year but an approximation to it."

The classes above quoted are arranged A, B, "very poor"; "lowest class"; and "casual earnings"; C and D, "poor," "irregular earnings," and "regular minimum"; E and F, "comfortable," "ordinary standard earnings," and "highly paid work," and G H, "well to do," "lower middle," and "upper middle."

The following statements further summarise the position:—A, B, C, D are "classes of poverty sinking to want"; E, F, G, H, "classes in comfort rising to affluence."[13] [13] I. p. 62.

"Omitting class A, which rather involves the question of disorder, we have in classes B, C and D the problem of poverty."[14] "B," the "very poor," are at all times [14] p. 131. more or less in "want," though only a percentage "not I think a large percentage," in distress. C and D "poor" "though they would be much the better for more of everything are not 'in want.' "They are neither ill-nourished nor ill-clad according to any standard that can reasonably be used."

The Data for the Classes Judged by the Individual Case.

17. As further definition of these classes thirty individual cases are quoted and classified. The test applied is expenditure. A "male adult" standard of consumption is assumed and the members of the family are calculated as equivalent to so many parts of a "male adult." Thus an adult equals 1, a woman ·75, and "in proportion for children."[15] So a family—man, wife, and three children [15] I. p. 132. —a son 18, two girls 8 and 6—make up 3·35[16]; the man [16] I. p. 136. being 1, the wife ·75, the son of 18—? ·75, the two girls ·85 between them.

In this "B" case the man is 38 "in poor health fresh from the infirmary"; his wife 43 is consumptive. The son of 18 earns 8s. a week as carman's boy. The neighbouring clergy send soup two or three times a week. The income is not stated but is supposed to be 10s. 3·4d. but the expenditure is stated to be 15s. 2½d. or 4s. 6d. a week per male adult.

In another "B" case (No. 4) in which the supposed income is 18s. 10·8d. per week, "the family run largely on credit." "The man, a bricklayer, gets something as caretaker, very little by his trade." The wife works as dressmaker and has to put out the washing. There are six children, thirteen, eleven, nine and a half, three and a half, two, and four months. The family is set down as equal to 3·7 in "male adult" measure and the expenditure at 24s. 5¼d. or per "male adult" 6s. 7½d. a week.[17] [17] I. p. 136, 143.

In a third, a "B" case on the line, it is stated, between "B" and "C" there are a man, wife, and five children (ten, nine, five, three, one). The man has irregular work as a wharf labourer at 20s. to 21s. a week. A girl not counted in the family is in service, and "still receives money and clothes from home." The wife earns 3s. 6d. a week by needlework. By steadiness on the man's part and good management on the part of the wife, they live as well as many families in "D." The "adult male" costs 6s. 8d. a week, the family 21s. 6d.—against a supposed weekly income of 22s.

In regard to these and the other cases quoted as typical the following points may be noted:—

In these cases no figures of actual income are given. There is an analysis of expenditure only. The supposed income is given and the amount expended very largely exceeds this. In case No. 1 (above), the expenditure is nearly a half more than the income. If this is to be taken as representing the normal condition of the family budget it is hardly to be accounted for by debt or the avoidance of payment, or by the thrice-weekly soup. Probably the real facts have not been disclosed. At least, the inference is strong that there must be some grave flaw in the statement.

In the second case the difference between supposed income and actual expenditure is almost as significant. The supposed weekly income is 18s. 6·8d.; the expenditure 24s. 5·75d. If the man's actual income were as great as his expenditure, he would assuredly not be in "B" but in "C," or even "D." Here again, the inference must be that the facts of the case as a whole have not been ascertained.

The third case is significant from another point of view. It is said to be on the line between "B" and "C," but so good is the use of means that the family "live as well as many families in 'D'." Here the income though irregular is taken at 22s. (20s. to 21s. irregular, earned by the man plus 3s. 6d. earned by the wife)—slightly more than the expenditure. The facts are probably better ascertained. Why then should not the family be placed in "D"? It is not wage, but the use of wage, that makes the difference between well-being and want. But, if that is true, as undoubtedly it is, this factor, socially a more important one than the amount of wage, would upset all these classifications. Here a family in which "poverty" is avoided by the good use of means is put in the same category with a family in which "poverty" is made acute by the thriftless use of means. The grouping indeed is arbitrary, unless this factor be taken into account and given its full value.

These cases then quoted by Mr. Booth to define his distinctions of classes more clearly, fail to do so. They suggest rather that the classes are practically arbitrary groupings based on no common principle. They show that the people have not really been classified by him into "poor" and "very poor."

The Data for Classes as Judged by the Street Notes.

[19] I., pp. 5, 7.

18. Turning to the streets given as specimens of classification [19], we find how large a number of families have to be passed over as unknown. In St. Hubert Street, for instance, an A street: fifty-seven of the families have children at school, and are therefore classified; forty-one having no children at school, and therefore unknown to the School Board visitors, are unclassified. This represents a very large proportion of whom nothing is known, and everything has to be inferred.

[20] I., p. 16

Or, if we take a C–D street, [20] we find the figures: four B families; four C; ten D; three E families; twenty-one in all; and fifteen families without children, and therefore unknown to the School Board visitors, unless possibly children in these families had recently left school. Another C–D street has four F families; twelve E; twelve D, and five C; thirty-three in all, and eighteen unclassified.

Thus the margin of the unknown is very considerable, and the grouping of the streets—so far as the statistics are concerned—based on something very like guess-work.

The details supplied in the specimen entries are extremely meagre, and many, if not most, of those required for a classification drawn on the lines of social science, are wanting.

If classes A, B, C, D are to be described as "classes sinking to want" this evidence is essential, for C–D, it is stated, are "neither ill-nourished nor ill-clad according to any standard that can reasonably be used; and D are, taken as a body, decent, steady men, paying their way and bringing up their children respectably." Such a conclusion evidently can only be based on comparative statistics, not a single day estimate. But it is impossible to regard all A, B, C, D as "sinking to want." The notes of cases (I., p. 140), for instance, suggest that this is hardly the fact.

A classification of families on the down grade or on the up grade might be of real service, but much fuller information would be required for that.

Are the Classes True Classes?

19. The first question is—are they true classes? A class is strictly a "summoning." Thus we might summon a class of the payers of income tax. Here the lines of division are accurately divided by specific count or admitted valuation. So in the case of paupers; all who apply for poor law relief can be counted accurately, and so from that point of view form a class, a "statistical" class, and in view of a particular law—the Poor Law—a social class. But in an investigation of society we can make no such clear counts or valuations; and our classification, to be valid as representing divisions of the population, must be a classification of people and their conditions in relation to social habits and organisation, or to put the question more broadly, the characteristics of the class must be characteristics which in the persons classified represent real factors in the formation of society.

Under the phrase "social habits," which are real factors in the formation of society, are included: habits of nurture and care of children; industry and use of leisure; income and its use; foresight; arrangements for the settlement of children in life, and their marriage.

Under the phrase "social organisation" are included the relation to religious bodies; to the elementary and other schools, school attendance, etc.; to trade unions or benefit societies; to charitable institutions and societies, and to the Poor Law, and so on.

All these questions have relation to society and social life, to whatever grade the individual and his family may belong. They are thus a basis for the classification of society as a whole; and it is obvious classes with different amounts of income might each be tested by such a rule of "social habit and organisation," though in different classes the actual methods of nurture and care of children, industry, etc., may vary considerably (as, for instance, in education in the elementary school and the public school, etc.), according to the purpose of life which each class has in view.

And classification of the people from the point of view of social life must then be founded on "real factors in the formation of society," real, that is to say, as representing the elements in society actually at work in making it.

20. The question is whether Mr. Booth's classification represents these real factors, or whether it represents rather more or less arbitrary groupings of the people.

If we take Class A we find it consists of those at the bottom of society whose "life is the life of savages with vicissitudes of extreme and occasional success." The measure here applied is strictly that of breach of social habit; and, if this be so, we want to know how many have broken loose from social habit. This is how the number is made up. (Vol. I. p. 37.)

This class Mr. Booth "puts at 11,000 or 1¼ of the population, but this is no more than a very rough estimate, as these people are beyond enumeration, and only a small proportion of them are on the School Board visitors' books. If I had been content to build up the total of the class from those of them who are parents of children at school in the same proportions as has been done with the other classes, the number indicated would not have greatly exceeded 3,000; but there is little family life among them, and the numbers in my tables are obtained by adding in an estimated number from the inmates of common lodging houses, and from the lowest streets. With these ought to be counted the homeless outcasts who, on any given night, take shelter where they can, and so may be supposed to be in part outside of any census. Those I have attempted to count consist mostly of casual labourers of low character and their families, together with those in a similar way of life, who pick up a living without labour of any kind."

In calculating the number in this class, then, practically the method of grouping on the basis of school information is set aside. On that basis the number would have been 3,000; the number actually set down is 11,000. What is the ground for this very large addition is not stated. "An estimated number has been added from the inmates of common lodging houses, from the lowest streets and from homeless outcasts." But the data for this estimate are not given.

Yet this class—an important point bearing on the question of degeneracy—is, it is stated, to a very considerable extent, hereditary. On the other hand, it is said also that it is recruited with adult men from all others, and that the number of the children in the class is portionately small—while that of young persons is "proportionately large." This analysis seems to be in itself contradictory. The class as described above may be considered a real class—uncivilised in a manner, living without the ordinary obligations of social habit. But comparative method is not applied to it. It is not possible therefore, from the data supplied, to say of what it is constituted or how far it is actually hereditary.

21. The second class is B—"Casual earnings"—"very poor." It contains 100,000 of 11·25 per cent. of the population.

In it are set 22,000 artisans, and 41,000 "casual labour."

"From whatever section (of employment) Class B is drawn, except among the sections of poor women, 19,095 in number (Table XVII.), there will be found many who, from shiftlessness, helplessness, idleness, or drink are inevitably poor."

This raises the question of the meaning of poor or "very poor"—whether they are or can be defined as a statistical class—or whether they are a social class and represent an aggregation of real conditions in relation to social habit and organisation.

23. Poor, like pauper originally, is "paucus" one of small means—a relative term.

It is clear that in the case of Class B as stated, many are shiftless, etc. The question with them is not of means, but the use of means.

174.

O 2

The general supposition is that the people are poor not from such causes, but from not having enough—they would, as Mr. Booth says of Classes C and D " be much the better for more of everything all round." And to carry out his programme Mr. Booth should have made a " depravity " class—thus defining at least two classes of poor—poor by literal want of means, and poor by " depravity." But the poor by literal want of means—who may be called perhaps the " economic " poor are those only who, having ability to use means with a view to wealth or well-being, do not possess them.

But the material for any such division, by which " poor " should present a social class is wanting. Really the term is used by Booth only in a quite general sense—not suitable for the definition either of a social or a statistical class. Poor by depravity, by inability to use means, by weakness of mind, or of body—all very different groups are merged in one mass—so that for any scientific purpose the term " poor " loses all value as a definition. Difficult as it is to give to so vague and relative a word any precise meaning by direct count, it loses all its meaning when it is used for a large mass of the people—without regard to causes. And on Mr. Booth's method the causes cannot be ascertained and they cannot, form, as they should, the ground of classification.

24. In the next class we have " C—intermittent wages " and above it, " D—small regular earnings." Here is an economic difference which is very marked. C are " men who usually work by the job or who are in or out of work according to the season or the nature of their employment. " Many of the irregularly employed could not keep a permanent job if they had it, and who must break out from time to time."

D are, as a body, decent steady men, paying their way and bringing up their children respectably.

Of C and D employment-sections are formed and the two thus find themselves in the same category—under, amongst other divisions, that of " irregular " in Section 3. They the " irregular " (Section 3) " have," it is said, " a very bad character for improvidence "—" wanting in ordinary prudence ; " but on the other hand Section 3 " for the most consists of hardworking struggling people, not worse morally than any other class, though thriftless and improvident."

Here are two classes of which one is found on the line of " intermittent wage " with either as result or as cause thriftlessness and improvidence, while the other D on the line of small regular earnings is the reverse. One asks whether the economic difference is the main factor. There is no evidence. Neither by examination of any large number of cases nor by any comparison of past and present is light thrown on the problem. But what is more, these two classes, economically so different, are merged and treated as one ; and in the extension of Mr. Booth's method from the East End and Hackney to London C and D as separate classes disappear. The class thus becomes no true class at all, either by economic difference or by social habit.

25. Reference has been made to the absence of any ascertainment of causes. A word more should be said as to this, for Mr. Booth submits a statement of causes, from notes of 4,000 families, and he thus classifies them :—

	A. B. Per cent.	C. D. Per cent.
Loafers - - - - - - -	4	0
Questions of employment ; casual, irregular work, etc. - - -	55	68
Questions of habit, drink, thriftlessness - - - - - - -	14	13
Question of circumstance ; large families, illness - - - -	27	19

It is difficult to reconcile these figures with the general statements of the classes given above ; nor do they furnish much information even as to causes.

Take employment. How far does this mean incapacity ? Mr. Booth says, " Incapacity of two sorts is no doubt common : that which leads to low pay and that which leads especially to irregularity of employment." But this does not help us at all. How far we ask does the factor of incapacity affect the question of employment—as a cause ? And there is no reply.

Here again also most diverse conditions are brought under one head. Mr. Booth does not give the data for his classification — indeed if we may judge from his entries, he has not the data. The classification fails at the critical point just when it is put to the test.

So if we ask why A, B—the outcast and residual classes—shows 14 per cent. in drink, and C with D the regular and steady class D 13, there is no evidence ; and the figures suggest insufficient knowledge.

Useful as school board information may be, it hardly touches the many who send their children without demur, and only require to be " scheduled."

The Remedy.

26. Then as to the remedy. It is to place B under State discipline, as it is a residual class of those who " from mental, moral and physical reasons are incapable of better work." " The life would not be attractive ; regular meals and fixed hours of work, which would not be short." This " would not encourage idleness or weaken the springs of energy. The difficulty lies solely in inducing or driving these people to accept a regulated life."

That is so indeed ; and the difficulty is not met by the suggestion.

(1) It is proposed that there should be a social standard of State dependence. Membership of class B would hardly suffice, as indicating a good standard for State intervention. A fairer one would be to bring under State supervision the people on the down-grade wherever they might be, if the project were in any way feasible.

(2) A new servile and dependent residual class would be created, but out of this mass of incapacity what hope is there of any ultimate reform ? Of the ssibility of driving them to the regulated life, no evidence is offered.

(3) In East London and Hackney the B class is estimated at 100,000, excluding C and its bad ones. How shall such a population be brought under State discipline, and what state is strong enough to educate such an amount of adult incapacity as the class represents by direct measures ? The grouping of the people in what are after all unreal classes aids not a whit in the solution of the problem. The remedy depends on the cause ; and the causes are lost in the classes ; and thus the class only confuses the issue.

(4) There is no evidence to show that such persons as are described as in Class B would be driven into regular employment. The evidence is the other way. Life itself is their taskmaster, and probably a more discerning one.

27. Instead of this system of investigation—it is to be hoped that at some time there may be made a real enquiry —based on closer and more complete observation, in which the enumeration will be actual and all the necessary facts verified. The conditions of families thus studied and grouped in types according to social habits might throw some real light on problems of poverty and deterioration.

Summary.

28. The process of this inquiry and its limitations are now clear. Certain data are estimated, counted, and described as brought under view at one time, but without regard to antecedent facts and conditions. Hence, while there is an immense elaboration of statistics there are no data for forming conclusions.

The data thus viewed at one time are partly founded on statements, but very largely based on inference only ; and they are far from complete.

Part of the data—that collected in regard to families with children at school in part of East London and Hackney is classified in numerical form and augmented by proportional additions of unknown members of families and in other ways, and thus made applicable to about half of the population of East London and Hackney, and then in turn these results are applied by a series of proportion sums and in other ways to the other, the unknown, half of the same population.

Then, lastly, the East London and Hackney figures are applied by yet other proportion sums and estimates to the metropolis as a whole.

In all these elaborate classifications, proportions, and adjustments there is an immense liability to error—and the errors cannot be checked in any sufficient manner. To *data* of the most general description precise figures are applied. Thus, facts by no means closely analysed and not really grouped at all (*cf.* the merging of classes C and D above) are shrouded in an appearance of precision which is very misleading. As has been evident over and over again, popular writers and speakers have forgotten the intricacies of the process of making the classes, and have used the figures only, as in themselves evidence, summary, conclusion, confession, and admission of wholesale want and distress. To this result many of the generalisations in the book lend themselves (*cf.* above: "classes of poverty sinking to want"); and to this use of the figures no protest is made; and unfortunately it is not possible to check errors and revise classifications in figures assigned to such conflicting and momentary data as are here brought together (though for the sake of the better guidance of public opinion it would indeed be desirable to do so).

The final consummation of the method is reached when by yet further proportion sums the figures of the metropolis are made applicable to the country at large, and the people discuss, "how the twelve millions on the verge of hunger may be fed."

Poverty is a relative term, a secondary or modifying cause. It may act as an obstacle or a spur according to the temperament and conditions on which it acts. It is applied here to classes more or less in want. But classes are not true classes *qua* poverty, but *qua* social habit. Unless in the lowest residual class to some extent, society is not a solid mass across which lines can be drawn horizontally as representing classes, but it is in a state of constant motion. To test social health, one should measure the upward and downward movements. Strictly, if we use the word at all, poverty, as a relative term accompanying certain social habits, would then represent ill-being, the downward movement, and wealth, or well-being, the upward movement.

To make these absolute classes many opposite and contradictory elements have to be forced into unity:

(*a*) It is taken for granted that there is an equation applicable in all classes (B to F) between (a) members of a family and (b) necessaries, *e.g.*, the man stands for 1 in the expenditure, the wife ·75, and so on.

(*b*) Every one is assumed—class by class—to be, though individual in ability and strength, subject to and acting in accordance with an average—the average which is assumed by placing quite different types in one class, *e.g.*, the careful, the drinking, and the thriftless. Thus the characteristics which are important in the making of good social habit are averaged away—indeed ignored. All that the use of money adds to the value of money is ignored and neutralised.

(*c*) In the same way all physical differences, even those of ordinary and everyday importance—age, sickness, feebleness, etc.—have to be ignored and averaged away.

Thus we have as a statistical creation a degraded society which is from the point of view of social science hardly less like nature than the "economic man" of some political economists.

"*Poverty.*"

The Method.

29. Mr. Rowntree takes the town of York as the subject of his inquiry. Mrs. Bernard Bosanquet has already criticised his method (and Mr. Charles Booth's) very effectively in a paper entitled "the Poverty Line," to which Mr. Rowntree has made some rejoinder, and in a paper in the *Contemporary Review* for January, 1904—"Physical deterioration and the poverty line." Many points dealt with in those papers may be passed over here.

Mr. Rowntree's aim is to ascertain the extent of destitution, *i.e.*, insufficiency of means to obtain necessaries and to learn how far this destitution is culpable—that is due to causes assumed to be within the control of the individual, drink, etc.

His method is this—

1. To count and estimate the number of people just as they are at a given time (as do Mr. Booth and Dr. Hunter) instead of using the comparative method.
2. To ascertain the income of the head of the family and of the children who earn and of the wife's.
3. To class families according to income, on the understanding that a man, wife, and two to four children are a "moderate" family, and that if a man has a larger family, he is placed in a lower income class and *vice versa*.
4. To fix a standard of necessary food and the cost of it and to add the latter *per individual* to the cost of rent, clothing, etc.
5. Then to divide the income class by class by the total cost of necessary living for the class, and thus ascertain how far the income suffices for that purpose.

The sum set out is this:—

$$\frac{\text{Moderate family} + \text{Cost of necessaries.}}{\text{Income.}}$$

The sufficiency of the method turns on: (*a*) the exactness of the investigation of the amount of income; (*b*) the reality of the classification of the families; (*c*) the standard fixed for necessaries and its applicability.

Amount of Income.

30. As to the first—the exactness of the investigation of income.

The income classes are—

A. under 18s. a week.
B. 18s. and under 21s.
C. 21s. and under 30s.
D. Over 30s.
E. Servants.
F. Servant keeping class.
G. In public Institutions.

The population dealt with is 75,000.

The inquiry refers to 46,754 people or 11,560 families.

The inquiry is made by house to house visitors, and the classes are settled after consultation with the clergy, etc., and as to wages, after general consultation with employers.

Everything depends on precision in regard to the p. 14.
income. This is what is said:—

The visitor "found that the people with few exceptions were willing to supply the information sought. In some cases there was a disposition to give incorrect information, but experience soon enabled him to distinguish between truth and falsehood, and in doubtful cases the facts stated p. 26
were checked by neighbours and others."

"The occupation of each of the workers was ascertained and sometimes also direct information regarding the wages earned. Where this information was not available the p. 27.
wage was *estimated*. In the case of skilled workers the wages *were assumed to be the average wage* which obtains in the district for the particular trade." These "it is believed, have been *estimated* with a large degree of accuracy." "In *all* cases, in assuming wages, allowance has been made for short time, including public holidays not paid for, also over-time, and cost of tools, etc." (The italics are mine.) This allowance is based on information from trade unions, masters, etc.

No uniform allowance has been made for loss of wage through illness, but in the estimates of short time some allowance under this head has been made.

A note on this—is given as a "rough indication"—2d. p. 27.
a week in a sick club gives 7s. 6d. for six weeks, and 3s. 6d. for another six weeks of sickness.

"Working upon these lines, the earnings of every wage earner have either been ascertained or carefully estimated."

Afterwards the word "estimate" is dropped, and the reader cannot but think he is dealing with actual earnings. On page 116 the words used are "the total earnings of every working class family were ascertained."

31. Now to make some comments:—

(*a*) The system was one of visitation without verification in the individual case, and as to the earnings of the man, the wife, and the children, this, even with the other sources of information mentioned, is quite insufficient—at least, it would be elsewhere.

Take at random a few families visited elsewhere by people of long experience—and then tested by verification:

(1) Man forty-five, dock labourer; stated earnings, 12s. to 18s.; ages and stated earnings of children; girl nineteen; boy fourteen, 12s.; total stated, earnings, 24s. to 30s.; total verified earnings, average 41s.

(2) Man thirty-five; railway labourer; stated earnings 23s.; wife 5s.; in all 28s.; total verified earnings 33s.; and so on.

Obviously visitation without verification in the individual case is no sufficient means for ascertaining the truth.

(*b*) Next the special examination of seemingly "doubtful" cases is a most misleading method—the inquirer is left in a position of thinking he knows he is right, without knowing where he may be wholly wrong.

(*c*) Next, the assumption that the skilled worker earned the average wage of the district is quite insufficient in compiling accurate data, which are to be applied to families whose incomes are to be treated as purchasing necessaries in very small amounts per consumer.

(*d*) Next, if deductions are made from the wage—these should be set out in their money value—that the sum may be before the reader, the notes quoted above are no real guide.

(*e*) Lastly, as to the earnings of children: "A girl or lad" it is stated "will pay 5s. or 9s. weekly, while a man will pay 9s. or 14s. to the household." But this is a very large variation surely, on which to base either estimate or average, apart from a knowledge of the actual facts—a difference of 55 per cent. Yet what the child actually gives may make all the difference in the family. The family is the social, and in great measure the industrial, unit. Many cases could be quoted in support of this.

The great variation in the sums available from children in particular cases can be shown by reference to many instances. The variations are hardly such as could be dealt with by general average.

32. In regard to the statements of wages Mr. Bowley's paper in the "Proceedings of the Royal Statistical Society," vol. lxv. of 30th June, 1902, page 359 may be considered. The paper is a note on wages in York, 1899 (B.S. Rowntree). After quoting from Mr. Rowntree's book passages respecting his method and the assessment of wages, etc., Mr. Bowley writes:—

"These figures are not published in detail in the book but merged in family incomes and grouped together. Mr. Rowntree, however, has caused the statements of all adult male wage earners to be extracted from his material." On his basis the wage given has only to be multiplied by fifty-two, to give the annual earning. "The figures are estimates, and in some cases estimates only to the nearest round number. Thus, in many of the skilled trades the average weekly wages approximate to 30s. in some cases being 1s. or 2s. more, in others 1s. or 2s. less, and in such cases the wages are entered at 30s." Hence "the marked grouping is very noticeable, indicating that the gap between skilled and unskilled labour is more perceptible in York than in London, or in the country as a whole." The table divided to show shilling differences is suggestive. The total number of wage earners dealt with is 9,544, the number set down as receiving a wage of 20s. is 1,532, those that receive a wage of 19s., only twenty-seven, those who receive a wage of 21s., only thirty-one. Some other figures stand thus:—

At 23s. - - -	499	At 30s. - - -	2,398
,, 24s. - - -	810	,, 35s. - - -	260
,, 25s. - - -	394	,, 40s. - - -	420
,, 28s. - - -	471		

The intervening entries of wage have but small numbers of earners entered against them.

Thus analysed, it is clear that the wage table suggests rounding such as one finds in census work in the case of ages. This rounding clearly seems due to the method adopted by Mr. Rowntree, and it appears to be a very great defect, especially as the question at issue is the ability to utilise a small income for the purchase of very small quantities of necessaries. It is not possible logically to connect large generalisations as to wage with minute investments of the generalised wages in a large number of particular cases.

Classification of the Families.

33. The second point is the sufficiency of the classification as to *families.** It stands thus:—

A.=Under 18s. for a moderate family.
B.=18s. and under 21s. for a moderate family.
C.=21s. and under 30s. for a moderate family.
D.=over 30s. for a moderate family.

34. Some points may be noted:

The great inequality of the classes as measured by income.

Class B stands at 18s. under 21s.

Class C stands at 21s. and under 30s.

Next a moderate family=two to four children—this is an element of great variability in making up the classes.

Next the effect of this adjustment of family scale to income scale is remarkable; thus, if the family income equals 22s. in the case of a parent and four children, the family is put in C, but if of five or more, it is pushed down at least into B. The difference of the one child puts the case within the range of a class earning 18s. and under 21s., instead of in the range of a class earning 21s. and under 30s., and strictly the case might be pushed even further down and become an A case

So again with a family—income 27s., two parents and two children are in C, but if they have only one child, then in D. This is quite arbitrary.

How the question stands is indicated by the following note on Class A. Dividing the income in the class made up in accordance with the rules stated above, we have in it families receiving from under 4s. 5d. a head to 2s. 3d. per head or less:—

Thus A equals moderate family with income *under* 18s. a week.

A moderate family equals two parents and two to four children. (2+2; 2+3; 2+4.)

Eighteen shillings or less is then divisible in a moderate family.

Therefore A equals any sum per head *under* the following figures:—

(1) A equals
18s. (or less)—
———
s.
2+2=4·5 }
or 2+3=3·6 } per head.
or 2+4=3·0 }

(2) If there is one parent and two children, or two parents and one child, A would equal any sum *under* the following figures:—
18s. (or less)—
——— = 6s. per head.
1+2
or 2+1

Unless these be pushed up into the 18s. to 21s. class—Class B.

(3) But, on the other hand, all the immoderate families in A must remain in A.

So A equals not only the above in (1) but often immoderate A families:—
18s. (or less).
———
2+2
or 2+3
or 2+4

But the following *as well* 2+5 } as divisers of the
2+6 } 18s. or less.

(4) And into A are also pushed all the immoderate families from B—earning above 18s. and under 21s., and including families falling to it from B to A. Taking shilling divisions the range covers the following cases:—

19s.
———
2 + 5 or 6, &c., children;
Which =2·7s. or 2·3s., &c., per head.

20s.
———
2 + 5 or 6, &c., children;
Which =2·8s. or 2·5s., &c.;

21s.
———
2 + 5 or 6, &c., children;
Which =3s. or 2·6s., &c.

Therefore A, including all the cases in (1) (3) and (4) above, covers any income from 4·5s. per head downwards.

This is obviously a very large range—a kind of sliding scale of income checked by family in which there is no underlying principle at all. For the question must be asked: Are these classes real classes? The answer must be in the negative. The classes are but groups more or less indefinitely formed in connection with differences of income and family. They are not based upon any real characteristics; nor would they be necessary for Mr. Rowntree's purpose, if he rested his case upon observation of particular instances and verified material only.

35. Another test might be applied on Mr. Rowntree's lines of averaging. I have taken fourteen families at hazard—not families in Mr. Rowntree's book, but families of persons who have applied for relief to a Charity Organisation Committee.

The average wage of the fourteen families as returned at 23·14s. In five of the fourteen families there is definite upward movement. These omitted, there remain nine families with an average wage of 9s. This *average* wage would bring all such families into Class A.

But as the following notes show, the relation between average wage and the condition of the families in the class is quite unreal. The cases are as follows:—

(*a*) Pension case, 12s. 6d., man and wife.
(*b*) Infirmary case.
(*c*) Heavy drinker.
(*d*) Discharged summarily: falsification of accounts.

(e) Bad eyes; large relief; begging; bad temper.

(f) Old, children well able to support, second marriage; family quarrel.

(g) Weakly; casual labourer; dependent largely on wife and children.

(h) Old, son "disappeared" but when found well able to help, rent 10s.

(i) Drink.

It will be noticed in how few of these cases the question of difficulty or "poverty" is one of income as against expenditure. By reference to Class A in Mr. Rowntree's sample of fifty-one entries on page 33 it will appear that in those cases too the true issue is not income as against expenditure but social habit. Thus in his fifty-one cases three are cadger cases, two are mentally deficient, in seven there is drink, five are entered as large families but not further investigated; and twenty-two are set down as poor law relief but not further investigated.

In the notes of these sample cases there is no statement of wages or of finance and no statement of ages. It seems doubtful indeed whether such evidence was obtained. It it were, it would tend greatly to modify Mr. Rowntree's own conclusions.

Standard fixed for Necessaries.

36. We turn next to the standard fixed for necessaries and its applicability. It is evident that the investigation, so far as it has yet been under consideration, does not afford a basis for precise numerical results. Yet it is upon this basis that the utility of the investigation depends. A food test is next applied. This is done in two ways.

The Local Government Board issued three years ago a dietary for people in the workhouses and infirmaries. There were three conditions mentioned in their circular letter—

(1) The inmates were examined by a medical officer.

(2) There were "plain" and "infirm" diets; and

(3) There were children's diets.

"Owing to the great difference of individual appetites m the case of children, it should be remembered that in their dietaries the prescribed rations represent the average allowance for a group of children and not the amount for a particular child."

(4) They note that the bread allowance is excessive; and;

(5) They note that there is a tendency to waste food which should be checked.

From these "plain" dietaries Mr. Rowntree has made a dietary and priced it according to the contract prices of the York Board of Guardians.

Next he has divided his families as grouped on the above plans between "Poverty lines."

Primary Poverty population, 7,230; families, 1,463.
Secondary Poverty (or self-caused), 13,072.

The "Primary Poverty" he has divided into six sections according to immediate cause.

Households affected.	Immediate cause.
403	Death of chief wage earner, 15·63.
146	Illness or old age of chief wage earner, 5·11.
38	Chief wage earner out of work, 2.31.
51	Irregularity of work, 2·83.
187	Large family, 22·16.
640	Regular work and low wages, 51·96.

The whole makes 9 per cent. of the population.

Taking these sections Mr. Rowntree makes budgets charging the cost of food as above, showing a deficiency for the section and then working it out per family. Each adult is taken as requiring an equal amount according to the standard of an average man.

37. To pass to comment. The result of all this cannot be otherwise than unreal, in spite of its being set down in seemingly precise statistics. Thus the consumption of the food depends not on an average of the consumption of adults but on the digestive powers of those who partake of it. Thus the old would take less, the young more; but no ages are given. Weight and size would tell; but the standard figures apply only to a man of eleven stone. The waste which is made in the workhouse need not be made in the home; but here it is not—on this method, cannot be, taken into account. Optional variations in diet, which are petty in themselves, but stand for much in individual cases, are also necessarily ignored.

The method might be of use if applied to individuals who had been medically examined, but here it is applied to the average of a class whose physical condition has not been examined, in conjunction with a statement of income which is after all only an estimate.

Then as to the standard itself. Mr. Rowntree requires in all classes 3,500 calories per diem, food being analysed into calories, *i.e.*, into degrees of heat productivity. The value of the food as a source of energy is measured by the amount of heat it produces on complete combustion. The insufficiency of this method has been recognised and it cannot be taken as finally settled—certainly it cannot be accepted without demur as here applied.

The experiments that have been made are mostly on inmates of Institutions under known conditions and living healthy lives. From these experiments it has been concluded that a healthy man of eleven stone weight and doing a moderate amount of muscular work must have from 3,000 to 3,500 calories if his needs are to be fully covered. "Standard dietaries," it is urged, "are only of limited use . . . . only drawn up to meet the needs of *typical* individuals living under known conditions and doing a moderate amount of muscular work." Further, the calories needed vary according to occupation: so a clerk at his desk is stated to require only 2,500 calories, a man at moderate work, such as house-painting, 2,631, a navvy at hard labour, a highly paid man, 3,513. A point like this makes the greatest difference in Mr. Rowntree's divisors; but for it he makes no allowance.

Many other points may be mentioned, but one will suffice. In the case of women Dr. Dunlop states that "their food requirements have received comparatively little attention," yet these elaborate calculations, with some modifications, are applied to them as they are to men. For all working male ordinary persons, except those of more than usual size, twelve stone in their clothes, and doing ordinary work, Dr. Dunlop furnishes dietaries amply sufficient which would cost say, 2d., a meal.

It may be concluded that Mr. Rowntree's method is extremely speculative, that the information on which it is based is far from sufficient, and that neither method nor information provide the material that would be required for drawing trustworthy "poverty lines," if, indeed, such a task could be accomplished at all with scientific scrutiny in reference to a large population. At least any inquiry should be restricted to observed data, to actual counts, and to classifications founded on social habit.

A Contrast.

38. One contrast may be added.

Mr. Charles Booth applying his method to London concludes that "after deducting loafers and criminals, 29·8 per cent. of the people are in perpetual poverty, owing to the family earnings being less than 21s. a week."

Mr. Rowntree, copying Mr. Booth's method, concludes that 9·91 per cent. of the population of York, which, he holds, is generally representative of urban population, have earnings insufficient for the maintenance of merely physical health.

Mr. Booth includes in his 29·8 per cent. those (*see* above) with 18s. to 21s. a week who are "neither ill-nourished nor ill-clad according to any standard that can reasonably be used." On the other hand, he "does not enter into questions of economical or wasteful expenditure." Accordingly Mr. Booth counts as in the same perpetual poverty alike the wasters and the well-nourished and well-clad.

Mr. Rowntree calculates, as he believes, those whose poverty is due to their own fault at 18·51, and those who have insufficient earnings etc., at 9·91, in all 28·42, and fortifies his conclusion by the similarity of his figures to Mr. Booth's.

It is plain that there is no real similarity. Mr. Charles Booth ignores economical or wasteful expenditure, and includes the well-clad and well-nourished and makes a percentage of 28·8.

Mr. Rowntree recognises wasteful expenditure or self-caused poverty, and finds none in his class provided with sufficient means to command the necessaries of life—none well-clad and well-nourished—and makes a percentage of 28·42.

How these percentages have been arrived at has been shown; and it is plain that they cannot be said to be consistent one with another, nor either of them based on sufficient information.

C. S. Loch.

Made in the USA
Las Vegas, NV
06 April 2026

2d930411-a3f4-49d3-945e-72f9ea38b438R01